A CLASSIFICATION OF THE LIVING MOLLUSCA

A CLASSIFICATION OF THE LIVING MOLLUSCA

Assembled by Kay Cunningham Vaught

Edited by
R. Tucker Abbott
and Kenneth J. Boss

American Malacologists, Inc.
Melbourne, Florida 32902, U.S.A.
1989

THIS BOOK IS
DEDICATED BY THE AUTHOR

To the late Dr. A. Myra Keen, whose early
encouragement has been a constant inspiration
during the several years it has taken
to compile this work

Edited by R. Tucker Abbott and Kenneth J. Boss.

Published by American Malacologists, Inc.
P.O. Box 2255
Melbourne, FL 32902-2255

Distributed by American Malacologists, Partnership
P.O. Box 1192
Burlington, MA 01803

Manufactured in the U.S.A. by BookCrafters, Chelsea, Michigan

ISBN 0-915826-21-6 Paperbound, plastic comb
ISBN 0-915826-22-4 Hardbound, sewn

Library of Congress Catalog Card Number: 88-83463

FIRST EDITION 1989

EDITORIAL PREFACE

With the proliferation of new taxa proposed over the last twenty years, there has long been a need for a current list of supraspecific names in the Mollusca. Several well-illustrated synoptic outlines appeared some years ago, such as Thiele's *Handbuch der systematischen Weichtierkunde* and the *Handbuch der Paläozoologie* by Wenz and Zilch. More recently, Franc in the *Traité de Zoologie*, Boss in the *Synopsis and Classification of Living Organisms*, and Vokes in his *Genera of the Bivalvia* have partially accomplished this task.

We can now thank an enthusiastic amateur malacologist for undertaking this useful and enormous compilation. Kay Cunningham Vaught, after ten years of privately funded research, has provided us with an up-to-date framework of the systematics of the Mollusca. Although this compilation covers all of the family and generic names with which most malacologists will ever deal, nonetheless there are many obscure names yet to be entered. Quite naturally, the opinions on the status of subjective synonyms is subject to change and interpretation by others. One should also be reminded that names limited to fossil forms are not included here.

The field of malacology and the community of private mollusk collectors owe Kay Vaught a debt for her contribution and praises for her accomplishment.

The Editors:

R. Tucker Abbott
American Malacologists, Inc.
Melbourne, Florida

Kenneth J. Boss
Museum of Comparative Zoology
Harvard University
Cambridge, Massachusetts

A CLASSIFICATION OF THE LIVING MOLLUSCA

(With listing of supraspecific scientific names)

INTRODUCTION

In recent years, many changes and additions have been made in the general classification of the Phylum Mollusca. Research is uncovering hitherto unsuspected affinities and rejecting others, not only in individual groups, but in the higher categories. Many groups have not as yet been investigated thoroughly, or are in the process of being re-examined in the light of new methods of identification and classification. Some of these changes have been under scrutiny for several years, and have been generally accepted. Others are more or less preliminary, and may fall under further research.

Since this is merely a compilation of the work of others, the writer is permitted only the function of applying the available data to a generally recognized framework in a manner consistent with the opinions of the majority. The literature on these changes is scattered throughout the world in various scientific journals and periodicals: this is an attempt (admittedly incomplete) to collect these recent arrangements and to reconcile them with the previously accepted works. Many references have been given, although space does not permit a complete listing.

It is hoped that with its accompanying index, this listing of known supraspecific scientific names (arranged in as logical an order as possible) will provide a useful tool for the advanced amateur, the student, and the professional malacologist.

The author particularly asks that any workers finding mistakes, omissions or emendations kindly communicate with her at the below address. Hopefully, a revised edition, possibly including names of fossil taxa, will eventually be issued.

ACKNOWLEDGMENTS

This information could not have been assembled without the assistance of many authorities, and the use of extensive library resources. Especially appreciated is the continued support and aid of Dr. William K. Emerson and Mr. Walter E. Sage III of the American Museum of Natural History in New York and Dr. Kenneth J. Boss of the Harvard Museum of Comparative Zoology, whose "Mollusca" section in "Synopsis and Classification of Living Organisms" has been of tremendous help.

Equally important are those who checked and corrected copy, and answered numerous questions: Mr. Anthony D'Attilio of the San Diego Museum, Dr. Alan Beu of the New Zealand Geological Survey; Dr. Rudiger Bieler, Dr. R. S. Houbrick and the late Dr. Joseph Rosewater of the National Museum of Natural History (Smithsonian Institution), Mr. Walter O. Cernohorsky of the Auckland Institute and Museum, Dr. George M. Davis of the Acadamy of Natural Sciences of Philadelphia, Dr. K. Elaine Hoagland of the Lehigh University Center for Marine Experimental Studies, Dr. Bruce A. Marshall of the National Museum of New Zealand, Mr. Richard E. Petit, Dr. Winston F. Ponder of the Australian Museum, Dr. Amelie Scheltema of the Woods Hole Oceanographic Institution, Dr. Alan Solem of the Field Museum of Natural History in Chicago, Dr. Richard A. Van Belle of the Institut Royal des Sciences Naturalles de Belgique, Dr. Emily H. Vokes of Tulane University, and Dr. Anders Waren of the Swedish Museum of Natural History in Stockholm. The contribution of each is gratefully acknowledged.

Kay Cunningham Vaught
8646 E. Paraiso Drive
Scottsdale, Arizona 85255
USA
November, 1988

A CLASSIFICATION OF THE LIVING MOLLUSCA

SYSTEMATIC NOTATION USED IN THIS OUTLINE:

KEY TO FORMAT:

Class (various endings)	Class: Gastropoda
Subclass (-a, -ia)	Subclass: Prosobranchia
Order (-a)	Order: Neogastropoda
Suborder (-a)	Suborder: Stenoglossa
SUPERFAMILY (-oidea)	**MURICOIDEA**
FAMILY (-idae)	MURICIDAE
SUBFAMILY (-inae)	MURICINAE
GENUS	MUREX Linnaeus, 1758
SUBGENUS	BOLINUS Pusch, 1837
Synonym	Purpura Jousseaume, 1880
(Homonym)	(non Bruguière, 1789)

NOTES: The rules of systematic notation have been violated in the interests of clarity as follows: No Italics have been used, due to printing restrictions.

Class, Order, SUPERFAMILY, FAMILY: Systematic order followed in the higher taxa.

SUBFAMILY: Systematic order used where an accepted arrangement is known. If not, the nominate subfamily is given first; the following in alphabetical order, all in capital letters.

GENUS: Nominate genus listed first; following genera in alphabetical order, in capitals.

SUBGENUS: Nominate Subgenus (sensu stricto) not listed. All subgenera in alphabetical order, in capital letters, indented below genus.

Synonyms, Homonyms: Listed in chronological order, in lower case letters.

INDEX: No family names listed; nominate genus gives the familial location in the outline. All classifications except synonyms and suprafamilial names in capital letters.

ABBREVIATIONS USED:

auctt. - auctorum: of authors; not as proposed originally.
em. - emendation: change in concept; intentional change made to correct an error.
err. - error: unintentional change.
ex: - taken from.
inv. - invalid: invalid name.
MS. - manuscript: unpublished work.
nom.cons. - nomen conservandum: otherwise unacceptable name.
nom.dub. - nomen dubium: uncertain name; insufficient evidence to permit identification.
nom.nov. - nomen novum: replacement for a preoccupied name.
nom.nd., nom.nud. - nomen nudum: invalid published name; technically unacceptable.
nom.nl., nom.null. - nomen nullum: invalid accidental change of name.
nom.oblit. - nomen oblitum: name unused in literature for over fifty years; rejected.
nom.rej. - nomen rejectum: name rejected permanently.
nom.van. - nomen vanus: invalid intentional change of name.
(non -) - indicates homonyms.
non-binom.: - invalid, non-binominal name.
part, pars: - part, in part.
pro., n.n. pro: - in place of.
s.l. - sensu lato: - in the broad sense.
s.s. - sensu stricto: - in the narrow sense.
?, (?): Questions placement in outline.

SYSTEMATIC ARRANGEMENT OF THE HIGHER TAXA USED IN THIS OUTLINE

PHYLUM MOLLUSCA
Class Aplacophora
Subclass Chaetodermomorpha
LIMIFOSSORIDAE
PROCHAETODERMATIDAE
CHAETODERMATIDAE
Subclass Solenogastres
Order Pholidoskepia
DONDERSIIDAE
SANDALOMENIIDAE
GYMNOMENIIDAE
LEPIDOMENIIDAE
MACELLOMENIIDAE
MEIOMENIIDAE
Order Neomeniamorpha
NEOMENIIDAE
HEMIMENIIDAE
Order Sterrofustia
PHYLLOMENIIDAE
IMEROHERPIIDAE
HETEROHERPIIDAE
Order Cavibelonia
PARARRHOPALIIDAE
RHOPALOMENIIDAE
ACANTHOMENIIDAE
AMPHIMENIIDAE
SIMROTHHIELLIDAE
DREPANOMENIIDAE
STROPHOMENIIDAE
PRONEOMENIIDAE
EPIMENIIDAE
SYNGENOHERPIIDAE
RHIPIDOHERPIDAE

Class Polyplacophora
Order Neoloricata
Suborder Lepidopleurina
LEPTOCHITONIDAE
HANLEYIDAE
AFOSSOCHITONIDAE
Suborder Ischnochitonina
ISCHNOCHITONIDAE
MOPALIIDAE
SCHIZOCHITONIDAE
CHITONIDAE
Suborder Acanthochitonina
CRYPTOPLACIDAE
ACANTHOCHITONIDAE

Class Monoplacophora
Order Tryblidiida
TRYBLIDIOIDEA
NEOPILINIDAE
VEMIDAE
MONOPLACOPHORIDAE
LAEVIPILINIDAE

Class Gastropoda
Subclass Prosobranchia
Order Archaeogastropoda
PLEUROTOMARIOIDEA
PLEUROTOMARIIDAE
SCISSURELLIDAE
HALIOTIDAE
NEOMPHALOIDEA
NEOMPHALIDAE
FISSURELLOIDEA
FISSURELLIDAE
PATELLOIDEA
ACMAEIDAE
LOTTIIDAE
PATELLIDAE
LEPETIDAE
LEPETELLOIDEA
LEPETELLIDAE
PSEUDOCOCCULINIDAE
PYROPELTIDAE
OSTEOPELTIDAE
COCCULINELLIDAE
ADDISONIIDAE
CHORISTELLIDAE
BATHYPHYTOPHILIDAE
COCCULINOIDEA
COCCULINIDAE
BATHYSCIADIIDAE
TROCHOIDEA
TROCHIDAE
PELYCIDIIDAE ?
STOMATELLIDAE
SKENEIDAE
CYCLOSTREMATIDAE
TURBINIDAE
PHASIANELLIDAE
TRICOLIIDAE
SEGUENZIOIDEA
SEGUENZIIDAE
NERITOIDEA
NERITOPSIDAE
NERITIDAE
SEPTARIIDAE
PHENACOLEPADIDAE
TITISCANIIDAE
HYDROCENOIDEA
HYDROCENIDAE
HELICINOIDEA
HELICINIDAE
PROSERPINIDAE
CERESIDAE
Order Mesogastropoda
CYCLOPHOROIDEA
CYCLOPHORIDAE
POTERIIDAE
MEGALOMASTOMIDAE
MAIZANIIDAE
DIPLOMMATINIDAE
PUPINIDAE
VIVIPAROIDEA
VIVIPARIDAE
AMPULLARIIDAE
VALVATOIDEA
VALVATIDAE
LITTORINOIDEA
LITTORINIDAE
EATONIELLIDAE
POMATIASIDAE
ANNULARIIDAE
ACICULIDAE
RISSOIDEA
HYDROBIIDAE
BYTHYNIIDAE
HYDROCOCCIDAE
POMATIOPSIDAE
TRUNCATELLIDAE
STENOTHYRIDAE
FALSICINGULIDAE ?
IRAVADIIDAE
BAICALIIDAE
BENEDICTIIDAE
MICROMELANIIDAE
RISSOIDAE
BARLEEIDAE
EMBLANDIDAE
EPIGRIDAE
ELACHISINIDAE
ASSIMINEIDAE
SKENIOPSIDAE
RISSOIDEA (Cont'd.)
TRACHYSMATIDAE
TROCHACLIDIDAE
CINGULOPSIDAE
ORBITESTELLIDAE
CAECIDAE
CTILOCERATIDAE
VITRINELLIDAE
TORNIDAE
RASTODENTIDAE
LOXONEMATOIDEA
ABYSSOCHRYSIDAE
CERITHIOIDEA
THIARIDAE
PLANAXIDAE
MELANOPSIDAE
MODULIDAE
CERITHIIDAE
DIALIDAE
LITIOPIDAE
CAMPANILIDAE
DIASTOMATIDAE
POTAMIDIDAE
PLEUROCERIDAE
FOSSARIDAE
SYRNOLOPSIDAE
TURRITELLIDAE
SILIQUARIIDAE
VERMETIDAE
STROMBOIDEA
STRUTHIOLARIIDAE
APORRHAIDAE
STROMBIDAE
HIPPONICOIDEA
HIPPONICIDAE
VANIKORIDAE
CALEDONIELLIDAE
CREPIDULOIDEA
CREPIDULIDAE
CAPULIDAE
TRICHOTROPIDAE
XENOPHOROIDEA
XENOPHORIDAE
CYPRAEOIDEA
CYPRAEIDAE
OVULIDAE
TRIVIIDAE
PEDICULARIIDAE
LAMELLARIIDAE
PSEUDOSACCULIDAE
HETEROPODA
ATLANTIDAE
CARINARIIDAE
PTEROTRACHEIDAE
NATICOIDEA
NATICIDAE
TONNOIDEA
TONNIDAE
FICIDAE
CASSIDAE
RANELLIDAE
BURSIDAE
Suborder Heteroglossa
CERITHIOPSOIDEA
CERITHIOPSIDAE
TRIFORIDAE
TRIPHOROIDEA
TRIPHORIDAE
EPITONIOIDEA
EPITONIIDAE
JANTHINIDAE
EULIMOIDEA
EULIMIDAE

EULIMOIDEA (Cont'd.)
STILIFERIDAE
ASTEROPHILIDAE
PAEDOPHOROPODIDAE
ENTOCONCHIDAE
THYCIDAE
ACLIDIDAE
Order Neogastropoda
MURICOIDEA
MURICIDAE
CORALLIOPHILIDAE
BUCCINIDAE
COLUMBELLIDAE
NASSARIIDAE
MELONGENIDAE
FASCIOLARIIDAE
VOLUTIDAE
HARPIDAE
VASIDAE
OLIVIDAE
MARGINELLIDAE
MITRIDAE
VOLUTOMITRIDAE
COSTELLARIIDAE
CANCELLARIOIDEA
CANCELLARIIDAE
CONOIDEA
CONIDAE
TURRIDAE
TEREBRIDAE
RISSOELLOIDEA
OMALOGYRIDAE
RISSOELLIDAE

Subclass Heterobranchia
Superorder Allogastropoda
ARCHITECTONICOIDEA
ARCHITECTONICIDAE
MATHILDIDAE
PYRAMIDELLOIDEA
PYRAMIDELLIDAE
AMATHINIDAE
Subclass Opisthbranchia
Order Cephalaspidea
PHILINOIDEA
ACTAEONIDAE
RINGICULIDAE
HYDATINIDAE
BULLINIDAE
SCAPHANDRIDAE
PHILINIDAE
AGLAJIDAE
GASTROPTERIDAE
DIAPHANIDAE
NOTODIAPHANIDAE
BULLIDAE
HAMINEIDAE
RETUSIDAE
RUNCINOIDEA
RUNCINIDAE
ILDICIDAE
Order Acochlidioidea
EUACOCHLIDIOIDEA
ACOCHLIDIIDAE
HEDYLOPSIDAE
MICROHEDYLIDAE
PLATYHEDYLIDAE
TANTULIDAE ?
PHILINOGLOSSOIDEA
PHILINOGLOSSIDAE
PLUSCULIDAE
Order Sacoglossa
CYLINDROBULLIDAE

Order Sacoglossa (Cont'd.)
VOLVATELLIDAE
JULIIDAE
OXYNOIDAE
GASCOIGNELLIDAE ?
CALIPHYLLIDAE
STILIGERIDAE
ELYSIIDAE
BOSELLIDAE
Order Anaspidea
AKERIDAE
Order Aplysiomorpha
APLYSIOIDEA
APLYSIIDAE
NOTARCHIDAE
Order Notasdpidea
UMBRACULOIDEA
UMBRACULIDAE
TYLODINIDAE
PLEUROBRANCHOIDEA
PLEUROBRANCHIDAE
Order Thecosomata
Suborder Euthecosomata
LIMACINIDAE
CAVOLINIIDAE
Suborder Pseudothecosomata
PERACLIDAE
CYMBULIIDAE
DESMOPTERIDAE
Order Gymnosomata
PNEUMODERMATIDAE
NOTOBRANCHAEIDAE
CLIOPSIDAE
CLIONIDAE
THLIPTODONTIDAE
HYDROMYLIDAE
LAGINIOPSIDAE
Order Nudibranchia
Suborder Doridoidea
GNATHODORIDOIDEA
DORIDOXIDAE
BATHYDORIDIDAE
ANADOROIDEA
CORAMBIDAE
GONIODORIDIDAE
OKENIIDAE
ONCHIDORIDIDAE
TRIOPHIDAE
NOTODORIDIDAE
POLYCERATIDAE
GYMNODORIDIDAE
VAYSSIEREIDAE
EUDORIDOIDEA
HEXABRANCHIDAE
CADLINIDAE
CHROMODORIDIDAE
ACTINOCYCLIDAE
MIAMIRIDAE
CONUALEVIIDAE
ALDISIDAE
ROSTANGIDAE
DORIDIDAE
ARCHIDORIDIDAE
GEITODORIDIDAE
HOMOIODORIDIDAE
DISCODORIDIDAE
KENTRODORIDIDAE
HALGERIDAE
PLATYDORIDIDAE
PHYLLIDIOIDEA
PHYLLIDIIDAE
DENDRODORIDIDAE
IDULIIDAE

Suborder Dendronotoidea
DENDRONOTOIDEA
TRITONIIDAE
MARIANINIDAE
LOMANOTIDAE
SCYLLAEIDAE
HANCOCKIIDAE
DENDRONOTIDAE
BORNELLIDAE
TETHYIDAE
DOTIDAE
PHYLLIROIDAE
Suborder Arminoidea
ARMINOIDEA
HETERODORIDAE
ARMINIDAE
LEMINDIDAE
METARMINIOIDEA
MADRELLIDAE
DIRONIDAE
ZEPHYRINIDAE
GONIAEOLIDIDAE
CHARCOTIIDAE
HEROIDAE
Suborder Aeolidoidea
EUEOLIDIOIDEA
NOTAEOLIDIIDAE
CORYPHELLIDAE
PARACORYPHELLIDAE
NOSSIDAE
FLABELLINIDAE
PROTAEOLIDIELLIDAE
PLEUROLIDIIDAE
EUBRANCHIDAE
PSEUDOVERMIDAE
TERGIPEDIDAE
FIONIDAE
CALMIDAE
GLAUCIDAE
FACELINIDAE
AEOLIDIIDAE
MYRRHINIDAE
SPURILLIDAE

Subclass Gymnomorpha
Order Systellommatophora
ONCHIDIOIDEA
ONCHIDIIDAE
Order Soleolifera
VERONICELLIDAE
RATHOUSIIDAE
RHODOPIDAE

Subclass Pulmonata
Order Archaeopulmonata
ELLOBIOIDEA
ELLOBIIDAE
OTINIDAE
Order Basommatophora
SIPHONARIOIDEA
TRIMUSCULIDAE
SIPHONARIIDAE
AMPHIBOLOIDEA
AMPHIBOLIDAE
CHILINOIDEA
CHILINIDAE
LATIIDAE
ACROLOXOIDEA
ACROLOXIDAE
LYMNOIDEA
LYMNAEIDAE
LANCIDAE

SYSTEMATIC ARRANGEMENT OF THE HIGHER TAXA USED IN THIS OUTLINE

GLACIDORBOIDEA
GLACIDORBIDAE
PHYSOIDEA
PHYSIDAE
PLANORBOIDEA
PLANORBIDAE
BULINIDAE
PROTANCYLIDAE
PATELLOPLANORBIDAE
NEOPLANORBIDAE
ANCYLIDAE

Order Stylommatophora
Suborder Orthurethra
ACHATINELLOIDEA
ACHATINELLIDAE
COCHLICOPOIDEA
COCHLICOPIDAE
AMASTRIDAE
PUPILLOIDEA
PUPILLIDAE
VALLONIIDAE
VERTIGINIDAE
ORCULIDAE
PLEURODISCIDAE
STROBILOPSIDAE
PYRAMIDULIDAE
CHONDRINIDAE
BULIMINOIDEA
BULIMINIDAE
CERASTUIDAE
Suborder Mesurethra
CLAUSILIOIDEA
CLAUSILIIDAE
Suborder Sigmurethra
PARTULOIDEA
PARTULIDAE
ORTHALICOIDEA
ORTHALICIDAE
ODONTOSTOMIDAE
AMPHIBULIMIDAE
URICOPTIDAE
CERIONIDAE
MEGASPIRIDAE (?)
COELOCIIDAE (?)
ACHATINOIDEA
FERUSSACIIDAE
SUBULINIDAE
ACHATINIDAE
COELAXIDAE
THYROPHORELLIDAE (?)
AILLYOIDEA
AILLYIDAE
OLEACINOIDEA
SPIRAXIDAE
OLEACINIDAE
TESTACELLIDAE
STREPTAXOIDEA
STREPTAXIDAE
STROPHOCHEILOIDEA
STROPHOCHEILIDAE
MEGALOBULIMIDAE
DORCASIIDAE
ACAVOIDEA
ACAVIDAE
CARYODIDAE
MACROCYCLIDAE
MEGOMPHICIDAE
RHYTIDOIDEA
RHYTIDIDAE
CHLAMYDEPHORIDAE
SYSTROPHIIDAE
HAPLOTREMATIDAE

PLECTOPYLIDOIDEA
SCULPTARIIDAE
PLECTOPYLIDIDAE
PUNCTOIDEA
ENDODONTIDAE
PUNCTIDAE
HELICODISCIDAE
CHAROPIDAE
OTOCONCHIDAE
DISCIDAE
OREOHELICIDAE
OOPELTIDAE
Suborder Elasmognatha
SUCCINEOIDEA
SUCCINEIDAE
ATHORACOPHOROIDEA
ATHORACOPHORIDAE
SAGDOIDEA
SAGDIDAE
GASTRODONTOIDEA
GASTRODONTIDAE
HELIXARIONOIDEA
EUCONULIDAE
HELIXARIONIDAE
ARIOPHANTIDAE
TROCHOMORPHIDAE
UROCYCLIDAE
CYSTOPELTIDAE
VITRINOIDEA
VITRINIDAE
ZONITIDAE
DAUDEBARDIIDAE
PARMACELLIDAE
MILACIDAE
LIMACOIDEA
LIMACIDAE
BOETTGERILLIDAE
AGRIOLIMACIDAE
TRIGONOCHLAMYDOIDEA
TRIGONOCHLAMYDIDAE
POLYGYROIDEA
POLYGYRIDAE
THYSANOPHORIDAE
CAMAENOIDEA
SOLAROPSIDAE
CAMAENIDAE
HELICOIDEA
SPHINCTEROCHILIDAE
HYGROMIIDAE
HELICIDAE
HELICELLIDAE
BRADYBAENIDAE
XANTHONYCHIDAE
ARIONOIDEA
ARIONIDAE
PHILOMYCIDAE

Class Cephalopoda
Subclass Nautiloidea
Order Nautilida
NAUTILIDAE

Subclass Coleoidea
Order Sepiida
SPIRULIDAE
SEPIIDAE
SEPIOLIDAE
SEPIADARIIDAE
IDIOSEPIIDAE
Order Teuthida
Suborder Myopsida
LOLIGINIDAE
PICKFORDIATEUTHIDAE

Suborder Oegopsida
LYCOTEUTHIDAE
ENOPLOTEUTHIDAE
OCTOPOTEUTHIDAE
ONYCHOTEUTHIDAE
GONATIDAE
CTENOPTERYGIDAE
BATHYTEUTHIDAE
HISTIOTEUTHIDAE
PSYCHROTEUTHIDAE
ARCHITEUTHIDAE
NEOTEUTHIDAE
WALVISTEUTHIDAE
BRACHIOTEUTHIDAE
OMMASTREPHIDAE
THYSANOTEUTHIDAE
LEPIDOTEUTHIDAE
BATOTEUTHIDAE
CYCLOTEUTHIDAE
CHIROTEUTHIDAE
MASTIGOTEUTHIDAE
JOUBINITEUTHIDAE
PROMACHOTEUTHIDAE
GRIMALDITEUTHIDAE
CRANCHIIDAE
Order Vampyromorpha
VAMPYROTEUTHIDAE
Order Octopoda
Suborder Cirrata
CIRROTEUTHIDAE
OPISTHOTEUTHIDAE
Suborder Incirrata
BOLITAENOIDEA
BOLITAENIDAE
AMPHITRETIDAE
IDIOCTOPODIDAE
OCTOPODOIDEA
OCTOPODIDAE
VITRELEDONELLIDAE
LAEMOTEUTHIDAE ?
ARGONAUTOIDEA
ARGONAUTIDAE
ALLOPOSIDAE
OCYTHOIDAE
TREMOCTOPODIDAE

Class Bivalvia
Subclass Protobranchia
Order Solemyoida
SOLEMYOIDEA
SOLEMYIDAE
NUCINELLOIDCEA
NUCINELLIDAE
Order Nuculoida
NUCULOIDEA
NUCULIDAE
PRISTOGLOMIDAE
NUCULANOIDEA
TINDARIIDAE
LAMETILIDAE
RADIIDENTIDAE
NUCULANIDAE
YOLDIIDAE
SILICULIDAE
MALLETIDAE
SATURNIIDAE
Subclass Pteriomorphia
Order Arcoida
ARCOIDEA
ARCIDAE
NOETIIDAE
PARALLELODONTIDAE
CUCULAEIDAE

SYSTEMATIC ARRANGEMENT OF THE HIGHER TAXA USED IN THIS OUTLINE

LIMOPSOIDEA
LIMOPSIDAE
GLYCYMERIDIDAE
PHYLOBRYIDAE
Order Mytiloida
MYTILOIDEA
MYTILIDAE
Order Pterioida
Suborder Pteriina
PTERIOIDEA
PTERIIDAE
MALLEIDAE
ISOGNOMONIDAE
PULVINITIDAE
Suborder Pinnina
PINNOIDEA
PINNIDAE
Order Limoida
LIMOIDEA
LIMIDAE
Order Ostreoida
Suborder Ostreina
OSTREOIDEA
OSTREIDAE
GRYPHAEIDAE
PLICATULOIDEA
PLICATULIDAE
DIMYIDAE
Suborder Pectinina
PECTINOIDEA
PECTINIDAE
SYNCYCLONEMIDAE
PROPEAMUSSIIDAE
SPONDYLIDAE
ANOMIOIDEA
ANOMIIDAE
PLACUNIDAE

Subclass Paleoheterodonta
Order Trigoinoida
TRIGONIOIDEA
TRIGONIIDAE
Order Unionoida
UNIONOIDEA
MARGARITIFERIDAE
UNIONIDAE
HYRIIDAE
MUTELOIDEA
MUTELIDAE
MYCETOPODIDAE
ETHERIDAE

Subclass Heterodonta
Order Veneroida
LUCINOIDEA
LUCINIDAE
FIMBRIIDAE
MACTROMYIDAE

LUCINOIDEA (Cont'd.)
THYASIRIDAE
UNGULINIDAE
CYRENOIDIDAE
GALEOMMATOIDEA
GALEOMMATIDAE
KELLIIDAE
LASAEIDAE
LEPTONIDAE
MONTACUTIDAE
CYAMIOIDEA
CYAMIIDAE
PERRIERINIDAE
SPORTELLIDAE
NEOLEPTONIDAE
GALATHEAVALVIDAE
BERNARDINIDAE
CHLAMYDOCONCHOIDEA
CHLAMYDOCONCHIDAE
CARDITOIDEA
CARDITIDAE
CONDYLOCARDIIDAE
CHAMOIDEA
CHAMIDAE
CRASSATELLOIDEA
CRASSATELLIDAE
ASTARTOIDEA
ASTARTIDAE
CARDINIIDAE
CARDIOIDEA
CARDIIDAE
HEMIDONACIDAE
LYMNOCARDIIDAE
TRIDACNOIDEA
TRIDACNIDAE
MACTROIDEA
MACTRIDAE
ANATINELLIDAE
CARDILIIDAE
MESODESMATIDAE
SOLENOIDEA
SOLENIDAE
CULTELLIDAE
TELLINOIDEA
TELLINIDAE
SEMELIDAE
PSAMMOBIIDAE
DONACIDAE
DREISSENOIDEA
DREISSENIDAE
GAIMARDIOIDEA
GAIMARDIIDAE
ARCTICOIDEA
ARCTICIDAE
TRAPEZIIDAE
KELLIELLIDAE
GLOSSOIDEA
GLOSSIDAE
VESICOMYIDAE

CORBICULOIDEA
CORBICULIDAE
PISIDIIDAE
VENEROIDEA
VENERIDAE
GLAUCONOMIDAE
TURTONIIDAE
COOPERELLIDAE
PETRICOLIDAE
Order Myoida
Suborder Myina
MYOIDEA
MYIDAE
CORBULIDAE
ERODONIDAE
SPHENIOPSIDAE
GASTROCHAENOIDEA
GASTROCHAENIDAE
HIATELLOIDEA
HIATELLIDAE
PHOLADOIDEA
PHOLADIDAE
XYLOPHAGAIDAE ?
TEREDINIDAE

Subclass Anomalodesmata
Order Pholadomyoida
PHOLADOMYOIDEA
PHOLADOMYIDAE
PARILIMYIDAE
PANDOROIDEA
THRACIIDAE
LATERNULIDAE
PERIPLOMATIDAE
LYONSIIDAE
PANDORIDAE
MYOCHAMIDAE
CLEIDOTHAERIDAE
CLAVAGELLOIDEA
CLAVAGELLIDAE
POROMYOIDEA
POROMYIDAE
CUSPIDARIIDAE
VERTICORDIIDAE

Class Scaphopoda
Order Dentaliida
DENTALIIDAE
LAEVIDENTALIIDAE
GADILINIDAE
OMNIGLYPTIDAE
Order Gadilida
ENTALINIDAE
PULSELLIDAE
SIPHONODENTALIIDAE
GADILIDAE
WEMERSONIELLIDAE

References: (Classification, Systematics)

Boss, K. J. "Mollusca" in S. Parker "Synopsis and Classification of Living Organisms" McGraw-Hill, New York pp.946-1166 - 1982

Golikov, A. N & Y. I. Starobogatov "Systematics of Prosobranch Gastropods" Malacologia 15(1):185-232 - 1975

Haszprunar, G. "The Heterobranchia; new concept of the phylogeny of the higher Gastropoda" Z.Zool.Syst.Evolut.forsch.23(1):15-37 -1985

Salvini-Plawen, L."Proposed Classification of : Mollusca" Malacologia 19(2):249-278 - 1980

---- & G. Haszprunar "The Vetigastropoda and the systematics of streptoneurous Gastropoda" J.Zool.Lond.211:747-770 - 1986

Taylor, D. W. & N. F. Sohl "An Outline of Gastropod Classification" Malacologia 1(1):7-32 - 1962

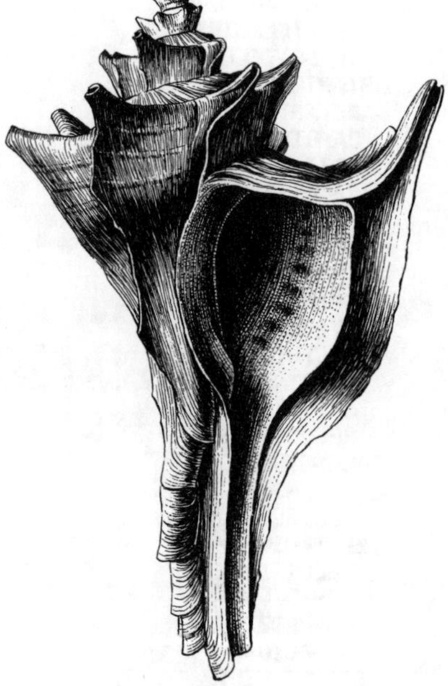

A CLASSIFICATION OF THE LIVING MOLLUSCA

(Listing of Supraspecific Names)

PHYLUM MOLLUSCA

Class Aplacophora
Subclass Chaetodermomorpha (Caudofoveata) (8)

LIMIFOSSORIDAE
LIMIFOSSOR Heath, 1904
METACHAETODERMA Thiele, 1913
PSILODENS Salvini-Plawen, 1977
SCUTOPUS Salvini-Plawen, 1968

PROCHAETODERMATIDAE (13)
PROCHAETODERMA Thiele, 1902
CHEVRODERMA Scheltema, 1985
SPATHODERMA Scheltema, 1985

CHAETODERMATIDAE
CHAETODERMA Loven, 1844
Crystallophrisson Mobius, 1874
FALCIDENS Salvini-Plawen, 1968

Subclass Solenogastres
Order Pholidoskepia

DONDERSIIDAE
DONDERSIA Hubrecht, 1888
HEATHIA Thiele, 1913 ?
Ichythomenia auctt.
(non Pilsbry, 1898)
ICHTHYOMENIA Pilsbry, 1898 ?
LYRATOHERPIA Salvini-Plawen, 1978
MICROMENIA LeLoup, 1948
Rupertomenia Schwabl, 1955
NEMATOMENIA Simroth, 1893
Myzomenia Simroth, 1893
Herpomenia Heath, 1911
STYLOMENIA Pruvot, 1899

SANDALOMENIIDAE
SANDALOMENIA Thiele, 1913

GYMNOMENIIDAE (Wireniidae)
GYMNOMENIA Odhner, 1919
GENITOCONIA Salvini-Plawen, 1967
WIRENIA Odhner, 1919

LEPIDOMENIIDAE
LEPIDOMENIA Kowalevsky, 1883
NIERSTRASZIA Heath, 1918

MACELLOMENIIDAE
MACELLOMENIA Simroth, 1893 ?

MEIOMENIIDAE
MEIOMENIA Morse, 1979
MEIOHERPIA Salvini-Plawen, 1985

Order Neomeniamorpha

NEOMENIIDAE
NEOMENIA Tullberg, 1875
HEATHIMENIA Salvini-Plawen, 1967

HEMIMENIIDAE
HEMIMENIA Nierstrasz, 1902
ARCHAEOMENIA Thiele, 1906

Unassigned genera:
PHOLIDOHERPIA Salvini-Plawen, 1978
NOTOMENIA Thiele, 1897

Order Sterrofustia

PHYLLOMENIIDAE
PHYLLOMENIA Thiele, 1913
HARPAGOHERPIA Salvini-Plawen, 1978
LITUIHERPIA Salvini-Plawen, 1978
OCHEYOHERPIA Salvini-Plawen, 1978

IMEROHERPIIDAE
IMEROHERPIA Salvini-Plawen, 1978

HETEROHERPIIDAE
HETEROHERPIA Salvini-Plawen, 1978
RHABDOHERPIA Salvini-Plawen, 1978

Order Cavibelonia

PARARRHOPALIIDAE
ELEUTHEROMENIINAE
ELEUTHEROMENIA Salvini-Plawen, 1967
GEPHYROHERPIA Salvini-Plawen, 1978

PARARRHOPALIINAE
PARARRHOPALIA Simroth, 1893
LABIDOHERPIA Salvini-Plawen, 1978
Pruvotina Cockerell, 1903 (part)
Eleutheromenia Salvini-Plawen (part)
PRUVOTINA Cockerell, 1903
Pro Paramenia Pruvot (part)
Perimenia Nierstrasz (part)

LOPHOMENIINAE
LOPHOMENIA Heath, 1911
FORCEPIMENIA Salvini-Plawen, 1968
HYPOMENIA Van Lummel, 1930
METAMENIA Thiele, 1913

HALOMENIINAE ?
HALOMENIA Heath, 1911

RHOPALOMENIIDAE
RHOPALOMENIA Simroth, 1893
Proneomenia Hubrecht, 1880 (part)
Entonomenia LeLoup, 1948
DINOMENIA Nierstrasz, 1902
DRIOMENIA Heath, 1911
PRUVOTIA Thiele, 1894

ACANTHOMENIIDAE
ACANTHOMENIA Thiele, 1913

AMPHIMENIIDAE
AMPHIMENIA Thiele, 1894
ALEXANDROMENIA Heath, 1911
MEROMENIA LeLoup, 1949
PACHYMENIA Heath, 1911
PARAGYMNOMENIA LeLoup, 1947
PLATHYMENIA Schwabl, 1961
PROPARAMENIA Nierstrasz, 1902

AMPHIMENIDAE (Cont'd.)
SPENGELOMENIA Heath, 1912
SPUTOHERPIA Salvini-Plawen, 1978
UTRALVOHERPIA Salvini-Plawen, 1979

SIMROTHIELLIDAE
SIMROTHIELLA Pilsbry, 1898
Solenopus Sars,1869 (part)
(non Shönherr, 1826
= Neomenia Tullberg)
Kruppomenia Nierstrasz, 1903
BIRASOHERPIA Salvini-Plawen, 1978
BISERRAMENIA Salvini-Plawen, 1967
SIALOHERPIA Salvini-Plawen, 1978
UNCIMENIA Nierstrasz, 1903

DREPANOMENIIDAE
DREPANOMENIA Heath, 1911

STROPHOMENIIDAE
STROPHOMENIA Pruvot, 1899
ANAMENIA Nierstrasz, 1908

PRONEOMENIIDAE
PRONEOMENIA Hubrecht, 1880
Rhopalomenia Simroth (part)
DORYMENIA Heath, 1911
Proneomenia Hubrecht (part)

EPIMENIIDAE
EPIMENIA Nierstrasz, 1908
Proneomenia Hubrecht (part)
Dinomenia Nierstrasz (part)

SYNGENOHERPIIDAE
SYNGENOHERPIA Salvini-Plawen, 1978

RHIPIDOHERPIIDAE
RHIPIDOHERPIA Salvini-Plawen, 1978

References:

Boss, K. J. in S. Parker (Ed.) "Synopsis and Classification of living Organisms" Mollusca, I. McGraw-Hill, N.Y. - 1982 (1)

Heath, H. "Solenogastres of the tropical Pacific" Mem.Mus.Comp.Zool.Harv.Coll. 45(1):1-182 - 1911 (2)

---- "Solenogastres of the East Coast of North America" Mem.Mus.Comp.Zool. Harvard Coll. 45(2):183-263 - 1918 (3)

LeLoup, "Solenogastres des campagnes de Prince Albert I. of Monaco" 1950 (4)

Nierstrasz, H. "The Solenogastres of the Siboga Expedition" Siboga-Expiditie Mon. 47:1-46 - 1902 (5)

Ponder, W.F. "A new Aplacophoran from New Zealand" Jour.Malac.Soc.Aust. 2(1):47-54 - 1970 (6)

Salvini-Plawen, L. von "Die Caudofoveata" Fifth Eur.Biolog.Symposium 4:27-51 - 1972 (7)

---- "Uber Lebendbeobachtungen an Caudofoveata, nebst Bermerkungen zum system der Klasse" Sarsia 31:105-126 - 1968 (8)

---- Sarsia 31:131-142 (Solenogastres) - 1968 (8a)

---- "Solenogastres and Caudofoveata; organisation und phlogenetische bedeuntung" Malacologia 9:191-216 - 1968 (9)

---- Mollusca-Caudofoveata "Marine Invertebrates of Scandinavia" 4:1-54 - 1975 (10)

---- "Antarctic and Subantarctic Solenogastres" Zoologica. Stuttgart 128:1-315 - 1978 (11)

Scheltema, A. H."Two new species of Chaetoderma from off West Africa" J.Moll.Stud. 42:223-234 - 1976 (12)

---- "The aplacophoran family Prochaetodermatidae in the North American Basin, including Chevroderma and Spathoderma, new genus" Biol.Bull.169:484-529 - 1985 (13)

Schwabl "Solenogastres" Pac.Sc. 17(3):261-281 - 1963 (14)

Thiele, J. "Solengastres" Das Tierreich (15) Berlin: Friedlander & Sohn, 55pp. - 1913

Class Polyplacophora (Placophora)
Order Neoloricata
Suborder Lepidopleurina

LEPTOCHITONIDAE
LEPTOCHITONINAE
LEPTOCHITONINA Strack, 1982
LEPIDOPLEURUS Leach in Risso, 1826
(non Carpenter in Dall, 1879)
LEPTOCHITON Gray, 1847
Deshayesiella Carpenter in Dall, 1879
Lophyropsis Thiele in Troschel, 1893
Beanella Thiele in Troschel, 1893
(non Dall, 1882)
Terenochiton Iredale, 1914
Xiphiozona Berry, 1919

LEPTOCHITON (Cont'd.)
PILSBRYELLA Nierstrasz, 1905
PARACHITON Thiele, 1909
OLDROYDIA Dall, 1894
HANLEYELLA Sirenko, 1973

HANLEYIDAE
HANLEYA Gray, 1857
Hanleia Carpenter, 1873 (nom.null.)
Laminoplax Ferreira, 1981
HEMIARTHRUM Carpenter in Dall, 1876

AFOSSOCHITONIDAE
CHORIPLAX Pilsbry, 1894
Microplax H. Adams & Angas, 1864

Suborder Ischnochitonina

ISCHNOCHITONIDAE

ISCHNOCHITONINAE

ISCHNOCHITON Gray, 1847
- Radsiella Thiele in Troschel, 1893 (non Pilsbry, 1892)
- Lophyriscus Thiele in Troschel, 1893
- Leptopleura Thiele in Troschel, 1893
- Rhodoplax Thiele in Troschel, 1893
- Strigichiton Hull, 1923
- Levicoplax Iredale & Hull, 1925
- Isochiton Ashby & Cotton, 1934
- Ovatoplax Cotton & Weeding, 1939

ISCHNOCHITON ss.
- Radsiella Pilsbry, 1892 (non Thiele, 1893)
- Anisoradsia Iredale & May, 1916
- Diktuonus Ashby, 1931

SIMPLISCHNOCHITON Van Belle, 1974

STENOSEMUS Middendorff, 1847
- Lophyrus G. O. Sars, 1878 (non Poli, 1791 - nom.null.)
- Chondropleura Thiele, 1906
- Lepidopleuroides Thiele, 1928
- Lophyrochiton Jakovleva, 1952

ISCHNORADSIA Shuttleworth, 1853 (non Carpenter in Dall, 1879)
- Lepidoradsia Carpenter in Dall, 1879

HETEROZONA Carpenter in Dall, 1879

HAPLOPLAX Pilsbry, 1894
- Chartoplax Iredale & Hull, 1924

AUTOCHITON Iredale & Hull, 1924
- Euporoplax Iredale & Hull, 1924
- Euretoplax Iredale & Hull, 1924

ISCHNOPLAX Carpenter in Dall, 1879 (2)
- Stereoplax Thiele in Troschel, 1893

STENOCHITON H. Adams & Angas, 1864
- Zostericola Ashby, 1919

STENOPLAX Carpenter in Dall, 1879

STENORADSIA Carpenter in Dall, 1879
- Maugerella Carpenter in Dall, 1879

LEPIDOZONA Pilsbry, 1892
- Lepidopleurus Carpenter in Dall, 1879 (non Leach in Risso 1826)
- Solivaga Iredale & Hull, 1925

TRIPOPLAX Berry, 1919
- Ischnoradsia Carpenter in Dall, 1879 (non Shuttleworth, 1853)
- Rhombochiton Berry, 1919
- Gurjanovillia Jakovleva, 1952
- Albrechtia Is. Taki, 1955

CONNEXOCHITON Kaas, 1979

CALLISTOPLACINAE

CALLISTOCHITON Carpenter in Dall, 1879
- Lophochiton Ashby, 1923 (non Berry, 1925)
- Callistelasma Iredale & Hull, 1925
- Callistassecla Iredale & Hull, 1925

CALLISTOPLAX Carpenter in Dall, 1879

CERATOZONA Dall, 1882
- Ceratophorus Carpenter in Dall, 1879 (nom.nud.) (non Shuckard, 1837, nor Gistl, 1848)
- Newcombia Carpenter in Dall, 1879 (nom.nud.)

CHAETOPLEURINAE

CHAETOPLEURA Shuttleworth, 1853 (2)
- Choetopleura Shuttleworth, 1856 (n.nl.)
- Lepidochiton Carpenter, 1857 (non Lepidochitona Gray, 1821)
- Acanthopleura Gray, 1857 (non Guilding, 1829)
- Pallochiton Dall, 1879
- Hemphillia Carpenter in Dall, 1879 (nom.nud.) (non Bland & Binney, 1872)
- Arthuria Carpenter in Dall, 1879 (n.nd.)
- Helioradsia Thiele in Troschel, 1893
- Helicoradsia Simroth, 1894 (nom.null.)
- Variolepis Plate, 1899
- Calloplax Thiele, 1909 ?
- Typhlochiton Dall, 1921
- Chetopleura Ashley, 1939 (nom.null.)

DINOPLAX Carpenter in Dall, 1882

NUTTALLOCHITON Plate, 1899
- Notochiton Thiele, 1906

CALLOCHITONINAE

CALLOCHITON Gray, 1847
- Clathropleura Tiberi, 1877 (non Thiele, 1893)
- Trachyradsia Carpenter in Dall, 1878
- Stereochiton Carpenter in Dall, 1882 (nom.nud., 1879)
- Icoplax Thiele in Troschel, 1893
- Scrobicoplax Iredale & Hull, 1929
- Paricoplax Iredale & Hull, 1929 ?
- Quaestiplax Iredale & Hull, 1929
- Ocellochiton Ashby, 1939

ACUTOPLAX Cotton & Weeding, 1939

EUDOXOCHITON Shuttleworth, 1853

EUDOXOPLAX Iredale & May, 1916

LEPIDOCHITONINAE

LEPIDOCHITONA Gray, 1821 (non Lepidochiton Carpenter, 1857)
- Trachydermon Carpenter, 1864
- Craspedochilus G. O. Sars, 1878
- Dawsonia Carpenter in Dall, 1879 (n.nd.) (non Hartt, 1868)
- Beania Carpenter in Dall, 1879 (n.nd.) (non Johnston, 1840)
- Middendorffia Carpenter in Dall, 1882
- pro Middendorfia Dall, 1879 & 1882
- Beanella Dall, 1882
- Cyanoplax Pilsbry, 1892
- Leptochitona Pilsbry, 1893 (nom.null.)
- Adriella Thiele in Troschel, 1893
- Mopaliopsis Thiele in Troschel, 1893
- Mopaliella Thiele, 1909
- Ploiochiton Berry, 1926
- pro Lophochiton Berry, 1925 (non Ashby, 1923)
- Trachochiton Risbec, 1946

SPONGIORADSIA Pilsbry, 1894

DENDROCHITON Berry, 1911
- Basiliochiton Berry, 1918

NUTTALLINA Carpenter in Dall, 1871

TONICELLA Carpenter, 1873
- Boreochiton G. O. Sars, 1878

TONICINA Thiele, 1906

SUBTERENOCHITON Iredale & Hull, 1924

JUVENICHITON Sirenko, 1975
- Micichiton Sirenko, 1975
- Nanichiton Sirenko, 1975

SCHIZOPLACINAE

SCHIZOPLAX Dall, 1878

MOPALIIDAE
MOPALIA Gray, 1847
Osteochiton Dall, 1886
PLAXIPHORA Gray, 1847
Euplaxiphora Shuttleworth, 1853
Guildingia Carpenter in Dall, 1882
Diaphoroplax, Maorichiton Iredale, 1914
Poneroplax Iredale, 1914
Semimopalia Dall, 1919
Aerilamma Hull, 1924
Vaferichiton Iredale & Hull, 1932
Hachijomopalia Is. Taki, 1954
FREMBLYA H. Adams, 1866
Streptochiton Carpenter in Dall, 1879 (nom.nud.)
Kopionella Ashby, 1919
MERCATORA Leloup, 1942
KATHARINA Gray, 1847
AMICULA Gray, 1847
Cryptochiton Gray, 1847 (non Middendorff, 1847)
Symmetrogephyrus Middendorff, 1847
Stimpsoniella Carpenter, 1873
Chlamydochiton Dall, 1878
Chlamydoconcha Pilsbry, 1893 (non Dall, 1884)
PLACIPHORELLA Carpenter in Dall, 1879
Langfordiella Dall, 1925
PLACOPHOROPSIS Pilsbry, 1893
Euplacophora Verrill & Smith, 1882 (nom.null.) pro Plaxiphora Gray, 1847
? Trochodochiton Rochebrune, 1884

SCHIZOCHITONIDAE Dall, 1889
SCHIZOCHITON Gray, 1847
LORICA H. & A. Adams, 1852
Aulacochiton Shuttleworth, 1853
Zelorica Finlay, 1927
LORICELLA Pilsbry, 1893
Squamophora Nierstrasz, 1905
Componochiton Milne, 1963

CHITONIDAE
CHITONINAE
CHITON Linnaeus, 1758
Scutigerulus Meuschen, 1787 (nom.nud.)
Trachyodon Dall, 1892
Amaurochiton, Chondroplax, Diochiton all Thiele in Troschel, 1893
Poeciloplax Thiele in Troschel, 1893
Sypharochiton Thiele in Troschel, 1893
Triboplax Thiele in Troschel, 1893
Georgus Thiele in Troschel, 1893
Delicatoplax Iredale & Hull, 1926
Tegulaplax Iredale & Hull, 1926
RADSIA Gray, 1847
RHYSSOPLAX Thiele in Troschel, 1893
Clathropleura Thiele in Troschel, 1893 (non Tiberi, 1877)
Anthochiton Thiele in Troschel, 1893
MUCROSQUAMA Iredale & Hull, 1926

ACANTHOPLEURINAE
ACANTHOPLEURA Guilding, 1829
Corephium Gray, 1847
Maugeria Gray, 1857
Francisia Carpenter in Dall, 1882 (nom.nud., 1879)
Rhopalopleura Thiele in Troschel, 1893
Amphitomura, Mesotomura Pilsbry, 1893
Acanthozostera Iredale & Hull, 1926
Planispina Iw. Taki in Is. Taki, 1962

ACANTHOPLEURINAE (Cont'd.)
ENOPLOCHITON Gray, 1847
LIOLOPHURA Pilsbry, 1893
CLAVARIZONA Hull, 1923
SQUAMOPLEURA Nierstrasz, 1905
Sclerochiton Carpenter in Dall, 1882

TONICIINAE
TONICIA Gray, 1847
Tonichia Gray, 1840 (nom.nud.)
Fannyia Gray, 1857 (non Fannia Robineau-Desvoidy, 1830)
Fannettia Dall, 1882
LUCILINA Dall, 1882
Lucia Gould, 1862
Toniciopsis Thiele in Troschel, 1893
ONITHOCHITON Gray, 1847
Pristochiton Clessin, 1904
Onithoplax Thiele, 1910
Onithella Mackay, 1933
Nodiplax Beau, 1967

Suborder Acanthochitonina

CRYPTOPLACIDAE
CRYPTOPLAX Blainville, 1818
Chitonellus Lamarck, 1819 (non Blainville, 1825)
Oscabrella Broderip, 1836
Chitoniscus Herrmannsen, 1846 (non Carpenter in Dall, 1882)
Phaenochiton Middendorff, 1847
Dichachiton Middendorff, 1847
Ametrogephyrus Middendorff, 1847

ACANTHOCHITONIDAE
ACANTHOCHITONINAE
CRYPTOCONCHUS Blainville in Burrow, 1815
ACANTHOCHITONA Gray, 1821
Gymnoplax Gray, 1821
Chitonellus Blainville, 1825 (non Lamarck, 1819)
Acanthochites Leach in Risso, 1826
Phakellopleura Guilding, 1829 (non Phacellopleura Carpenter in Dall, 1879)
Hamachiton Middendorff, 1847
Platysemus Middendorff, 1847
Strechochiton A. Adams in Tapparone-Canefri, 1874
Stectoplax Carpenter in Dall, 1882 (nom.nud., 1879)
Meturoplax Pilsbry, 1894
CRASPEDOCHITON Shuttleworth, 1853
Angasia Carpenter in Dall, 1882 (non White, 1863, nor Bale, 1864)
Phacellozona Pilsbry, 1894
Craspedoplax Iredale & Hull, 1925
Amblyplax Ashby, 1926
Lophoplax Ashby, 1926
NOTOPLAX H. Adams, 1861
Phacellopleura Carpenter in Dall, 1879 (non Phakellopleura Guilding, 1829)
Macandrellus Carpenter in Dall, 1879
Leptoplax Carpenter in Dall, 1882 (nom.nud.)
Loboplax Pilsbry, 1893
Mecynoplax Thiele in Troschel, 1893
Aristochiton Thiele, 1909
Pseudotonicia Ashby, 1928
Ikedaella Is. & Iw. Taki, 1929
Crocochiton Cotton & Weeding, 1939

NOTOPLAX (Cont'd.)

SPONGIOCHITON Carpenter in Dall, 1882
Thaumastochiton Thiele, 1909
BASSETHULLIA Pilsbry, 1928
Glyptelasma Iredale & Hull, 1925 (non Pilsbry, 1907)

CHONEPLAX Carpenter in Dall, 1882
Chitoniscus Carpenter in Dall, 1882 (nom.nud.) (non Herrmannsen, 1846)

CRYPTOCHITONINAE
CRYPTOCHITON von Middendorff, 1847 (non Gray, 1847)

References:

Van Belle, R. A. "Systematic Classification of the Chitons" in "Informations" de la Societe Belge de Malacologie, Serie 11(1-3):1-178 1983 (1)

---- Addenda I "Informations" Serie 13:2 - 1985

Ferreira, A. J. "The genus Chaetopleura" Veliger 25(3):203-224 - 1983 (2)

Class Monoplacophora
Order Tryblidiida

TRYBLIDIOIDEA
NEOPILINIDAE (2)
NEOPILINA Lemche, 1957
LEMCHEPHILA Moskalev, Starobogatov & Filatove, 1983 (5)
Lemchephyala auctt. (?)

VEMIDAE
VEMA Clark & Menzies, 1959

MONOPLACOPHORIDAE
MONOPLACOPHORUS Moskalev, Starobogatov & Filatova, 1983

LAEVIPILINIDAE
LAEVIPILINA McLean, 1979

Class Gastropoda
Subclass Prosobranchia
Order Archaeogastropoda

PLEUROTOMARIOIDEA
PLEUROTOMARIIDAE
ENTEMNOTROCHUS Fischer, 1885
MIKADOTROCHUS Lindholm, 1927
PEROTROCHUS Fischer, 1885

SCISSURELLIDAE (1)
SCISSURELLA Orbigny, 1824
Schismope Jeffreys, 1856
Woodwardia Crosse & Fischer, 1861
SCISSURONA Iredale, 1924
ANATOMA Woodward, 1859 (3a)
Schizotrochus Monterosato, 1877
INCISURA Hedley, 1904

SCISSURELLIDAE (Cont'd.)
SINEZONA Finlay, 1927
Schismope auctt. (non Jeffreys)
Woodwardia auctt.(non Crosse & Fischer)
SUKASHITROCHUS Habe & Kosuge, 1964

HALIOTIDAE (4, 6)
HALIOTIS Linnaeus, 1758
Teinotus H. & A. Adams, 1854
Haleotis Binkhorst, 1861
Tinotis Fischer, 1885
EUHALIOTIS Wenz, 1938
EUROTIS Habe & Kosuge, 1964
EXOHALIOTIS Cotton & Godfrey, 1933
MARINAURIS Iredale, 1927
NORDOTIS Habe & Kosuge, 1964
NOTOHALIOTIS Cotton & Godfrey, 1933
OVINOTIS Cotton, 1943
PADOLLUS Montfort, 1810
Neohaliotis Cotton & Godfrey, 1933
PAUA Fleming, 1952
SANHALIOTIS Iredale, 1929
SCHISMOTIS Gray, 1856
SULCULUS H. & A. Adams, 1854
USAHALIOTIS Habe & Kosuge, 1964

Suborder Euomphalina (3)
NEOMPHALOIDEA
NEOMPHALIDAE (3)
NEOMPHALUS McLean, 1981

LEPETODRILOIDEA (Placement ?) (3b)
LEPETODRILIDAE
LEPETODRILUS McLean, 1988

GORGOLEPTIDAE
GORGOLEPTIS McLean, 1988

References:
McLean, J. H. "West American Scissurellidae" Veliger 9(4):404-410 - 1967 (1)

---- "New monoplacophoran limpet: Continental shelf off Southern California" Cont.Sci.Nat.Hist.Mus. Los Angeles County 307:1-19 - 1979 (2)

---- "The Galapagos Rift limpet Neomphalus" Malacologia, 21(1-2)291-336 - 1981 (3)

---- "New Species: N.E. Pacific Archaeogastropoda" Veliger 26(3):233-239 - 1984 (3a)

---- "New Archaeogastropod limpets from hydrothermal vents: Lepetodrilacea" Phil.Trans.R.Soc.Lond.B.319:1-32 - 1988 (3b)

Moore, R. C. (Ed.) "Treatise on Invertebrate Paleontology" I.Mollusca 1:221-223 -1960 (4)

Moskalev, L I., Y.A. Starobogatov & Z.A. Filatova "New data: abyssal Monoplacophora" Zool.Zh. 62(4):981-996 Journ.Acad.Sci.USSR - 1983 (5)

Pickery, R. "Haliotidae" Gloria Maris 19(2):23-32 - 1980 (6)

FISSURELLOIDEA

FISSURELLIDAE
EMARGINULINAE
EMARGINULA Lamarck, 1801
Emarginulus Montfort, 1810
Imarginula Gray, 1821
SEMPERIA Crosse, 1867
SUBZEIDORA Iredale, 1924
CLYPIDINA Gray, 1847
EMARGINELLA Pilsbry, 1890
FISSURISEPTA Seguenza, 1853 (3)
HEMITOMA Swainson, 1840
Subemarginula Gray, 1847
(non Hagenow, 1851)
Siphonella Issel, 1869
MONTFORTIA Récluz, 1843
MONTFORTISTA Iredale, 1929
LAEVIEMARGINULA Habe in Kuroda, 1953
NESTA H. Adams, 1870
LAEVINESTA Pilsbry & McGinty, 1952
(Placement ?) (3)
NOTOMELLA Cotton, 1957
Entomella Cotton, 1945
(non Cossmann, 1888)
PUNCTURELLA Lowe, 1827
Sipho Brown, 1827
(non Fabricius, 1823)
Sypho Brown, 1833
Cremoria Gray, 1842
Cemoria Leach, 1852
(non Risso, 1826)
CRANOPSIS A. Adams, 1860
RIMULANAX Iredale, 1924 (2)
Kira Habe, 1951
RIXA Iredale, 1924
VACERRENA Iredale, 1958 (2)
pro Vacerra Iredale, 1924
(non Godman, 1900)
RIMULA Defrance, 1827
Rimularia Bronn, 1838
SCUTUS Montfort, 1810
Parmophorus Blainville, 1817
Parmaphora Bowdich, 1822
Parmophora Desmarest, 1859
Scutum Sowerby, 1842
(non Schumacher, 1817)
Aviscutum Iredale, 1940
NANNOSCUTUM Iredale, 1937
TUGALI Gray in Dieffenbach, 1843
Tugalia Gray, 1847
PARMOPHORIDEA Wenz, 1938
Parmophorella Strebel, 1907
(non Matthew, 1886)
TUGALINA Habe in Kuroda, 1953

EMARGINULINAE (Cont'd.)
ZEIDORA A. Adams, 1860 (Zidora Fischer, 1885; em.)
Crepiemarginula Seguenza, 1880
Legrandia Beddome, 1833
(non Hanley, 1872)

DIODORINAE
DIODORA Gray, 1821
Glyphis Carpenter, 1857
(non Agassiz, 1843)
Capiluna Gray, 1857
Monodilepas Finlay, 1927
AUSTROGLYPHIS Cotton & Godfrey, 1934
ELEGIDION Iredale, 1924
Eligidion Cotton & Godfrey, 1945
FISSURIDEA Swainson, 1840
MEGATHURA Pilsbry, 1890
Macrochasma Dall, 1915
STROMBOLI Berry, 1954

FISSURELLINAE
FISSURELLA Bruguière, 1798
Fissurellus Montfort, 1810
Balboaina Peréz-Farfante, 1943 (4b)
Balvoina, Balboina auctt. (err.)
Carcellesia Peréz-Farfante, 1952
Corinna Christiaens, 1973
CLYPIDELLA Swainson, 1840
CREMIDES H. & A. Adams, 1854
AMBLYCHILEPAS Pilsbry, 1890
Sophismalepas Iredale, 1924
COSMETALEPAS Iredale, 1924
Profissurellidea Wenz, 1938
DENDROFISSURELLA McLean & Kilburn, 1986
LUCAPINA Sowerby, 1835
Foraminella Guilding in Sowerby, 1835
MACROCHISMA Sowerby, 1839
Macroschisma Agassiz, 1846
DOLICHISCHISMA Iredale, 1940
Dolichoschisma Iredale, 1940 (err.)
FOROLEPAS Iredale, 1940
MEDUSAFISSURELLA McLean & Kilburn, 1986
MONTFORTULA Iredale, 1915
MONTFORTULANA Habe, 1961

FISSURELLIDINAE
FISSURELLIDEA Orbigny, 1841 (4a)
Megatebennus Pilsbry, 1890
LEUROLEPAS McLean, 1970 (4)
LUCAPINELLA Pilsbry, 1890
Chlamydoglyphis Pilsbry, 1890
PUPILLAEA Gray in Sowerby, 1835
Pupillia Gray, 1840
BUCHANANIA Lesson, 1830 (4a) (No shell-adults)
Ephrada Gistl, 1848

References:

Farfante, I.P. "Fissurella, Lucapina, Lucapinella: Western Atlantic" Johnsonia 1(10):9-20 - 1943
—— "The genus Diodora in the Western Atlantic" Johnsonia 1(11):1-20 - 1943 (1a)
—— "The genus Zeidora, Nesta, etc." Johnsonia 2(24):93-148 - 1947 (1b)

Herbert, D. G. & R. N. Kilburn "Taxonomic studies: Emarginulinae of South Africa and Mozambique" So.Afr.J.Zool.21(1):1-27 - 1986 (2)

Hickman, C. S. "Radular Patterns, Systematics, Diversity and Ecology of Deep-Sea Limpets" Veliger 26(2):73-92 - 1983 (3)

McLean, J. H. "Description of new genus; Eastern Pacific fissurellid" Veliger 12(3):362-367 - 1970 (4)
—— "Shell reduction and loss in the Fissurellidae: a review of the genera and species in Fissurellidea" Am.Malac.Bull.2 - 1984 (4a)
—— "Systematics of Fissurella in the Peruvian and Magellanic faunal provinces" - 1984 Cont.Sci.Nat.Hist.Mus.L.A.354:1-70 (4b)

—— & R.N. Kilburn "Propodial Elaboration in So. African and Indian Ocean Fissurellidae" (4c) Nat.Hist.Mus.LA.Co.Cont.Sci.379:1-12 -1986

PATELLOIDEA
ACMAEIDAE (6b)
ACMAEINAE
ACMAEA Eschscholtz, 1833
Radiacmea Iredale, 1915
Naccula Iredale, 1924
Actinoleuca Oliver, 1926
Asteracmea Oliver, 1926
Thalassacmea Oliver, 1926
Nomaeopelta Berry, 1958
Collisellacmaea Christiaens, 1973

PECTINODONTINAE
PECTINODONTA Dall, 1882

LOTTIIDAE (6b)
LOTTIINAE
Tribe Lottiini
LOTTIA Gray, 1833
Tecturella Carpenter, 1860
(non Stimpson, 1853)
Tecturina Carpenter, 1861 (nom.nud.)
Lecania Carpenter, 1866 (nom.nud.)
(non Macquart, 1839)
Collisella Dall, 1871
Conoidacmea Habe, 1944
Kikukozara Habe, 1944
Simplacmaea Christiaens, 1975
ERGINUS Jeffreys, 1877
? Problacmaea Golikov & Kussikin,1972 (6)
NIVEOTECTURA Habe, 1944
TECTURA Gray, 1847 (6b)
Atalacmea, Notoacmea,
Parvacmea all Iredale, 1915
Conacmea, Subacmea Oliver, 1926
? Testudinalia Moskalev, 1966
Tribe Scurriini (6c)
SCURRIA Gray, 1847
DISCURRIA Lindberg, 1988

PATELLOIDINAE (Oliver, 1926)
PATELLOIDA Quoy & Gaimard, 1834
Collisellina Dall, 1871
Chiazacmea Oliver, 1926
POTAMACMAEA Peile, 1922

RHODOPETALINAE (6a)
RHODOPETALA Dall, 1921

PATELLIDAE
PATELLINAE
PATELLA Linnaeus, 1758 (2)
Patellaria Gmelin, 1793
Patellus Montfort, 1810
Patellopsis Thiele in Troschel, 1891
(non Nobre, 1896)
Costatopatella Pallary, 1912
Granopatella Pallary, 1920
ANCISTROMESUS Dall, 1871
CYMBULA H. & A. Adams, 1854
ELIANA Christiaens, 1973
LAEVIPATELLA Pallary, 1920
OLANA H. & A. Adams, 1854
PATELLASTRA Monterosato, 1884
PATELLIDEA Thiele in Troschel, 1891
PATELLONA Thiele in Troschel, 1891
PENEPATELLA Iredale, 1929
SCUTELLASTRA H. & A. Adams, 1854
Patellanax Iredale, 1924
HELCION Montfort, 1810
ANSATES Sowerby, 1839
Patina Leach in Gray, 1847
PATINASTRA Thiele in Troschel, 1891

NACELLINAE
NACELLA Schumacher, 1817
PATINIGERA Dall, 1905
Patinella Dall, 1871
(non Gray, 1848)
CELLANA H. Adams, 1869
Helcioniscus Dall, 1871

LEPETIDAE (5)
LEPETA Gray, 1847
BATHYLEPETA Moskalev, 1977
IOTHIA Forbes, 1849 (Jothia)
Pilidium Forbes & Hanley, 1849
(non Müller, 1846)
LIMALEPETA Moskalev, 1978
MAORICRATER Dell, 1956
CRYPTOBRANCHIA Middendorff, 1851
Cryptoctenidia Dall, 1918
PROPILIDIUM Forbes and Hanley, 1849
Rostrisepta Seguenza, 1866
SAGAMILEPETA Okutani, 1987 (10)

LEPETELLOIDEA
LEPETELLIDAE (9b,c)
LEPETELLA Verrill, 1880
TECTICRATER Dell, 1956
TECTISUMEN Finlay, 1927

PSEUDOCOCCULINIDAE (7a)
PSEUDOCOCCULININAE
PSEUDOCOCCULINA Schepman, 1908
BANDABYSSIA Moskalev, 1976
KURILABYSSIA Moskalev, 1976
MESOPELEX Marshall, 1986
NOTOCRATER Finlay, 1927
Punctolepeta Habe, 1958
TENTAOCULUS Moskalev, 1976

CAYMANABYSSIINAE (4)
CAYMANABYSSIA Moskalev, 1976
AMPHIPLICA Haszprunar, 1988
COLOTRACHELUS Marshall, 1986
COPULABYSSIA Haszprunar, 1988
YAQUINABYSSIA Haszprunar, 1988

PYROPELTIDAE (8a)
PYROPELTA McLean & Haszprunar, 1987

OSTEOPELTIDAE (7b)
OSTEOPELTA Marshall, 1987

COCCULINELLIDAE (7)
COCCULINELLA Thiele, 1909

ADDISONIIDAE (8)
ADDISONIA Dall, 1882

CHORISTELLIDAE (1)
CHORISTELLA Bush, 1897
Choristes Auctt. (non Carpenter, 1872)

BATHYPHYTOPHILIDAE (?)
BATHYPHYTOPHILUS Moskalev, 1978
AENIGMABONUS Moskalev, 1978

COCCULINOIDEA (8a)
COCCULINIDAE (7)
COCCULINA Dall, 1882
? Dallia Jeffreys, 1883 (non Bean, 1878)
COCCOCRATER Haszprunar, 1987
COCCOPIGYA Marshall, 1986
pro Coccopygia Dall, 1889
(non Reichenbach, 1882)

COCCULINIDAE (Cont'd.)
FEDIKOVELLA Moskalev, 1976
PARACOCCULINA Haszprunar, 1987
TEUTHIROSTRIA Moskalev, 1976

BATHYSCIADIIDAE (9)
BATHYSCIADIUM Dautzenberg & Fischer, 1901
BONUS Moskalev, 1973
BATHYPELTA Moskalev, 1971

References: Patelloidea
Bouchet P. & A. Waren "Abyssal molluscan fauna of the Norwegian Sea and its relation to other fauna" Sarsia 64:211-243 - 1979 (1)

Christiaens "Revision du genre Patella" - 1973 Bull.Mus.Nat.Hist.Natur.Zool.121:1305-1392 (2)

Golikov, A.N. & O.G. Kussakin "Tecturidae; position systematique" Malac.11(2):287-294 - 1972 (3)

Haszprunar,G."Anatomy, affinities: pseudococculinid Limpets" I. Zool.Scrip.16(14):305-324 - 1987
II. Ibid. 17(2):161-179 - 1988 (4)

Hickman, C. S."Radular patterns; deep sea limpets" Veliger 26(2):73-92 - 1983 (5)

Lindberg, D. R. "Problacmaea moskalevi, a new addition to the Eastern Pacific limpet fauna" Veliger 22(1):57-60 - 1979 (6)
—— "Rhodopetalinae, a new subfamily of Acmeidae, boreal Pacific" Malac.20(2):291-303.1981 (6a)
—— "Name changes in Acmaeidae" Veliger 29(2):142-148 - 1986 (6b)
—— "Systematics of the Scurriini —" Ibid.30(4):387-394 - 1988 (6c)

Marshall, B. A. "The family Cocculinellidae" Nat.Mus.N.Z.Records 2(12):139-143 - 1983 (7)
—— "Cocculinidae, Pseudococculinidae:New Zealand, New South Wales" N.Z.Journ.Zool.12 -1986 (7a)
—— "Osteopeltidae, new family - limpets; whale bone;deep-sea"J.Moll.Stud.53:121-127-1987(7b)

McLean, J. H. "The archaeogastropod; Addisoniidae" Veliger 28(1):99-108 - 1985 (8)
—— & G. Haszprunar "Pyropeltidae, new family —" Veliger 30(2):196-205 - 1987 (8a)

Moskalev, L. I. "Pacific Ocean Bathysciadidae —" Zool.Zh. 52(9):1297-1303 - 1973 (9)
—— "On the generic Classification; Cocculinidae" Tr.In-Ta.Inst.Ok.AN.USSR 99:59-70 - 1976 (9a)
—— "Revision of the Lepetellidae of the Pacific" Tr.In-Ta.Inst.Okean.108:52-78 - 1977 (9b)
—— "The Lepetellidae & forms similar to them" Tr.In-Ta Inst.Okean.113:132-146- 1978 (9c)

Okutani, T. "Sagamilepeta, new genus for Lepeta —" Venus 46(2):127-130 - 1987 (10)

Ponder, W.F. & R.G. Creese "Revision: Australian Notoacmea, Collisella & Patelloida" J.Malac.Soc.Aust.4(4):167-208 - 1980 (11)

TROCHACEA
TROCHIDAE
MARGARITINAE
MARGARITES Gray, 1847
Margarita Leach in Ross, 1819 (non 1814)
Eumargarita Fischer, 1885
Valvatella Melvill, 1897 (non Gray, 1857)
BATHYMOPHILA Dall, 1881
CANTHARIDOSCOPS Galkin, 1955
MARGARELLA Thiele in Troschel, 1893
Margaritella Thiele in Troschel, 1891 (non Meek & Hayden, 1860)
Promargarita, Submargarita Strebel, 1908
MARGARITOPSIS Thiele, 1906
NEOMARGARITES Golikov & Gulbin, 1984
OMPHALOMARGARITES Habe & Ito, 1965
PUPILLARIA Dall, 1909
AGATHODONTA Cossmann, 1918 (8)
Agnathodonta Wenz, 1938
ANTIMARGARITA Powell, 1951
ANXIETAS Iredale, 1917 (11)
BATHYBEMBIX Crosse, 1892
Bembix Watson, 1879 (non Koninck, 1844)
Ginebis Otuka, 1942
CALLIOTROPIS Seguenza, 1903 (13)
Solariellopsis Schepman, 1908 (non de Gregorio, 1886)
SOLARICIDA Dall, 1919
DANILIA Brusina, 1865
Craspedotus Philippi, 1847 (non Schönherr, 1844)
Heliciella O. G. Costa, 1861
ECHINOGURGES Quinn, 1979
EUCHELUS Philippi, 1847
Aradasia Gray, 1850
Tallorbis G. & H. Nevill, 1869

EUCHELUS (Cont'd.)
Huttonia Kirk, 1882 (non Pickard Cambridge, 1880)
ANTILLACHELUS Woodring, 1928
HERPETOPOMA Pilsbry, 1890
NEVILLIA H. Adams, 1868
VACEUCHELUS Iredale, 1929
GRANATA Cotton, 1957
HYBOCHELUS Pilsbry, 1899
LISCHKEIA Fischer in Kiener & Fischer, 1880
ADAMSENIDA Habe, 1957
CIDARINA Dall, 1909
TURCICULA Dall, 1881
MIRACHELUS Woodring, 1928
OLIVIA Cantraine, 1835 (non Bertolini, 1810)
PUTZEYSIA Sulliotti, 1889
Gemmula Seguenza, 1876 (non Weinkauff, 1875)
TIBATROCHUS Nomura, 1940
TROPIDOMARGA Powell, 1951
TURCICA H. & A. Adams, 1854
Ptychostylus Gabb, 1866
PERRINIA H. & A. Adams, 1854 ?

PLANITROCHINAE
PLANITROCHUS Perner in Barrande, 1903

MONODONTINAE
MONODONTA Lamarck, 1799
Monodontes Montfort, 1810
Labio Oken, 1815
Odontis Sowerby, 1825
Trochidon Swainson, 1840
Pimpellies Gistl, 1848
AUSTROCOCHLEA Fischer, 1885
Austrochlea Pilsbry, 1889 (err.)

MONODONTA (Cont'd.)
NEOMONODONTA Kuroda, Habe & Oyama, 1971
OSILINUS Philippi, 1847
Gibbium Gray, 1847 (non Scopoli, 1777)
Trochius, Neptheusa Gray, 1852
Trochocochlea Mörch, 1852
Caragolus, Nephteusa Monterosato, 1884
Mutilastra, Pseudosilinus Nordsieck,1975
BANKIVIA Beck in Krauss, 1849
LEIOPYRGA H. & A. Adams, 1863
CANTHARIDUS Montfort, 1810
Cantharis Férussac, 1821
(non Linnaeus, 1758)
Elenchus Swainson,1840 (non Curtis, 1831)
Cantharidium Schaufuss, 1869
IWAKAWATROCHUS Kuroda & Habe, 1954
MICRELENCHUS Finlay, 1927
PLUMBELENCHUS Finlay, 1927
PHASIANOTROCHUS Fischer, 1885
CHRYSOSTOMA Swainson, 1840
DILOMA Philippi, 1845
Zediloma Finlay, 1927
CAVODILOMA Finlay, 1927
FRACTARMILLA Finlay, 1927
CHLORODILOMA Pilsbry, 1889
Latona Hutton,1884 (non Schumacher, 1817)
MELAGRAPHIA Gray, 1847
Anisodiloma Finlay, 1927
Neodiloma Fischer, 1885
OXYSTELE Philippi, 1847
PICTODILOMA Habe, 1946
JUJUBINUS Monterosato, 1884
Clelandella Winkworth, 1932
Gravijubinus F. Nordsieck, 1975
Pictijubinus F. Nordsieck, 1975
MIRULINUS Monterosato, 1917
STRIGOSELLA Sacco, 1896
TEGULA Lesson, 1835
AGATHISTOMA Olsson & Harbison, 1953
CHLOROSTOMA Swainson, 1840
OMPHALIUS Philippi, 1847
Neomphalius Fischer, 1885
PROMARTYNIA Dall, 1909
STEARNSIUM S. S. Berry, 1958
THALOTIA Gray, 1840
ALCYNA A. Adams, 1860
CALTHALOTIA Iredale, 1929
ODONTOTROCHUS Fischer in Kiener & Fischer,1880
PROTHALOTIA Thiele, 1930

CATAEGINAE (10a)
CATAEGIS McLean & Quinn, 1987

GIBBULINAE
GIBBULA Risso, 1826
Magulus, Puteolus Monterosato, 1888
Conotrochus Pilsbry, 1889
(non Seguenza, 1864)
Phorculellus Sacco, 1896
pro Phorculus Monterosato, 1888
(non Cossmann, 1888)
Phorculorbis Cossmann, 1918
ADRIARIA Pallary, 1917
CALLIOTROCHUS Fischer in Kiener & Fischer,1880
EURYTROCHUS Fischer in Kiener & Fischer,1880
CANTHARIDELLA Pilsbry, 1889
COLLICULUS Monterosato, 1888
Glomulus Monterosato, 1888
Glossulus Pallary, 1938
ENIDA A. Adams, 1860
FORSKALENA Iredale, 1918
Forskalia H. & A. Adams, 1854
(non Koelliker, 1853)

GIBBULA (Cont'd.)
FORSKALIOPSIS Coen, 1931
(non Haeckel, 1888)
HISSEYAGIBBULA Kershaw, 1955
NOTOGIBBULA Iredale, 1924
PHORCUS Risso, 1826
STEROMPHALA Gray, 1847
Korenia Friele, 1877
Gibbulastra Monterosato, 1884
Gibbuloidella Sacco, 1896
TUMULUS Monterosato, 1888
CITTARIUM Philippi, 1847 (3)
Meleagris Montfort, 1810
(non Linnaeus, 1758)
Livona Gray, 1847
FOSSARINA A. Adams & Angas, 1864
Minos Hutton, 1884
CLYDONOCHILUS Fischer, 1890
MINOPA Iredale, 1924
NANULA Thiele, 1924
NORRISIA Bayle, 1880
Trochiscus Sowerby, 1838
(non Heyden, 1826)
PHORCULUS Cossmann, 1888
TROCHINELLA Iredale, 1937

CALLIOSTOMATINAE
CALLIOSTOMA Swainson, 1840
Conulus Nardo,1841
(non Leske, 1778)
Callistoma Agassiz, 1846
Ziziphinus Gray, 1840 (nom.nud.)
Montagua Gray, 1852 (non Leach, 1814)
Stylotrochus Seguenza, 1876
(non Haeckel, 1862)
Fluxina Dall, 1881
Manotrochus Fischer, 1885
Jacinthinus Monterosato, 1889
Eucasta Dall, 1889
Ampullotrochus Monterosato, 1890
Dymares Schwengel, 1942
ALERTALEX Dell, 1956
BENTHASTELENE Iredale, 1936
ELMERLINIA Clench & Turner, 1960
FAUTOR Iredale, 1924
Salsipotens Iredale, 1924
KOMBOLOGION Clench & Turner, 1960
LAETIFAUTOR Iredale, 1929
MAUREA Oliver, 1926 (Calliostoma Syn.)
Mauriella Oliver, 1926
Calliotropis Oliver, 1926
(non Seguenza, 1903)
Mucrinops, Venustas Finlay, 1927
Calotropis Thiele, 1929
OTUKAIA Ikebe, 1942
SINUTOR Cotton & Godfrey, 1935
SPIKATOR Cotton & Godfrey, 1935
TRISTICHOTROCHUS Ikebe, 1942
ZIZIPHINUS Gray in Dieffenbach, 1843
ASTELE Swainson, 1855
Leiotrochus Conrad, 1862
Eutrochus A. Adams, 1864
ASTELENA Iredale, 1924
CALLISTELE Cotton & Godfrey, 1935
CORALASTELE Iredale, 1930
EURASTELE Coen, 1946
MAZASTELE Iredale, 1936
OMPHALOTUKAIA Yoshida, 1948
PULCHRASTELE Iredale, 1929
SCROBICULINUS Monterosato, 1889
DENTISTYLA Dall, 1889
FALSIMARGARITA Powell, 1951
PHOTINASTOMA Powell, 1951

CALLIOSTOMATINAE (Cont'd.)
PHOTINULA H. & A. Adams, 1854
Photina H.& A. Adams in A. Adams,1853
(non Burmeister, 1838)
Kingotrochus Ihering, 1902
VENUSTATROCHUS Powell, 1951

TROCHINAE
TROCHUS Linnaeus, 1758
Polydonta Schumacher, 1817
Lamprostoma Swainson, 1840
AKOYA Habe, 1961
AWAJITROCHUS Habe, 1958
BELANGERIA Fischer in Kiener & Fischer, 1880
COELOTROCHUS Fischer
in Kiener & Fischer, 1880
Neozelandia Cossmann, 1918
INFUNDIBULOPS Pilsbry, 1889
INFUNDIBULUM Montfort, 1810
Carinidea Swainson, 1840
KANEKOTROCHUS Habe, 1958
PRAECIA Gray, 1857
THORISTA Iredale, 1915
Anthora Gray, 1857
(non Doubleday, 1844)
THORISTELLA Iredale, 1915
CLANCULUS Montfort, 1810
Otavia Risso, 1826
Fragella Swainson, 1840
CAMITIA Gray, 1842
CLANCULOPSIS Monterosato, 1888
Clanculella Sacco, 1896
EUCHELICLANCULUS Kuroda, Habe & Oyama, 1971
EUCLANCULUS, EURICLANCULUS, ISOCLANCULUS,
MACROCLANCULUS, MICROCLANCULUS
all Cotton & Godfrey, 1934
MESOCLANCULUS Iredale, 1924
PARACLANCULUS Finlay, 1927
OLIGOMERIA Galkin & Golikov, 1985 (?)
PSEUDOTALOPIA Habe, 1961
TALOPENA Iredale, 1918
TECTUS Montfort, 1810
Pyramis Schumacher, 1817
(non Röding, 1798)
Pyramidea Swainson, 1840
CARDINALIA Gray, 1842
ROCHIA Gray, 1857

UMBONIINAE
UMBONIUM Link, 1807
Globulus Schumacher, 1817
Rotella Lamarck, 1822
ETHALIA H. & A. Adams, 1854
Liotrochus Fischer in
Kiener & Fischer, 1880
ETHALIELLA Pilsbry, 1905
SUCHIUM Makiyama, 1924
ZETHALIA Finlay, 1927
Ethaliopsis Cossmann, 1918
(non Schepman, 1908)
ANTISOLARIUM Finlay, 1927
CALLUMBONELLA Thiele, 1924
Umbotrochus Thiele, 1914
(non Perner, 1903)
ETHMINOLIA Iredale, 1924
Sericominolia Kuroda & Habe, 1952
GAZA Watson, 1879 (13)
CALLOGAZA Dall, 1881
ISANDA H. & A. Adams, 1854
ARCHIMINOLIA, CONOTALOPIA, PARMINOLIA,
VANITROCHUS all Iredale, 1929
UMBONELLA A. Adams, 1863
LIRULARIA Dall, 1909 (9)

UMBONIINAE (Cont'd.)
MONILEA Swainson, 1840
Talopia Gray, 1842 (nom.nud.)
PRIOTROCHUS Fischer in
Kiener & Fischer, 1880
Aphanotrochus Von Martens, 1880
ROSSITERIA Brazier, 1895
Solanderia Fischer in Kiener
and Fischer, 1880
(non Duchassaing & Michelin, 1846)

SOLARIELLINAE (13)
SOLARIELLA Wood, 1842
Machaeroplax Friele, 1877
ETHALIOPSIS Schepman, 1908
MICROPILISCUS Dall, 1927
SUAVOTROCHUS Dall, 1924
ZETELA Finlay, 1927
ILANGA Herbert, 1987 (5)
MICROGAZA Dall, 1881
Bowdengaza Woodring, 1928
MINOLIA A. Adams, 1860
Minosia Dunker, 1882
MINOLOPS Iredale, 1929
SPECTAMEN Iredale, 1924
Zeminolia Finlay, 1927

HALISTYLINAE
HALISTYLUS Dall, 1890
BOTELLOIDES Strand, 1928 (11a)
pro Botellus Iredale, 1924
(non Moniez, 1887)
CHARISMA Hedley, 1915 (11a)
Cavostella, Cavotera Laseron, 1954

PELYCIDIIDAE ? (12)
PELYCIDION Fischer in Fischer & Perrier, 1873
Nannoteretispira Habe, 1961

STOMATELLIDAE
STOMATELLA Lamarck, 1816
Phymotis Rafinesque, 1815 (nom.nud.)
Gena Gray 1842 (16)
Plocamotis Fischer, 1885
BRODERIPIA Gray, 1847
PSEUDOSTOMATELLA Thiele, 1921
STOMATOLINA Iredale, 1937
ROYA Iredale, 1912
STOMATIA Helbling, 1779
Stomax Montfort, 1810
MICROTIS A. Adams, 1850
Microtina A. Adams in Sowerby, 1854
MIRACONCHA Bergh in Schepman, 1908
NIPHONIA A. Adams, 1860
SYNAPTOCOCHLEA Pilsbry, 1890

SKENEIDAE
SKENEA Fleming, 1825
Delphinoidea T. Brown, 1827
Delphionoidea T. Brown, 1844
CAPORBIS Bartsch, 1915
CIRSONELLA Angas, 1877
CROSSEA A. Adams, 1865
Crosseia Fischer, 1885
CROSSEOLA Iredale, 1924
DOLICROSSEA Iredale, 1924
Doliocrossea Wenz, 1938 (error)
CYCLOSTREMISCUS Pilsbry & Olsson, 1945
MIRALABRUM Pilsbry & Olsson, 1945
PACHYSTREMISCUS Olsson & McGinty, 1942
PONOCYCLUS Pilsbry, 1953
DARONIA A. Adams, 1861
EUDARONIA Cotton, 1945

SKENEIDAE (Cont'd.)
DILLWYNELLA Dall, 1889
GANESA Jeffreys, 1883
GRANIGYRA Dall, 1889
LISSOSPIRA Bush, 1897
HAPLOCOCHLIAS Carpenter, 1864
LEPTOGYRA Bush, 1897
LEUCORHYNCHIA Crosse, 1867
LISSOTESTA Iredale, 1915
LISSOTESTELLA Powell, 1946
LODDERENA Iredale, 1924
LODDERIA Tate, 1899
Cyclostremella Tate, 1898
(non Bush, 1897)
LOPHOCOCHLIAS Pilsbry, 1921
MOELLERIOPSIS Bush, 1897
PARVITURBO Pilsbry & McGinty, 1945
PHILORENE Oliver, 1915
PONDORBIS Bartsch, 1915
PUTILLA A. Adams, 1867 (11)
Notosetia Iredale, 1915 ?
Conicella Laseron, 1954 ?
WANGANELLA Laseron, 1954 (?)
RHODINOLIOTIA Tomlin & Shackleford, 1915
SPINOBROOKULA Lozouet, 1982
SRILANKAIA Mienis, 1974 (?)
THARSIELLA Bush, 1897
Tharsis Jeffreys, 1883
(non Giebel, 1847)
Porcupinia Cossmann, 1900
(non Haeckel, 1887)
THOLOPEX DiGeronimo, 1974
TUBIOLA A. Adams, 1863
PARTUBIOLA Iredale, 1936

CYCLOSTREMATIDAE
CYCLOSTREMA Marryat, 1818
Cyclotrema Gray, 1840 (nom.nud.)
Pseudoliotina Cossmann, 1925
BROOKULA Iredale, 1912
Vetulonia Dall, 1913
AEQUISPIRELLA Finlay, 1924
CHUNULA Thiele, 1925 ?
LIOCARINIA Laseron, 1954
LIOTELLA Iredale, 1915
CONJECTURA Finlay, 1927
CORONADOA Bartsch, 1946
ELACHORBIS Iredale, 1915
MUNDITIA Finlay, 1927
PTERARENE Sakurai & Habe, 1977
TASMOCROSSEA Dell, 1952
ZALIPAIS Iredale, 1915

TURBINIIDAE
LIOTIINAE (10b)
LIOTIA Gray, 1842
ANADEMA H. & A. Adams, 1854
ARENE H. & A. Adams, 1854
Otollonia Woodring, 1928
Marevalvata Olsson & Harbison, 1953
CINYSCA Kilburn, 1970
pro Cynisca H. & A. Adams, 1854
(non Gray, 1844)
ILAIRA H. & A. Adams, 1854
Liotiaxis Iredale, 1936
LIOTINA Fischer, 1885
AUSTROLIOTIA Cotton, 1948
BATHYLIOTINA Habe, 1961
DENTARENE Iredale, 1929
GLOBARENE Iredale, 1929
LIOTINARIA Habe, 1955
MACRARENE Hertlein & Strong, 1951

ANGARIINAE (10b)
ANGARIA Röding, 1798
Delphinula Lamarck, 1804
Delphinulus Montfort, 1810
Praxidice Rafinesque, 1815
Scalator Gistl, 1848
Angarus Gray, 1857

TURBININAE
TURBO Linnaeus, 1758
Laeviturbo Cossmann, 1918
Bothropoma Thiele, 1924
Amphiboliturbo Magne, 1940
Neocollonia Kuroda & Habe, 1952
BATILLUS Schumacher, 1817
Angarina Bayle, 1878
Delphinulopsis Wright, 1878
(non Laube, 1870)
CALLOPOMA Gray, 1850
CARSWELLENA Iredale, 1931
CHAENOTURBO McLean, 1970
DINASSOVICA Iredale, 1937
EUNINELLA Cotton, 1939
HALOPSEPHUS Rehder, 1943
LUNATICA Röding, 1798
Olearia Herrmannsen, 1847
LUNELLA Röding, 1798
Marmarostoma Swainson, 1840
(non Swainson, 1829)
Ocana H. Adams, 1861
MARMAROSTOMA Swainson, 1829
Senectus Swainson, 1840
MODELIA Gray, 1850
NINELLA Gray, 1850
SARMATICUS Gray, 1847
Cidaris Swainson, 1840 (pars)
(non Röding, 1798)
SUBNINELLA Thiele, 1929
TAENIATURBO Woodring, 1928
PRISOGASTER Mörch, 1850
Amyxa Troschel, 1852

ASTRAEINAE
ASTRAEA Röding, 1798
Imperator Montfort, 1810
Canthorbis Swainson, 1840
Macropelmus Gistl, 1848
BELLASTRAEA Iredale, 1924
MEGASTRAEA J. H. McLean, 1970
OKINAWASTRAEA Habe & Okutani, 1981
UVANILLA Gray, 1850
ASTRALIUM Link, 1807
Calcar Montfort, 1810
Cyclocantha Swainson, 1840
Stella Herrmannsen, 1849
Distellifer Iredale, 1937
Pagocalcar Iredale, 1937
Rugastella Iredale, 1937
Micrastraea Cotton, 1939
FOLIASTRALIUM Habe & Okutani, 1980 ?
BOLMA Risso, 1826 (2)
Oobolma Sacco, 1896 ?
Pseudastralium Schepman, 1908
Incilaster Finlay, 1926
Galeoastraea Kuroda & Habe, 1958
Harisazaea Habe, 1958
SENOBOLMA Okutani, 1964 ?
COOKIA Lesson, 1832
Tubicanthus Swainson, 1840 (pars)
GUILDFORDIA Gray, 1850
LITHOPOMA Gray, 1850
Pachypoma Gray, 1850
POMAULAX Gray, 1850

ACREMODONTINAE (6a)
ACREMODONTA Marshall, 1983

HOMALOPOMATINAE
HOMALOPOMA Carpenter, 1864
Leptonyx Carpenter, 1864
(non Swainson, 1833)
ARGALISTA Iredale, 1915
CANTRAINEA Jeffreys, 1883
Cantraineia Fischer, 1885
COLLONISTA Iredale, 1918
LEPTOTHYROPSIS Woodring, 1928
PHANEROLEPIDA Dall, 1907
CIRSOCHILUS Cossmann, 1888
LEPTOCOLLONIA Powell, 1951
LEPTOTHYRA Pease, 1869
Petropoma Gabb, 1877
MOELLERIA Jeffreys, 1865 (1)
SPIROMOELLERIA Baxter & McLean, 1984
PANOCOCHLEA Dall, 1908

TURBINIDAE s.l.
GABRIELONA Iredale, 1917 (19)

PHASIANELLIDAE (17)
PHASIANELLA Lamarck, 1804
Phasianus de Montfort, 1810
(non Linnaeus, 1758)
Bolina Rafinesque, 1815 (nom.nud.)
? Eutropia Humphrey in
Swainson, 1840
Orthopnoea Gistl, 1847
Mimelenchus Iredale, 1924

TRICOLIIDAE
TRICOLIA Risso, 1826 (19)
Eudora Leach in Gray, 1852
(non Péron & Lesueur, 1810)
Chromotis A. Adams, 1863
Eucosmia Carpenter, 1864
(non Stephens, 1831)
Tricoliella Monterosato, 1884
Steganomphalus Harris & Burrows, 1891
Eulithidium Pilsbry, 1898
Usatricolia Habe, 1956
Eutricolia, Tricolietta F. Nordsieck,1975
AIZYELLA Cossmann, 1889
EOTRICOLIA Kuroda & Habe, 1952
HILOA Pilsbry, 1917
ORTHOMESUS Pilsbry, 1888

SEGUENZIOIDEA
SEGUENZIIDAE (6, 14)
SEGUENZIA Jeffreys, 1876
Mioseguenzia F. Nordsieck, 1976 ?
ANCISTROBASIS Dall, 1889
ASTHELYS Quinn, 1987 (14a)
BASILISSA Watson, 1879
BASILISSOPSIS Dautzenberg & Fischer, 1897
CALLIOBASIS Marshall, 1983
CARENZIA Quinn, 1983
FLUXINELLA Marshall, 1983
Fluxiella Okutani, 1968, 1974 (n.nd.)
GUTTULA Schepman, 1908
HADROCONUS Quinn, 1987
ROTELLENZIA Quinn, 1987
SEGUENZIELLA Marshall, 1983
SEGUENZIOPSIS Marshall, 1983
THELYSSA F. M. Bayer, 1971
THELYSSINA Marshall, 1983

References:

Baxter, R. & J. H. McLean "The genus Moelleria and Spiromoelleria in the Northern Pacific" Veliger 27(2):219-226 - 1984 (1)

Beu, A. G. & W. F. Ponder "Revision of genus Bolma" Rec.Aust.Mus.32(1-3):1-68 - 1979 (2)

Guidastri, R.I.et al "Systematic position: Trochus wiseri" Arch.Moll.143:125-136 - 1984 (3)

Herbert, D. G. "Revision: Solariellinae - Southern Africa" Ann.Natal.Mus.28(2):283-382 -1987 (5)

Marshall, B. A. "Recent and Tertiary Seguenziidae from the New Zealand region." N.Z.Jour.Zool.,Vol.10:235-262 - 1983 (6)
—— "Acremodonta, new subfamily of Trochidae" Rec.Nat.Mus.N.Z. 2(10):127-130 - 1983 (6a)

McLean, J. H. "The families Liotidae and Skeneidae in the Eastern Pacific" Echo 2:18 - 1969 (7)
—— "Agathodonta nortoni, living member of a lower Cretaceous Trochid genus" Nautilus 98(3):121-123 - 1984 (8)
—— "New species of Northeastern Pacific (9) Archaeogastropoda" Veliger 26(3):237 - 1984a
—— "The importance of gill structure in molluscan classification" West.Soc.Mal.Ann.Rep. 14:10 - 1985 (10)
—— "Angariinae & Liotiinae" (10b) Ann.Rept.Western Soc.Malacol.19:16 - 1987
—— & J. F. Quinn "Cataegis, n. gen.; Continental Slope" Nautilus 101(3):111-116 - 1987 (10a)

Ponder, W. F. "Review of the genera of Rissoidae" Rec.Aust.Mus.Supp.4:1-221 - 1985 (11)

—— "Revision: genus Botelloides" 1985 (11a) Sp.Pub.S.Aust.Dept.Mines,Energy.5:301-327

—— & S. J. Hall "Pelycidiidae, a new family of Archaeogastropod mollusks" Nautilus 97(1):30-35 - 1983 (12)

Quinn, J. F. Jr. "Biological results; University of Miami deep-sea expedition: systematics of Trochidae collected in Straits of Florida" Malacologia 19(1):1-62 - 1979 (13)
—— "Revision of the Seguenzia, summary and evaluation of the Superfamily" Proc.Biol.Soc.Wash.96(3)355-364 -1983 (14)
—— "Revision of the Seguenzia II: new genera" Nautilus 101(2):51-58 - 1987 (14a)

Rehder, H. A. "The genus Turcicula" Proc.Malac.Soc.London 31:222-225 -1955 (15)
—— "The marine mollusks of Easter Island" Smiths.Contr.Zool.289:1-166 - 1980 (16)

Robertson, R. "The Family Phasianellidae in the West Atlantic" Johnsonia 3(37):245-284 - 1958 (17)
—— "The Genus Gabrielona in the Indo-Pacific and the West Indies" Indo-Pac. Mollusca 3(14):41-61 - 1973 (18)
—— "Archaeogastropod biology, systematics of the genus Tricolia: Indo-West Pacific" Monographs of Marine Mollusca 3:1-1985(19)

Suborder Neritimorpha

NERITOIDEA

NERITOPSIDAE

NERITOPSIS Grateloup, 1832
Radula Gray, 1842

NERITIDAE

NERITINAE

NERITA Linnaeus, 1758
Neritarius Duméril, 1806
Dontostoma Herrmannsen, 1847
Tenare Gray, 1858
Peloronta Oken, 1815 ?
AMPHINERITA von Martens, 1887
Odontostoma Mörch, 1852
(non Turton, 1829)
FLUVINERITA Pilsbry, 1932 (freshwater)
HEMINERITA von Martens, 1887
LINNERITA Vermeij, 1984
MELANERITA von Martens, 1889
PUPERITA Gray, 1857
RITENA Gray, 1858
Pila Mörch, 1852
(non Röding, 1798)
Cymostyla von Martens, 1887
THELIOSTYLA Mörch, 1852
Natere Gray, 1858
Ilynerita von Martens, 1887
NERITINA Lamarck, 1816
Laphrostoma Rafinesque, 1815
Lamphrostoma auctt. (err.)
Lamprostoma auctt. (err.)
Chernites Gistl, 1848
Neritella Gray, 1848
Clypeolum Récluz, 1850 (non 1842)
Labialia, Onychina Scudder, 1882
CLYPEOLUM Récluz, 1842
Alina Récluz, 1842
(non Risso, 1826)
DOSTIA Gray, 1847
Mitrula Récluz, 1850
(non Gray, 1821)
NEREINA Cristofori & Jan, 1832
NERIPTERON Lesson, 1830
Neritopteron Fischer, 1885
NERITONA von Martens, 1869
PROVITTOIDA H. B. Baker, 1923
Provittoidea Wenz, 1938
PSEUDONERITA H. B. Baker, 1923
VITTA Mörch, 1852
Scapha Mörch, 1852
(non Récluz, 1841)
VITTINA H. B. Baker, 1923
Paranerita Bourne, 1909
(non Hampson, 1901)
VITTOIDA H. B. Baker, 1923
Vittoidea Wenz, 1938
NERITODRYAS von Martens, 1869
THEODOXUS Montfort, 1810
Theodoxis Montfort, 1810
Elea Fitzinger, 1833
Neritoglobus Kobelt, 1871
Theodoxia Bourguignat, 1877
Theodora, Theodorus auctt. (err.)
ALINOCLITHON H. B. Baker, 1923
CLITHON Montfort, 1810
Cliton Lesson, 1830
Corona Récluz, 1850
(non Albers, 1850)
Urceus Bruguière, 1792 ?
NERITAEA Roth, 1855
Neritoconus Kobelt, 1871

THEODOXUS (Cont'd.)
NERITOCLITHON H. B. Baker, 1923
NINNIA Westerlund, 1903
PICTONERITINA Iredale, 1936
VITTOCLITHON H.B. Baker, 1923
TANZANIELLA Lupu, 1979 ?

NERITILIINAE (Freshwater)

NERITILIA von Martens, 1879
SEPTARIELLINA Bequaert & Clench, 1936

SMARAGDIINAE

SMARAGDIA Issel, 1869
Gaillardotia Bourguignat, 1877
SMARAGDELLA H. B. Baker, 1923
SMARAGDISTA Iredale, 1936
MAGADIS Melville & Standen, 1899
PISULINA G. & H. Nevill, 1869

SEPTARIIDAE

SEPTARIA Férussac, 1807
Cimber Montfort, 1810
Septarius Gray, 1821
Catillus Gray, 1847
(non Brongniart, 1822)
Cibota H. & A. Adams, 1858
(non Herrmannsen, 1852)
Laodia Gray, 1868
Paria Gray, 1868
NAVICELLA Lamarck, 1816
Navicellus Gray, 1821
Scapha Récluz, 1841
Stenopoma Gray, 1868
PARASEPTARIA Risbec, 1942
ORTHOPOMA Gray, 1867 (Thiele) ?

PHENACOLEPADIDAE (Scutellinidae)

PHENACOLEPAS Pilsbry, 1891
Scutella Broderip, 1834
(non Lamarck, 1816)
Scutellina Gray, 1847
(non Agassiz, 1841)
Scutulina Cossmann, 1912
AMAPILEUS Iredale, 1929
CINNALEPETA Iredale, 1929
ZACALANTICA Iredale, 1921
PLESIOTHYREUS Cossmann, 1888

TITISCANIIDAE (no shell)

TITISCANIA Bergh, 1890

HYDROCENOIDEA (8)

HYDROCENIDAE (Terrestrial) (4)

HYDROCENA Pfeiffer, 1847
Monterissa Iredale, 1944
GEORISSA Blanford, 1864 ?
Omphalorissa Iredale, 1933 (nom.nud.)
Petrorissa Kuroda & Habe, 1945
CHONDRELLA Pease, 1871
GEORISSOPS Pilsbry & Hirase, 1908
LAPIDARIA Kang, 1986

HELICINOIDEA (8)

HELICINIDAE

HELICININAE

HELICINA Lamarck, 1799
Pitonnellus Montfort, 1810 ?
Pitonnillus Montfort, 1810 ?
Colyma Rafinesque, 1815
Pachytoma Swainson, 1840
Euneritella Wagner, 1905
Ampliata Wagner, 1907
Festiva Wagner, 1910

HELICINA (Cont'd.)
OXYRHOMBUS Fischer & Crosse, 1893
Angulata Wagner, 1905
Tamsiana Wagner, 1907
PSEUDOLIGYRA H. B. Baker, 1954
TRISTRAMIA Crosse, 1863
Caloplisma Fischer & Crosse, 1893
Retorquata Wagner, 1905
ALCADIA Gray, 1840 (2)
Isoltia Guppy, 1859
Eualcadia Wagner, 1907
Eucaladia auctt. (err.)
Incrustata Wagner, 1907
Intusplicata Wagner, 1907
Palliata Wagner, 1907
Sericea Wagner, 1907
ANALCADIA Wagner, 1907
Diaphana Guppy, 1895
(non Brown, 1827)
EMODA H. & A. Adams, 1856
Glyptemoda Clench & Aguayo, 1950
GLYPTALCADIA Boss & Jacobson, 1973
HJALMARSONA Baker, 1954
IDESA H. & A. Adams, 1856
Schrammia Guppy, 1895
Leialcadia Wagner, 1907
Mammilla Wagner, 1907
(non Tryon, 1883)
Weinlandella H. B. Baker, 1954
PENISOLTIA H. B. Baker, 1954
Pro Hispida Wagner, 1907
(non Bate, 1868)
STRIATEMODA H. B. Baker, 1940
BOURCIERA Pfeiffer, 1852
Pseudohelicina Sykes, 1907
CERATOPOMA Möllendorff, 1893
Diversicolor Wagner, 1905
Negopenia Iredale, 1941 (nom.nud.)
PALAEOHELICINA Wagner, 1905
Kalokonia Wagner, 1905
Rhabdokonia Wagner, 1905
HEMIPOMA Wagner, 1905
OGASAWARANA Wagner, 1905 (5)
OLYGYRA Say, 1818
Oligyra Say, 1819
Orbiculata Wagner, 1905
Succincta Wagner, 1905
Subglobulosa Wagner, 1905
OROBOPHANA Wagner, 1905
PLEUROPOMA Möllendorff, 1893
Albocincta Wagner, 1908
APHANOCONIA Wagner, 1905
Pachystoma Wagner, 1905
(non Guilding, 1828)
Reticulata Wagner, 1905
Sculpta Wagner, 1905
Sphaeroconia Wagner, 1909 ?
SCHASICHEILA Shuttleworth, 1852
Schasichila Fischer, 1885
Atoyac H. B. Baker, 1928
MISANTLA H. B. Baker, 1928
NECAXA H. B. Baker, 1928
STOASTOMA C. B. Adams, 1849 (Stoastomatinae)
Hemicyclostoma Pfeiffer, 1865
STOASTOMOPS H. B. Baker, 1924
SWIFTELLA H. B. Baker, 1941
STURANYA Wagner, 1905
Sturanyella Pilsbry & Cooke, 1934
Sturyanella auctt.
SULFURINA Möllendorff, 1893
Hypostrongyla Tomlin, 1930
Pestomena Iredale, 1941 (nom.nud.)
KOSMETOPOMA Wagner, 1905

CERATODISCINAE (Helicininae) (?)
CERATODISCUS Simpson & Henderson, 1901
FADYENIA Chitty, 1857
Lindsleya Chitty, 1857
Metcalfeia Chitty, 1857
BLANDIA Chitty, 1857
Petitia Chitty, 1857
Wilkinsonaea Chitty, 1857
LEWISIA Chitty, 1857

HENDERSONIINAE
HENDERSONIA Wagner, 1905
WALDEMARIA Wagner, 1905

VIANINAE
VIANA H. & A. Adams, 1856
Ampullina Blainville, 1824
(non Bowdich, 1822)
Hapata Gray, 1856
Rhynchocheila Shuttleworth, 1878
Fitzia Guppy, 1895
Krebsia Guppy, 1895
(non Mörch, 1877)
Excavata Wagner, 1907
CALIDVIANA H. B. Baker, 1954 ?
pro Callida Wagner, 1908
(non Agassiz, 1846)
Bakerviana Aguayo & Jaume, 1957
USTRONIA Wagner, 1908
CALYBIUM Morelet, 1891
GEOTROCHATELLA Fischer, 1891
EUTROCHATELLA Fischer, 1885
Trochatella Swainson, 1840
(non Lesson, 1830)
CUBAVIANA H. B. Baker, 1922
MICROVIANA H. B. Baker, 1928
PRIOTROCHATELLA Fischer, 1893
SEMITROCHATELLA Aguayo & Jaume, 1958
TORREVIANA Aguayo, 1943
TROSCHELVIANA H. B. Baker, 1922
GEOPHORUS Fischer, 1885
Pecoviana Iredale, 1941 (nom.nud.)
DIPLOPINAX Bartsch, 1921
SCHISTOPINAX Bartsch, 1921
HEUDEIA Crosse, 1885
MILUNA Wagner, 1905
LUCIDELLA Swainson, 1840
Prosopis Weinland, 1862
(non Fabricius, 1804)
PERENNA Guppy, 1867
POENIA H. & A. Adams, 1856
Urichia Guppy, 1895
POENIELLA H. B. Baker, 1923
PSEUDOTROCHATELLA Nevill, 1891
PYRGODOMUS Fischer & Crosse, 1893
Artecallosa Wagner, 1908

PROSERPINIDAE (8)
PROSERPINA G. B. Sowerby, 1839
Despoena Newton, 1891
DESPOENELLA H. B. Baker, 1923
pro Odontostoma Orbigny, 1842
(non Turton, 1830)

CERESIDAE (8) (Proserpinellinae)
CERES Gray, 1856
ARCHECHARAX Thompson, 1980
Cyane H. Adams, 1870 (8)
(non Felder, 1861)
LINIDIELLA Jousseaume, 1889
Chersodespoena Sykes, 1900
PROSERPINELLA Bland, 1865
STAFFOLA Dall, 1905

References; Neritoidea, Hydrocenoidea, Helicinoidea:
Boss, K. J., in S. F. Parker "Synopsis and Classification of Living Organisms" Mollusca Vol.1:946-1166 - 1982 (1)

—— & M. K. Jacobson "Monograph: Alcaldia - Cuba" Bull.Mus.Comp.Zool.145(7):311-358 - 1973 (2)

Clench, W. J. & M. K. Jacobson "Gen. Calidviana, Ustronia, Troschelviana, Semitrochatella: Cuba" Bull.Mus.Comp.Zool.Harv.141:403-463 - 1971 (3)

Kang, Z. B. "Description of new freshwater snails: China" Arch.Moll.117(1-3):73-91 - 1986 (4)

Minato, H. "Land shells of Miyako Islands" Venus 39(2):77-97 - 1980 (5)

Moore, R. C. (Ed.) "Treatise on Invertebrate Paleontology" I.Mollusca 1 - 1960 (6)

Russell, H. D. "Recent mollusks of the family Neritidea in the Western Atlantic" Bull.Mus.Comp.Zool.88(4):345-404 - 1941 (7)

Thompson, F. G. "Proserpinoid Land Snails: their relationships within the Archaeogastropoda" Malacologia 20(1):1-33 - 1980 (8)

Order Mesogastropoda

CYCLOPHOROIDEA (Terrestrial)
CYCLOPHORIDAE
CYCLOPHORINAE
CYCLOPHORUS Montfort, 1810
Eucyclophorus Kobelt & Möllendorff,1897
? Salpingophorus Kobelt & Möllendorff,1897
CRICOPHORUS Kobelt & Möllendorff, 1897
CYCLOHELIX Mörch, 1852
GLOSSOSTYLUS Kobelt & Möllendorff, 1897
LITOSTYLUS Kobelt & Möllendorff, 1897
PAPUOCYCLUS Ancey, 1895
Rhytidorhaphe Kobelt & Möllendorff,1897
AFRODITROPIS Bequaert & Clench, 1936
AULOPOMA Troschel, 1847
CHONDROCYCLUS Ancey, 1898
CRASPEDOTROPIS Blanford, 1864
CROSSOPOMA von Martens, 1891
CYATHOPOMA Blanford, 1861
JERDONIA Blanford, 1861
PSEUDOJERDONIA Kobelt, 1902
CYCLOSURUS Morelet, 1881
CYCLOTUS Swainson, 1840
Eucyclotus Möllendorff, 1890
AULACOPOMA Möllendorff, 1897
NAKADAELLA Ancey, 1904
OPISTHOPORUS Pfeiffer, 1851
PROCYCLOTUS Fischer, 1891
PSEUDOCYCLOPHORUS Möllendorff, 1890
SIPHONOCYCLUS Möllendorff, 1900
DITROPIS Blanford, 1869
DITROPOPSIS E. Smith, 1897
DIASPIRA Soos, 1911 ?
DITROPISENA Iredale, 1933 ?
ELGONOCYCLUS Verdcourt, 1982
JAPONIA Gould, 1859
LAGOCHILUS Blanford,1864 (Lagocheilus - Err.)
MYLICOTROCHUS P. & F. Sarasin, 1899
LAOTIA Saurin, 1953 (?)
LEPTOPOMA Pfeiffer, 1847
Dermatocera H. & A. Adams, 1855 (part)
ENTOCHILUS Kobelt & Möllendorff, 1897
LEUCOPTYCHIA Crosse, 1878
TROCHOLEPTOPOMA Kobelt & Möllendorff, 1897
LEPTOPOMOIDES G. Nevill, 1878
MARELEPTOPOMA Moolenbeek & Faber, 1984
MYCHOPOMA Blanford, 1869 ?
MICRAULAX Theobald, 1876
MYXOSTOMA Troschel, 1847
PINTERIA Varga, 1972 ?
PTYCHOPOMA Möllendorff, 1885
SCABRINA Blanford, 1863
Dasytherion Mabille, 1887
THEOBALDIUS G. Nevill, 1878
TOFFOLETTIA Giusti, 1971

Placement ? (10)
BARBACYCLOTUS, CALACYLOTUS, FILOCYCLUS, LAGOCYCLUS, MEGACYCLOTUS all Bartsch, 1942

SPIROSTOMATINAE
SPIROSTOMA Heude, 1885
Coelopoma A. Adams, 1867
(non Agassiz, 1839)
Spiropoma Kobelt & Möllendorff, 1900

ALYCAEINAE (12)
ALYCAEUS Gray, 1850
Orthalycaeus Pfeiffer, 1876 (part)
Awalycaeus Kuroda, 1956 (5)
PINCERNA Preston, 1907 ?
CHAMALYCAEUS Kobelt & Möllendorff, 1897
DICHARAX Kobelt & Möllendorff, 1900
Charax Benson, 1859
(non Risso, 1826)
METALYCAEUS Pilsbry, 1900
RAPTOMPHALUS Godwin-Austen, 1914
CYCLORYX Godwin-Austen, 1914 ?
DIORYX Benson, 1859

PTEROCYCLINAE ?
PTEROCYCLUS Benson, 1832 (Pterocyclos - err.)
Spiraculum Pearson, 1833 (part)
Steganotoma Troschel, 1837
PEARSONIA Kobelt, 1902
PSEUDOSPIRACULUM Kobelt, 1902
Spiraculum Pearson, 1833 (part)
Diplopterum Kobelt & Möllendorff, 1897
(non Swainson, 1839)
PLATYRHAPHE Möllendorff, 1890 ?
Cryptaulus Bavay, 1903
RHIOSTOMA Benson, 1860

POTERIIDAE (Amphicyclotidae)
POTERIA Gray, 1850
Platystoma Mörch, 1860
(non Meigen, 1803)
Ptychocochlis Simpson, 1894
Plectocyclotus Kobelt & Möllendorff,1897
CYCLOBAKERIA Bartsch, 1942
AMPHICYCLOTUS Crosse & Fischer, 1879 (6)
Cyclopoma Troschel, 1847
(non Agassiz, 1833)
CYCLOBLANDIA Bartsch, 1942
BUCKLEYIA Higgins, 1872
Buckleya Mousson, 1873 (err.)
CALAPEROSTOMA Pilsbry, 1935 ?
CROCIDOPOMA Shuttleworth, 1857 (11)
CYCLOCUBANA Bartsch, 1942 (1)
CYRTOTOMA Mörch, 1852
Aperostoma Troschel, 1847 (part)
Habropoma Fischer & Crosse, 1880

POTERIIDAE (Cont'd.)
GASSIESIA Clench, 1949
GONATORAPHE Möllendorff, 1898 (ex Rissoidea)
MEXCYCLOTUS Bartsch, 1942
NEOCYCLOTUS Fischer & Crosse, 1886
OSTODES Gould, 1862 (5)
DUBLONIA Clench, 1949

Placement ? (6)
FISIOPOMA, KONDORHAPHE Clench, 1949
PARAMIELLA Clench, 1949
pro Paramia Clench, 1949
(non Bleeker, 1863)

MEGALOMASTOMIDAE (11)
MEGALOMASTOMA Swainson, 1840
MEGALOMASTOMOIDES Bartsch, 1942
NEOPUPINA Kobelt, 1902
Megaloma Woodward, 1854
(non Hall, 1852
Lomastoma Woodward, 1856
(non Rafinesque, 1819)
APEROSTOMA Troschel, 1847 (10)
AUSTROCYCLOTUS, CYCLADAMSIA, CYCLOHIDALGOA, INCIDOSTOMA all Bartsch, 1942
CYCLOJAMAICIA Bartsch, 1942
CYCLOPILSBRYA Bartsch, 1942
CYCLOCAYMANIA Bartsch, 1942
CYCLOVENDREYSIA Bartsch, 1942
FARCIMEN Troschel, 1847
FARCIMOIDES Bartsch, 1942
LIRACYCLOTUS Bartsch, 1942
TOMOCYCLUS Crosse & Fischer, 1872

HAINESIINAE ?
HAINESIA Pfeiffer, 1856
Dacrystoma Crosse & Fischer, 1871
Mascaria Angas, 1878
ACROPTYCHIA Crosse & Fischer, 1877
Euptychia Crosse & Fischer, 1873
(non Hübner, 1816)
ANCEYIELLA Mabille, 1886

MAIZANIIDAE
MAIZANIA Bourguignat, 1889
Aferulus von Martens, 1897
Hijabia, Natalia Godwin-Austen, 1897
Austrocyclus Ancey, 1898
Cyclophoropsis Dautzenberg, 1908
CRASPEDOPOMA Pfeiffer, 1847
Bolania Gray, 1840 (nom.nud.)
Hygrobium Lowe, 1852
MAIZANIELLA Bequaert & Clench, 1936 ?
MICROMAIZANIELLA, PTEROMAIZANIELLA, SPIRULOZANIA all Van Bruggen, 1982
NEOMAIZANIA Van Bruggen, 1985
THOMEOMAIZANIA Bequaert & Clench, 1936

DIPLOMMATINIDAE
DIPLOMMATININAE
DIPLOMMATINA Benson, 1849
Eudiplommatina Kobelt & Möllendorff,1898
ANGIGASTER Pilsbry & Hirase, 1904
DIPLOPTYCHIA, METADIANCTA, SINICA all Möllendorff, 1885
MOUSSONIA Semper, 1865
ADELOPOMA Döring, 1884
Cylindropalaina, Eupalaina Kobelt & Möllendorff, 1898 ?
ANOSTOMELLA von Martens, 1867
ARINIA H. & A. Adams, 1856
Euarinia Kobelt & Möllendorff, 1898
LEUCARINIA Möllendorff, 1893

DIPLOMMATININAE (Cont'd.)
CLOSTOPHIS Benson, 1860
DIANCTA von Martens, 1867
Eudiancta Kobelt & Möllendorff, 1898
PARADIANCTA Möllendorff, 1895
GASTROPTYCHIA Kobelt & Möllendorff, 1900
Paxillus H. & A. Adams, 1851
HELICOMORPHA Möllendorff, 1890
MESSAGERIA Bavay & Dautzenberg, 1903
HUNGERFORDIA Beddome, 1889
NICIDA W. Blanford, 1868
OPISTHOSTOMA W. & H. Blanford, 1860
Gyrostropha Ancey, 1887
GEOTHAUMA Crosse, 1892
PLECTOSTOMA H. Adams, 1865
PALAINA Semper, 1865
CYLINDROPALAINA Möllendorff, 1897
MACROPALAINA Möllendorff,1897
PSEUDOPALAINA Möllendorff, 1898
ECLOGARINIA Iredale, 1933
VELEPALAINA Iredale, 1937

COCHLOSTOMATINAE
COCHLOSTOMA Jan, 1830
Pomatias Hartmann, 1821
(non Stüder,1789)
Hartmannia R. B. Newton, 1891
Maculatus Westerlund, 1883
Scalarinella Clessin, 1889
(non Döring in Dohrn, 1874)
Strobelia Clessin, 1889
Eupomatias A. J. Wagner, 1897
AURITUS Westerlund, 1883
Pomatiella Clessin, 1889 (part)
HOLCOPOMA Kobelt & Möllendorff, 1899
Pleuropoma A. J. Wagner, 1897
(non Möllendorff, 1893)
OBSCURELLA Clessin, 1889
Anotus Westerlund, 1883 (part)
(non Agassiz, 1846)
Rhabdotakra A. J. Wagner, 1897
PERSONATUS Westerlund, 1883
Stereopoma A. J. Wagner, 1897
TITANOPOMA A. J. Wagner, 1897
TURRITUS Westerlund, 1883

POLLICARIINAE ? (12)
POLLICARIA Gould, 1856
Hainesia L. Pfeiffer, 1856 (part)
Hybocystis Benson, 1859

PUPINIDAE (Terrestrial)
PUPININAE
PUPINA Vignard, 1829
Regestoma Hasselt, 1823 ?
Registoma Gray, 1842
AMBIPUPINA, DIPLOPUPINA, DOLIPUPINA, ESTOPUPINA all Iredale, 1937
SIPHONOSTYLA Kobelt, 1897
TYLOTOECHUS Kobelt & Möllendorff, 1897
Eupupina L. Pfeiffer, 1876 (part)
Mesostoma Heude, 1886
(non Ehrenberg, 1836)
SUAVOCALLIA Iredale, 1933
CALLIANELLA R. B. Newton, 1891
Callia Gray, 1841 (non Serville, 1835)
HARGRAVESIA H. Adams, 1870
Hyalopsis Pease, 1871
HEDLEYA Cox, 1892 ?
HILDAPUPINA, LOPUPINA, NECOPUPINA all Iredale, 1937
SIGNEPUPINA Iredale, 1937
Parpupina Iredale, 1937

PUPININAE (Cont'd.)
MOULINSIA Grateloup, 1840
POROCALLIA Möllendorff, 1893

PUPINELLINAE
PUPINELLA Gray, 1850
PUPINOPSIS H. Adams, 1866
BELLARDIELLA Tapparone-Canefri, 1883
NODOPOMATIAS Gude, 1921
Eupomatias Godwin-Austen, 1917
(non A. J. Wagner, 1897)
PSEUDOPOMATIAS Möllendorff, 1885
Fargesia Heude, 1886
RAPHAULUS Pfeiffer, 1856
Anaulus Pfeiffer, 1855
(non Ehrenberg, 1844)
SCHISTOLOMA Kobelt, 1902
Coptocheilus Gould, 1862
(non Amyot & Serville, 1843)
STREPTAULUS Benson, 1857
TORTULOSA Gray, 1847
Cataulus Pfeiffer, 1851
EUCATAULUS Kobelt, 1902

LIAREIDAE (4) (? Realiidae)
LIARIA Gray, 1852
Realia (Gray, 1849) Pfeiffer, 1851
CYTORA Kobelt & Möllendorff, 1897
Murdochia Ancey, 1901

VIVIPAROIDEA (mainly freshwater) (2)
VIVIPARIDAE
VIVIPARINAE
VIVIPARUS Montfort, 1810
Vivipara Sowerby, 1813
Viviparella Rafinesque, 1815
? Paludina Lamarck, 1816
(? non Férussac, 1812)
? Centrapala, Notopala Cotton, 1964
CALLINA Hannibal, 1912 (non Lowe, 1854 ?)
DACTYLOCHLAMYS Rao, 1925
HETEROGEN Annandale, 1921
LECYTHOCONCHA Annandale, 1920
LAGUNCULA Benson, 1842 ?
RECTIVIVIPARUS Pilsbry & Bequaert, 1927 ?
MARGARYA Nevill, 1877
RIVULARIA Huede, 1890
TAIA Annandale, 1918 ?
TEMNOTAIA Annandale, 1919
TULOTOMA Haldeman, 1840 (8)
Tylotoma Fischer, 1885

BELLAMYINAE (2)
BELLAMYA Jousseaume, 1886
ANULOTAIA Brandt, 1968
CIPANGOPALUDINA Hannibal, 1912 (10)
EYRIESIA Fischer, 1885
FILOPALUDINA Habe, 1964
IDIOPOMA Pilsbry, 1901

BELLAMYINAE (Cont'd.)
MEKONGIA Crosse & Fischer, 1876
NEOTHAUMA E. Smith, 1880
SIAMOPALUDINA Brandt, 1968
SINOTAIA Haas, 1939
TROCHOTAIA Brandt, 1974

CAMPELOMINAE (Lioplacinae - 8)
CAMPELOMA Rafinesque, 1819
Ambloxis Rafinesque, 1818 ?
Melantho Bowdich, 1822
LIOPLAX Troschel, 1856

AMPULLARIIDAE (Pilidae) (8)
AMPULLARIA Montfort, 1810
Conchylium Cuvier, 1816
CERATODES Guilding, 1828
FELIPPONEA Dall, 1919
POMELLA Gray, 1847 ?
AFROPOMUS Pilsbry & Bequaert, 1927
ASOLENE Orbigny, 1837
Ampulloidea Orbigny, 1840
LIMNOPOMUS Dall, 1904
LANISTES Montfort, 1810
LEROYA Grandidier, 1887
MELADOMUS Swainson, 1840
MARISA Gray, 1824 ?
PILA Röding, 1798
Pomus Humphrey, 1797
Ampullaria Lamarck, 1799 ?
Pachystoma Guilding, 1828
(non Latreille, 1809)
Pachylabra Swainson, 1840
TURBINCOLA Annandale & Prashad, 1921
POMACEA Perry, 1811
SAULEA Gray, 1867

VALVATOIDEA (Freshwater; brackish)
VALVATIDAE
VALVATA O. F. Müller, 1774
Gyrorbis Fitzinger, 1833
Planella Schlüter, 1838
Planorbitina Betta, 1870
ATROPIDINA Lindholm, 1906
LIRATINA Lindholm, 1906
COSTOVALVATA Polinski (1929) 1932
MICROCONCINNA Starobogatov, 1983
TROPIDINA H. & A. Adams, 1854
VANCLEAVEIA F. C. Baker, 1930 ?
ANDRUSOVIA Brusina, 1903
BORYSTHENIA Lindholm, 1913
Jelksia Bourguignat, 1877
(non Taczanowski, 1871)
CINCINNA (Hübner, 1810) Férussac, 1821
Valvatinella Betta, 1870
PAMIROCINCINNA Starobogatov & Sitnikova,1983
SIBEROVALVATA Starobogatov & Zatravkin, 1985
MEGALOVALVATA Lindholm, 1906
BIWAKOVALVATA Starobogatov, 1983 ?

References:
Alcalde, A. & M.K. Jacobson "New operculate snail: Jamaica" Nautilus 72(4):111-115 - 1959 (1)

Brandt, R.A.M. "The non-marine aquatic Mollusca: Thailand" Arch.Moll.105(1-4):1-423-1974 (2)

Burch, J.B. "Outline: classification; Aust. terr. molluscs" J.Mal.Soc.Aust.3(3):127-156.1976 (3)

Climo, F.M."Systematic position: Cytora, Liarea"(4) Trans.Roy.Soc.N.Z.Biol.Sci.12(18)209-212 -1970

Girardi, E. L. "The Samoan land snail Ostodes" Veliger 20(3):191-250 - 1978 (5)

Morrison, J.P.E."American cyclophorid land snails" J.Wash.Acad.Sci. 45(5):149-162 - 1955 (6)

Minato H. & T. Habe "Land shell fauna of Ujigunto, Kusakakigunto I"Venus 41(2):124-140-1982 (7)

Pain, T. "The Ampullariidae: historical survey" Journ.Conch.27(7):453-462 - 1972 (8)

References: (Mesogastropoda, Cont'd.)

Patterson, C. M. "Chromosomes: Tulotoma angulata" Malacologia 2(2):259-265 - 1965 (8x)

Solem, A. "Systematics of the land and freshwater Mollusca of the New Hebrides" Fieldiana: Zool.43(1):1-359 - 1959 (9)

Taylor, D. W. "Remarkable snail fauna: Coahuila, Mexico" Veliger 9(2):152-228 - 1966 (10)

de la Torre, C., P. Bartsch & J. P. E. Morrison "The Cyclophorid operculate land mollusks of America" Bull.U.S.Nat.Mus.181:1-306-1942(11)

Thompson, F. G. "Some Mexican and Cent. American land snails of the family Cyclophoridae" Zoologica 54(2):35-77 - 1969 (13)

Van Bruggen, A.C. "Further notes on Afrotropical prosobranch land molluscs."Proc.K.Ned.Acad. Wet (Biol.Med.Sci.89(4):357-358 - 1986 (14)

LITTORINOIDEA
LITTORINIDAE
LACUNINAE (5)
LACUNA Turton, 1827
Temana Leach in Gray, 1847
BOETICA Dall, 1918
EPHERIA Leach in Gray, 1847
Epheriella Pallary, 1914
TEMANELLA Rovereto, 1899
Medoria Leach in Gray, 1847
(non Robineau-Desvoidy, 1830)
AQUILONARIA Dall, 1886
CARINOLACUNA Thiele, 1929
HALOCONCHA Dall, 1886
Lacunaria Dall, 1884
(non Conrad, 1866)
Lacunella Dall, 1884
(non Deshayes, 1861)
LACUNITUNICA Golikov & Gulbin, 1984
STENOTIS A. Adams, 1863
Sublacuna Pilsbry, 1895

LITTORININAE
LITTORINA Férussac, 1822
Neritoides T. Brown, 1827
(non Meuschen, 1779)
Litorina Menke, 1828
Litormia Mörch, 1849
Neritrema Récluz, 1869
Litonia Weinkauff, 1878
Isonema Hall in Provancher, 1891
(non Meek & Worthern, 1886)
Algaroda, Littorivaga Dall, 1918
Neritotrema Wenz, 1939
Ezolittorina Habe, 1958
ALGAMORDA Dall, 1918
AUSTROLITTORINA Rosewater, 1970 (8)
LITTORINOPSIS Mörch, 1876
Lamellitorina Tryon, 1887
MELARHAPHE Menke, 1828
Melaraphis Philippi, 1836
LAEVILACUNARIA Powell, 1951
PELLILACUNELLA Powell, 1951
LAEVILITORINA Pfeffer, 1886
CORNEOLITORINA Powell, 1951
LITTORARIA Griffith & Pidgeon, 1834
PALUSTORINA Reid, 1986 (6)
MACQUARIELLA Finlay, 1926
MAINWARINGIA Nevill, 1884 (6a)
NODILITTORINA von Martens, 1897 (1)
Echinolittorina Habe, 1956
GRANULILITTORINA Habe & Kosuge, 1966 (8)
LIRALITTORINA Rosewater, 1981
TECTININUS Clench & Abbott, 1942
PEASIELLA Nevill, 1884
PELLILITORINA Pfeffer, 1886
PROBLITORA Iredale, 1931 (?)
RISELLOPSIS Kesteven, 1902
RISSOLITORINA Ponder, 1966

TECTARIINAE
TECTARIUS Valenciennes, (1832)
Cidaris Röding, 1798
(non Leske, 1778)
Echinella Swainson, 1840
(non Bory St. Vincent, 1824)
Pagodus Gray, 1839
Pagodella Swainson, 1840
Echinellopsis Rovereto, 1899
(non Bory St. Vincent, 1824)
Hamus Klein in H. & A. Adams, 1858
(non Klein in Watson, 1886)
Tectarium P. Fischer, 1885
CENCHRITIS Von Martens, 1900

ECHINININAE
ECHININUS Clench & Abbott, 1942
Echinella Swainson in H.& A. Adams,1854
(non Swainson, 1840)
Nina Gray, 1850 (non Horsfield, 1829)

BEMBICIINAE
BEMBICIUM Philippi, 1846
Risella Gray, 1847
CREMNOCONCHUS Blanford, 1869
Cremnobates Blanford, 1868
(non Swainson, 1855)
LISSOCONCHUS Thiele, 1929

EATONIELLIDAE (5a)
EATONIELLA Dall, 1876
Eatonia E. Smith, 1875
(non Hall, 1857)
Dardanula Iredale, 1915
pro Dardania Hutton, 1882
(non Stål, 1860)
Cerostraca Oliver, 1915
Abscindostoma, Albitoniella,
Dardaniopsis Ponder, 1965
ALBOSABULA Ponder, 1965
CAVEATONIELLA Ponder, 1965
PELLAX Finlay, 1927
CRASSITONIELLA Ponder, 1965
LIRATONIELLA Ponder, 1965
PUPATONIA Ponder, 1965

POMATIASIDAE (Terrestrial)
POMATIASINAE
POMATIAS Stüder, 1789
Cyclostoma Draparnaud, 1801
(non Lamarck, 1799)
Cyclostomus Montfort, 1810
Erica Moquin-Tandon, 1848
ARCHITEA Da Costa, 1869 (3)
Archytaea Fischer, 1885 (err.)
GEORGIA Bourguignat, 1882
Revoilia Bourguignat, 1882 ?
ARABIA Fischer-Piette ?
GUILLAINIA Crosse, 1884

POMATIASINAE (Cont'd.)
LEONIA Gray, 1850
LITHODON Gray, 1850
OTOPOMA Gray, 1850
Socotora Pallary, 1925
TROPIDOPHORA Troschel, 1847
Eutropidophora Kobelt & Möllendorff,1898
LIGATELLA von Martens, 1880
Rochebrunia Bourguignat, 1882
TUDORELLA Fischer, 1885

CYCLOTOPSINAE
CYCLOTOPSIS Blanford, 1864

ANNULARIIDAE (9)
ANNULARIINAE (Choanopomatinae)
ANNULARIA Schumacher, 1817
Choanopoma Pfeiffer, 1848
Annularodes, Annularosa, Annularita, all Henderson & Bartsch, 1920
ANNULARELLA Henderson & Bartsch, 1920 (4x)
ANNULAROPS Henderson & Bartsch, 1920
ANNULARODISCA Henderson & Bartsch, 1920
DIPLOPOMA Pfeiffer, 1859
SUBANNULARIA Jaume, 1984
ABBOTTELLA Henderson & Bartsch, 1920
BLAESOSPIRA Crosse, 1890
CHRISTOPHIPOMA Bartsch, 1946
COLONELLA Bartsch, 1946
EYERDAMIA Bartsch, 1946
EUTUDORA Henderson & Bartsch, 1920
Eutudorisca, Eutudorella, Eutudorops all Henderson & Bartsch, 1920
JAMAICIA C. B. Adams, 1850
LAGOPOMA Bartsch, 1946
MEGANIPHA Thompson, 1978 (9)
PETASIPOMA Bartsch, 1946
RAMSDENIA Preston, 1913
ROLLEIA Crosse, 1891
TUDORA Gray, 1850
Tudorellata, Tudorisca, Tudorops all Henderson & Bartsch, 1920
COLOBOSTYLUS Crosse & Fischer, 1888 (4)
Colobostyloides Bartsch, 1946
BONAIREA, TUDORATA H. B. Baker, 1924
WEINLANDIPOMA Bartsch, 1946

ADAMSIELLINAE
ADAMSIELLA Pfeiffer, 1851
ADAMSIELLOPS Henderson & Bartsch, 1920

CHONDROPOMINAE (Chondropomatinae)
CHONDROPOMA Pfeiffer, 1847
Chondropomatus, Chondropometes, Chondropomartes Henderson & Bartsch,1920
Chondrops, Levipoma Bartsch, 1946
Chondropomisca de la Torre & Bartsch,1938
ARTICULOPOMA Bartsch, 1946
CHONDROPOMELLA Bartsch, 1932
CHONDROPOMIUM Henderson & Bartsch, 1920

CHONDROPOMA (Cont'd.)
CHONDROPOMONUS Henderson & Bartsch, 1920
GUTIERREZIUM Jaume, 1984
PARACHONDRODES Henderson & Bartsch, 1920
PARACHONDROPS Henderson & Bartsch, 1920
CHONDROTHYRA Henderson & Bartsch, 1920
Chondrothyretes, Chondrothyrium, Chondrothyroma Henderson & Bartsch,1920
PLICATHYRA Jaume, 1984

CISTULOPSINAE
CISTULOPS H. B. Baker, 1924
Cistula Gray, 1847 (4) (non Say, 1825)
Citula (Humphreys, 1797) (rej.)
LICINA Gray, 1847
CHOANOPOMOPS H. B. Baker, 1924
TROSCHELVINDEX H. B. Baker, 1924

RHYTIDOPOMATINAE
RHYTIDOPOMA Sykes, 1901
Pro Ctenopoma Pfeiffer, 1856 (non Peters, 1844)
Crossepoma, Hispanipoma, Orcuttipoma, Colonina all Bartsch, 1946
TORRELLA Henderson & Bartsch, 1920
TORRELLISCA Henderson & Bartsch, 1920
CLYDONOPOMA Pilsbry, 1933
ECCRITOPOMA Pilsbry, 1933
OPISTHOSIPHON Dall, 1905
Opisthosiphona Henderson & Bartsch,1920
Leptopisthosiphon Bartsch, 1946
PARACHONDRIA Dall, 1905
Parachondrisca, Parachondrella, Henderson & Bartsch, 1920
Clenchipoma, Haitipoma, Kisslingia, Klattea, Sallepoma all Bartsch,1946
RHYTIDOTHYRA Henderson & Bartsch, 1920
XENOPOMA Crosse, 1890

ACICULIDAE (4) (Rissoacea ?)
ACICULA Hartman, 1821
Auricella Jurine, 1817 (part)
Acme Hartman, 1821
Acutalia Schaufuss, 1869
Pupula Charpentier, 1837
PLATYLA Moquin-Tandon, 1855
HYALACME P. Hesse, 1917
CAZIOTIA Pollonera, 1905
MENKIA Boeters, Gittenberger & Subai, 1985 (2)
RENEA Nevill, 1880
PLEURACME Kobelt, 1894
Megalacme Kobelt & Möllendorff, 1897

Placement ?
ABCHASOHELA Hüdec & Lezhawa,1971 (Helicidae ?)
pro Hela Hüdec & Lezhawa, 1970 (non Münster, 1830)
KAINARELLA Starobogatov, 1972
WATSONIA Fischer, 1971 (non de Folin, 1879)

References:
Bandel, K. & D. Kadolsky "Western Atlantic Nodilittorina" Veliger 25(1):1-42 - 1982 (1)

Boeters, H. D., E. Gittenberger Y. P. Subai "Eine neue gattung der Aciculidae" Basteria 49 (1-3):59-63 - 1985 (2)

Boss,K.J. & K.S. Merrill "Architea costa not an Architonicid but a Pomatiasid" Nautilus 98(2):77-79 - 1984 (3)

Fretter, V. & A. Graham "The Prosobranch Mollusks of Great Britain and Denmark" Part III J.Moll.Stud. Supp.5 - 1978 (4)

Jaume, M.L. "Rectifications: Molluscos terrestres" Misc.Zool.la Habana 20(1-4) - 1984 (4x)

Ponder, W.F. "Lacuninae: Subfamily of Littorinidae" Malac.Review - 1976 (5)

References: (Littorinoidea - Cont'd.)
Ponder, W.F. & E.K. Yoo "Revision: Eatoniellidae of Aust." Rec.Aust.Mus.31:606-658 -1977(1978)(5a)

—— & J. Rosewater "Rectifications: Littorinidae" Proc.Biol.Soc.Wash.92(4):773-782 - 1979 (5b)

Reid, D.G. "Littorinid mollusks: mangroves, Indo-Pac."Br.Mus.(Nat.Hist.)978:222 pp. 1986(6)

—— "Mainwaringia; littorinid genus from Asia" Jour.Moll.Stud.52(3):225-242 - 1986 (6a)

Rosewater, J. & D. Kadolsky "Rectifications: II" Proc.Biol.Soc.Wash.94(4):1233-1236 1981 (8)

Thompson, F.G."New genus:comments; Annulariidae" Nautilus 92(1):41-54 - 1978 (9)

RISSOIDEA
HYDROBIIDAE (Hydrobioidea - 19c)
HYDROBIINAE
HYDROBIA Hartmann, 1921
Paludestrina Orbigny, 1839
Littorinella Braun, 1842 ?
Subulina Schmidt, 1851
(non Beck, 1837)
Ecrobia Stimpson, 1865
Eupaludestrina, Pseudopaludinella,
Thalassiobia all Mabille, 1877
SABINEA Sowerby, 1842
Peringia Paladilhe, 1874 ?
ADRIOHYDROBIA Radoman, 1973
ANTROSELATES Hubricht, 1963
AROAPYRGUS H. B. Baker, 1931
pro Aroa H. B. Baker, 1930
(non Walker, 1855)
CATAPYRGUS Climo, 1974
FLUVIDONA Iredale ?
FLUVIOPUPA Pilsbry, 1911
HADOPYRGUS Climo, 1974
HEMISTOMIA Crosse, 1872
HETEROCYCLUS Crosse, 1872
HOYIA F.C. Baker, 1926
LITHOGLYPHOIDES Sturany & Wagner, 1914
Zagrabica Brusina, 1884
LOBOGENES Pilsbry & Bequaert, 1927
OBROVIA Radoman, 1973
OPACUINCOLA Ponder, 1966
PALUDINELLA Pfeiffer, 1841 ? (20)
Solenomphala Boettger, 1887
Assiminella Monterosato, 1906
LEUCOSTELE Semper in Thiele, 1927 ? (23)
RUPACILLA Thiele, 1927 ? (23)
SCHUTTIELLA Brandt, 1974
PAXILLOSTIUM Gardner, 1970
PETTERDIANA Brazier, 1896
Brazieria Petterd, 1889
(non Ancey, 1887)
Pseudoampullaria Ancey, 1898
TASMANIELLA Ancey, 1898
Beddomeia Petterd, 1889
(non Nevill, 1878)
PHREATODROBIA Hershler & Longley, 1986
PHREATOCERAS Hershler & Longley, 1987 (14b)
pro Hadoceras Hershler & Longley, 1986
(non Strand, 1934)
POTAMOLITHUS Pilsbry, 1896 (22a)
LITHOCOCCUS Pilsbry, 1911
PROBYTHINELLA Thiele, 1928
PSEUDAMNICOLA Paulucci, 1878
ABESKUNUS Kolesnikov, 1976
PTERIDES Pilsbry, 1909
PYRGOPHORUS Ancey, 1888 ?
SOAPITIA Binder, 1961
STAADTIA Schlickum, 1961
SYROFONTANA Schütt, 1983
TATEA Tenison-Woods, 1879
TRICHONIA Schütt, 1980
VENTROSIA Radoman, 1977

AMNICOLINAE (22a; 5a)
AMNICOLA Gould & Haldeman, 1841
Euamnicola Crosse & Fischer, 1891
LYOGYRUS Gill, 1863

BELGRANDIELLINAE ? (2b)
BELGRANDIELLA Wagner, 1927
Microna Clessin, 1890
(non Ziegler in Frauenfeld, 1863)
Graziana Radoman, 1975
pro Frauenfeldia Clessin, 1878
(non Egger, 1863)
LITHABITELLA Boeters, 1970
TURCORIENTALIA Radoman, 1973
MARTENSAMNICOLA Izzatullaev et al, 1985 (15x)
NUREKIA Izzatullaev et al, 1985
TURKMENAMNICOLA Izzatullaev et al, 1985
VALVATAMNICOLA Izzatullaev et al, 1985

BYTHINELLINAE ? (11)
BYTHINELLA Moquin-Tandon, 1855
Microna Ziegler, 1852 ?
Maresia Bourguignat, 1877
MARSTONIOPSIS Altena, 1936
PARABYTHINELLA Radoman, 1973

COCHLIOPINAE (21)
COCHLIOPA Stimpson, 1865
CLENCHIELLA Abbott, 1948 (4)
Carulinorbis Yen, 1979
pro Carinorbis Yen, 1946
(non Conrad, 1862)
LIMNOTHAUMA Haas, 1955
NANIVITREA Thiele, 1927 ?
SUBCOCHLIOPA Morrison, 1946

FONTIGENTINAE (5)
FONTIGENS Pilsbry, 1933
Stimpsonia Clessin, 1878
(non Girard, 1853)

GRAECOANATOLICINAE (19x)
GRAECOANATOLICA Radoman, 1973

HORATIINAE (11; 2)
HORATIA Bourguignat, 1887
DAPHNIOLA Radoman, 1973
DAUDEBARDIELLA Boettger, 1905
ARGANIELLA Giusti & Pezzoli, 1984
BOLEANA Radoman, 1973
BYTHIOSPEUM Bourguignat, 1882
CAVERNISA Radoman, 1978
ERYTHROPOMATIANA Radoman, 1978
FISSURIA Boeters, 1981
GOCEA Hadzisce, 1956
Ohridohoratia Hadzisce, 1956
HADZIELLA Kuscer, 1932 (11c)
HAUFFENIA Pollonera, 1898 (? Pseudohoratiinae)
NEOHORATIA Schütt, 1961
VRANIA Radoman, 1978
KERKIA Radoman, 1978

HORATIINAE (Cont'd.)
IGLICA Wagner, 1927 (11b)
RHAPHICA Schütt, 1975
LANZAIA Brusina, 1906 (19x)
Costellina Kuscer, 1933
MOITESSIERIA Bourguignat, 1863 (2a)
OHRIGOCEA Hadzisce, 1956
Karevia Hadzisce, 1956
PALADILHIA Bourguignat, 1865
LARTETIA Bourguignat, 1869
Vitrella Clessin, 1877
(non Swainson, 1840)
LHOTELLERIA Bourguignat, 1877
Locardia de Folin, 1880
? PALADILHIOPSIS Pavlovic, 1913 (20)
Insignia Angelov, 1972
SPIRALIX Boeters, 1972
PEZZOLIA Bodon & Giusti, 1986
PLAGIGEYERIA Tomlin, 1930 (19x)
pro Geyeria Wagner, 1914
(non Buchecker, 1876)
Microsalpinx Kuscer, 1932
PONTOBELGRANDIELLA Radoman, 1978
SAGANOA Kuroda & Habe, 1958 (Suganoa - err.)
? Phreatica Velkovrh, 1970 (Genus ?)
Kuschelita Climo, 1974
SAXURINATOR Schütt, 1960
SOGDAMNICOLA Izzatullaev, Sitnokova
& Starobogatov, 1984
STRUGIA Radoman, 1973

ISLAMIINAE (11c)
ISLAMIA Radoman, 1973
AVENIONIA Nicolas, 1882 (11a)
Paulia Bourguignat, 1882
(non Gray,1841)
PAULUCCIA Giusti & Pezzoli, 1980

LITHOGLYPHINAE (22a)
LITHOGLYPHUS Hartmann, 1821
CLAPPIA Walker, 1909
ANTROBIA Hubricht, 1971 (15)
FLUMINICOLA Stimpson, 1865
Heathilla Hannibal, 1912
HYPSOBIA Heude, 1889 (23)
LEPYRIUM Pilsbry & Olsson, 1951
GILLIA Stimpson, 1865
LITHOGLYPHOPSIS Thiele, 1928
MANNINGIELLA Brandt, 1970
PARAPYRGULA Annandale & Prashad, 1919
PILSBRYUS Yen, 1944 (fossil ?)
SADUNIELLA Brandt, 1970
SHADINIA Akramovskij, 1976
SOMATOGYRUS Gill, 1863
WALKERILLA Thiele, 1928
TAIHUA Annandale, 1924
WYKOFFIA Brandt, 1968 ?

? LITHOGLYPHULINAE (19x)
LITHOGLYPHULUS Schlickum & Schütt, 1971

LITTORIDININAE (9b; 14)
LITTORIDINA Eydoux & Souleyet, 1852
Heleobia, Littorinida Stimpson, 1865
TEXADINA Abbott & Ladd, 1951
APHAOSTRACON Thompson, 1968 (9b)
BALCONORBIS Hershler & Longley, 1986
BRACHYPIRGULINA Haas, 1955
COAHUILIX Taylor, 1966 (21)
COCHLIOPINA Morrison, 1946 (14)
ECPOMASTRUM Haas, 1957
DURANGONELLA Morrison, 1945
HELIGMOPOMA Haas, 1955

LITTORIDININAE (Cont'd.)
HELEOBOPS Thompson, 1969 (9b)
LITTORIDINOPS Pilsbry, 1952
LYRODES Döring, 1884
MEXIPYRGUS Taylor, 1966
MEXITHAUMA Taylor, 1966
PALUDISCALA Taylor, 1966
RHAMPHOPOMA Haas, 1955
RUFOLACUNA Ponder, 1976
SPURWINKIA Davis, Mazurkiewicz
& Mandracchia, 1982 (9b)
STROMBOPOMA Haas, 1955
STYGOPYRGUS Hershler & Longley, 1986
TRYONIA Stimpson, 1865
Isaea Conrad, 1871
(non Edwards, 1830)
Dryis, Liris Conrad, 1871
Conradia Wenz, 1925
(non A. Adams, 1863)
Hyalopyrgus Thompson, 1968 (14c)
PAUPERTRYONIA Taylor, 1987
ZETEKINA Morrison, 1947
Zetekella Morrison, 1946
(non Drake, 1944)

NYMPHOPHILINAE (Orientaliidae) (22)
NYMPHOPHILUS Taylor, 1966
ALBANIANA Radoman, 1973
ALZONIELLA Giusti & Bodon, 1984 (11a)
ANAGASTINA Radoman, 1978
pro Anagasta Radoman, 1973
(non Heinrich, 1956)
ANTIBARIA Radoman, 1973
APACHECOCCUS Taylor, 1987 (21b)
BIRGELLA F. C. Baker, 1926 (22a)
BRACENICA Radoman, 1973
CINCINNATIA Pilsbry, 1891
DALMATINELLA Radoman, 1973 (?)
DOLAPIA Radoman, 1973
FALSIBELGRANDIELLA Radoman, 1973
GRAECORIENTALIA Radoman, 1973
GROSSUANA Radoman, 1973
HEIDEELLA Backhuys & Boeters, 1974 (1)
ISTRIANA Velkovrh, 1971
MALAPRESIA Radoman, 1973
MERCURIA Boeters, 1971 (?)
NARENTIANA Radoman, 1973
NAUMIA Radoman, 1973
NOTOGILLIA Pilsbry, 1953
OHRIDOHAUFFENIA Hadzisce, 1956
OHRIDOSTURANYA Radoman, 1973
ORIENTALINA Radoman, 1978
pro Orientalia Radoman, 1972
(non Bykova, 1947)
POLINSKIOLA Radoman, 1974
pro Sturanyia Radoman, 1962
(non Wagner, 1905)
PRESPIANA Radoman, 1973
PRESPOLITOREA Radoman, 1973
PSEUDAVIONIA Bodon & Giusti, 1982
PSEUDORIENTALIA Radoman, 1973
PYRGULOPSIS Call & Pilsbry, 1886 (14c)
Marstonia F. C. Baker, 1926
Fontelicella, Microamnicola,
Natricola all Gregg & Taylor, 1965
Mexistiobia Hershler, 1985
RAPHINEMA Thompson, 1969
ROTONDIA Radoman, 1964
SPILOCHLAMYS Thompson, 1968
STIOBIA Thompson & McCaleb, 1978
TERRANIGRA Radoman, 1978
TRICHONIA Radoman, 1973
VINODOLIA Radoman, 1973

NYMPHOPHILINAE (Cont'd.)
YAQUICOCCUS Taylor, 1987 (21b)
ZAUMIA Radoman, 1973
ZAVALIA Radoman, 1973

POTAMOPYRGINAE (2b)
POTAMOPYRGUS Stimpson, 1865
Huttonia Johnston, 1891
(non Kirk, 1882)
Austropyrgus Cotton, 1942 (23)
INDOPYRGUS Thiele, 1928

PSEUDOHORATIINAE (19x)
PSEUDOHORATIA Radoman, 1967
LYHNIDIA Hadzisce, 1956
STRUGIA Radoman, 1973

PYRGORIENTALIINAE (19b)
PYRGORIENTALIA Radoman, 1973
KIRELIA Radoman, 1973

REHDERIELLINAE (4)
REHDERIELLA Brandt, 1974

SADLERIANINAE (19x)
SADLERIANA Clessin, 1887
BELGRANDIA Bourguignat, 1869

SEMISALSINAE (11, 19x)
SEMISALSA Radoman, 1974
Thermohydrobia Paulucci, 1878
Falsihydrobia Chukhchin, 1975

HYDROBIIDAE (Placement ?)
AKIYOSHIA Kuroda & Habe, 1954
CASIOHYDROBIA Starobogatov, 1978
CORROSELLA Boeters, 1970
CILGIA Schütt, 1968
DABRIANA Radoman, 1974
DALMATELLA Velkovrh, 1970
MACEDOPYRGULA Radoman, 1974
pro Trachypyrgula Radoman, 1955
(non Cossmann, 1921)
ONOBOPS Thompson, 1969
ORYGOCERAS Brusina, 1882 (Orygoceratidae 21a)
PSEUDOBYTHINELLA Liu & Zhang, 1982
PSEUDOISLAMIA Radoman, 1979
SCHUETTEMMERICIA Schlickum, 1961
STROBELIELLA Cazzaniga, 1981
WUCONCHONA Kang-Zaibin, 1983

BITHYNIIDAE
BITHYNIA Leach, 1818 (Bythinia auctt. - err.)
Bulimus Scopoli, 1777 (part)
Bithinia Gray, 1821 (nom.nud.)
Elona Moquin-Tandon, 1855
(non H. & A. Adams, 1855 ?)
DIGONIOSTOMA Annandale, 1920
Sataria Annandale, 1924
GABBIA Tryon, 1865 (?)
Digyrcidum Letourneux in Locard, 1882
Alocinma Annadale & Prashad, 1919
EMMERICIOPSIS Thiele, 1928
PARABITHYNIA Pilsbry, 1928
Paranerita Annandale, 1920
(non Bourne, 1909)
SIERRAIA Connoly, 1929
HYDROBIOIDES Nevill, 1884
Parafossarulus Annandale, 1924
KOLHYMAMNICOLA Starobogatov, 1976
LIMNITESTA Mandahl-Barth, 1974
PARODIZIA Medina, 1959 (?)
PETROGLYPHUS Möllendorff, 1894 ?

BYTHINIIDAE (Cont'd.)
MYSORELLA Annandale, 1919
Mysoria Godwin-Austen, 1918
(non E. Y. Watson, 1893)
PSEUDOVIVIPARA Annandale, 1918
WATTEBLEDIA Crosse, 1886

HYDROCOCCIDAE
HYDROCOCCUS Thiele, 1928

POMATIOPSIDAE (Freshwater) (9)
POMATIOPSINAE
Tribe Pomatiopsini
POMATIOPSIS Tryon, 1862
Chilocyclus Gill, 1863
BLANFORDIA A. Adams, 1863
Vicina Pilsbry, 1924
CECINA A. Adams, 1861
COXIELLA E. Smith, 1894
FLORIDISCROBS Pilsbry & McGinty, 1949 (18f)
FUKUIA Abbott & Hunter, 1949
IDIOPYRGUS Pilsbry, 1911
Aquidauania Davis, 1979
Hydracme Haas, 1938
ONCOMELANIA Gredler, 1881
Hemibia Heude, 1889
Katayama Robson, 1915
Schistosomophora Bartsch, 1936
TOMICHIA Benson, 1851
Tribe Erhaiini
ERHAIA Davis & Kuo, 1985 (9d)

TRICULINAE (9, 9d)
Tribe Triculini
TRICULA Benson, 1843
HALEWISIA Davis, 1979
PACHYDROBIA Crosse & Fischer, 1876
ROBERTSIELLA Davis & Greer, 1980
LACUNOPSIS Deshayes, 1876
NEOTRICULA Davis, 1986
FENOUILIA Heude, 1889
DELAVAYA Heude, 1889
PARAPROSOSTHENIA Annandale, 1919 (fossil)
SIBEROBYTHYNELLA Ioganzen, 1982 (?)
Tribe Julieniini
JULLIENIA Crosse & Fischer, 1876
KUNMINGIA Davis & Kuo, 1984
KARELAINIA Davis, 1979
HYDRORISSOIA Bavay, 1895
PACHYDROBIELLA Thiele, 1918
NEOPROSOSTHENIA Davis & Kuo, 1984
HUBENDICKIA Brandt, 1968

TRUNCATELLIDAE
TRUNCATELLA Risso, 1826 (Brackish water)
Fidelis Risso, 1826 (juv.)
Choristoma Cristofori & Jan, 1832
(non Hartman, 1840)
Erpetometra Lowe, 1830 (1832 ?)
Herpetometra Lowe in Gray, 1856 (err.)
Truncatula Leach in Gray, 1847
Truncatula Risso in Caziot, 1910 (err.)
(non Leach, 1847)
Glaucothoe, Zeanoe Leach, 1852
Albertisia Issell, 1880
TAHEITIA H. & A. Adams, 1863 ?
Blandiella Guppy, 1871
TOMLINITELLA Clench & Turner, 1948 ?
Tomlinella Clench & Turner, 1938
(non Viader, 1938)
GEOMELANIA Pfeiffer, 1845 (Terrestrial)
SCALATELLA von Martens, 1860
CHITTYA H. & A. Adams, 1858 (Chittia - err.)

STENOTHYRIDAE (Fresh, brackish)
STENOTHYRA Benson, 1856
pro Nematura Benson, 1836
(non Fischer, 1813)
Nematuna auctt. (err.)
Obesitena Iredale, 1943
GANGETIA Ancey, 1890
Astenothyra Annandale & Prashad, 1921
IRANOTHYRA Schütt, 1973

FALSICINGULIDAE ? (18f)
FALSICINGULA Habe, 1958

IRAVADIIDAE (Freshwater) (18e)
IRAVADIA Blanford, 1867
Irawadia auctt. (err.)
FLUVIOCINGULA Kuroda & Habe, 1954
Mesodestea Laseron, 1956
PSEUDOMERELINA Ponder, 1984
PSEUDONOBA Boettger, 1902
Paranoba Laseron, 1950
Dipsotoma, Lucidinella Laseron, 1956
Iraquirissoa Dance & Eames, 1966
ACLICERATIA Ponder, 1984
CERATIA H. & A. Adams, 1852
Hebetaclis Pilsbry, 1922
CHEVALLIERIA Cossmann, 1888
Nanadoma Laseron, 1956
FAIRBANKIA Blanford in Stoliczka, 1868
Pellamora Iredale, 1943
Wakauraia Kuroda & Habe, 1954
HYALA H. & A. Adams, 1852
LIROCERATIA Ponder, 1984
NOZEBA Iredale, 1915
Pasitheola Cossmann, 1896 (unnec.)
pro Pasithea Lea, 1833
Antinodulus Cossmann, in Cossmann
& Peyrot, 1919
Syntharella Laseron, 1955
RHOMBOSTOMA Seguenza, 1876 (fossil ?)
RISSOPSIS Garrett, 1873

BAICALIIDAE
BAICALIA von Martens, 1876
Limnorea Dybowski, 1875
(non Péron & Lesueur, 1809)
Ligea Dybowski, 1875
Trachybaicalia von Martens, 1876
BAICALIELLA Lindholm, 1909
EUBAICALIA Lindholm, 1924
DYBOWSKIOLA Lindholm, 1913
Dybowskia Dall, 1877
(non Jakolev, 1876)
GERSTFELDTIA Clessin, 1880
GODLEWSKIA Crosse & Fischer, 1879
LIOBAICALIA von Martens, 1876 ?
MAACKIA Clessin, 1880
PARABAICALIA Lindholm, 1909
PSEUDOBAICALIA Lindholm, 1909
TERATOBAICALIA Lindholm, 1909
TRICHIOBAICALIA Lindholm, 1909
PYRGOBAICALIA Starobogatov, 1972

BENEDICTIIDAE
BENEDICTIA Dybowski, 1875
BAICALCOCHLEA Lindholm, 1927
DALAINORIA Lindholm, 1927
KOBELTOCOCHLEA Lindholm, 1909

MICROMELANIIDAE
MICROMELANIINAE
MICROMELANIA Brusina, 1874
Goniochilus Sandberger, 1874

? PYRGULINAE (Pyrgulidae, 19x, 11c)
PYRGULA Cristofori & Jan, 1832
Brachypyrgula Polinski, 1932
Xestopyrgula Polinski,1932
Iljinella Roshka, 1978
ADRIOINSULANA Radoman, 1978
CHILOPYRGULA Brusina, 1896
DIANELLA Gude, 1913
Diana Clessin, 1878 (non Risso, 1826)
FALSIPYRGULA Radoman, 1973
GINAIA Brusina, 1896
MACEDOPYRGULA Radoman, 1973
pro Trachypyrgula Radoman, 1955
(non Cossmann, 1921)
MICROPYRGULA Polinski, 1932
NEOFOSSARULUS Polinski, 1932
OCHRIDOPYRGULA Radoman, 1955
PRESPOPYRGULA Radoman, 1973
PSEUDOCASPIA Starobogatov, 1972
PYRGOHYDROBIA Radoman, 1955
STANKOVICIA Polinski, 1932
TURRICASPIA Dybowski & Grochmalicki, 1917
TRACHYCASPIA Dybowski & Grochmalicki, 1917
TRACHYOCHRIDIA Polinski, 1932 ?

CONOCASPIINAE
CASPIA Dybowski, 1888
CASPIELLA Thiele, 1928
CLESSINIOLA Lindholm, 1924
Clessinia Dybowski, 1888
(non Döring, 1874)
NEMATURELLA Sandberger, 1874
MARTINIETTA Schlickum, 1974
pro Martiniella Schlickum & Moayedpour,1973
(non Grabau & Tieu, 1931)

EMMERICIINAE
EMMERICIA Brusina, 1870
Choerina, Tournoueria Brusina, 1870
EMMERICIELLA Pilsbry, 1909 ?

EKADANTINAE ? (23)
EKADANTA Rao, 1928

RISSOIDAE (18-f)
RISSOINAE
RISSOA Freminville in Desmarest, 1814
Rissoia (invalid emend.)
Loxostoma Bivona-Bernardi, 1838
Loxotoma auctt. (error)
Goniostoma Megerle in Villa, 1841
Zippora Leach in Gray, 1847 (nom.nud.)
Turboella Leach in Gray, 1847 (Turbella)
Anatasia, Apanthausa Gistl,1848 (n.nud.)
Persephona Leach, 1852 (non 1817)
Lamarckia Leach, 1852
Rissostomia Sars, 1878
Apicularia Monterosato, 1884
Sabanea Leach in Monterosato, 1884
(non Leach in Gray, 1847)
Schwartzia Bucquoy, Dautzenberg &
Dollfus, 1884
Elatiella, Gueriniana, Lilacinia,
Liavenustia, Nitituboella,
Sfaxiella all F. Nordsieck, 1972
ALVANIA Risso, 1826
Acinus, Actonia, Acinopsis
all Monterosato, 1884
Turbona Leach in Gray, 1847
Zacanthusa Leach, 1854
Massotia Bucquoy, Dautzenberg
& Dollfus, 1884
Arsenia Monterosato, 1891

ALVANIA (Cont'd.)
Thapsiella Fischer, 1885
pro Thapsia Monterosato, 1884
(non Albers, 1860)
Alvaniella Monterosato in Sacco, 1895
Acinulus Seguenza, 1903
Alcidiella Monterosato in Cossmann,1921
pro Alcidia Monterosato, 1890
(non Westwood, 1879)
Conalvinia Ponder, 1967
Andrewiella, Alvanolira, Coronalvania,
Lanciella all F. Nordsieck, 1972
Profundialvania Taviani, 1975
Deliciosalvania Piani, 1980 (nom.nud.)
CRISILLA Monterosato, 1917
Crisella auctt. (err.)
"Cingula Monterosato" auctt.
(non Fleming, 1828)
Crisillosetia F. Nordsieck, 1972
GALEODINA Monterosato, 1884
PUNCTULUM Jeffreys, 1884
LINEMERA Finlay, 1924
Ihungia Marwick, 1931 (fossil) ?
AMPHIRISSOA Dautzenberg & Fischer, 1897
ATTENUATA Hedley, 1918
Nobolira Finlay, 1926
Adolphinoba Powell, 1930
BENTHONELLA Dall, 1889
Hela Jeffreys, 1870
(non Münster, 1830)
Assiminopsis Locard, 1897
BOREOCINGULA Golikov & Kussakin, 1974
CINGULA Fleming, 1818
Sabanaea Leach in Gray, 1847
Cingilla Monterosato, 1884
LIROCINGULA Ponder, 1985
FRIGIDOALVANIA Warén, 1974
Arctonia Golikov, 1977
LIRONOBA Iredale, 1915
LUCIDESTEA Laseron, 1956
Falsisetia Golikov & Kussakin, 1967
MANZONIA Brusina, 1870
ALVINIA Monterosato, 1884
Moniziella, Flemellia F. Nordsieck, 1972
pro Seguenziella Sacco, 1904
(non Neviani, 1901)
pro Taramellia Seguenza, 1903
(non Campana, 1903)
pro Flemingia Jeffreys, 1884
(non Johnston, 1832)
SIMULAMERELINA Ponder, 1985
MERELINA Iredale, 1915
Promerelina Powell, 1926
Mereliniopsis Ponder, 1967
OBTUSELLA Cossmann, 1921
pro Cingulina Monterosato, 1884
(non A. Adams, 1860)
ONOBA H. & A. Adams, 1852
pro Turbonilla Leach in Gray, 1847
(non Risso, 1826)
Subonoba Iredale, 1915
Austronoba Powell, 1927
Willetia Gordon, 1939
SUBESTEA Cotton, 1944
Liroculma Ponder, 1967
OVIRISSOA Hedley, 1916
Badenia Finlay, 1930
pro Powellia Finlay, 1927
(non Maskell, 1879)
MANAWATAWHIA Powell, 1937
PARASHIELA Laseron, 1956
PERINGIELLA Monterosato, 1878
POWELLISETIA Ponder, 1965

RISSOINAE (Cont'd.)
PUSILLINA Monterosato, 1884
Auriconoba, Benzia, Multiturboella,
Radiata, Varisetia F. Nordsieck, 1972
Turgidina Verduin, 1979
HAURAKIA Iredale, 1915
Haurakiopsis Powell, 1937
Vitricithna Laseron, 1956
AMERIRISSOA Ponder, 1985
VICINIRISSOA Ponder, 1985
RACHIPTERON Thompson, 1964 ? (21)
SETIA H. & A. Adams, 1852
Parvisetia Monterosato, 1884 ?
Rudolphosetia Monterosato, 1917
PSEUDOSETIA Monterosato, 1884
STRIATESTEA Powell, 1927
VOORWINDIA Ponder, 1985

RISSOININAE (18f)
RISSOINA Orbigny, 1840
Zebinella Mörch, 1876
Moerchiella Nevill, 1885
Parazebinella Boettger, 1893
Stiva Hedley, 1904
Chiliostigma Melvill, 1918
Caporista Iredale, 1955
Austrosina, Condylicia, Peripetella,
Pleneconea, Zymalata all Laseron, 1956
Laseronia Cotton, 1959
Sulcorissoina Kosuge, 1965
Ailinzebina Ladd, 1966
RISSOLINA Gould, 1861
Costalynia, Fractoralla Laseron, 1956
PHOSINELLA Mörch, 1876
Phintorena Iredale, 1955
Planapexia Laseron, 1956
APATAXIA Laseron, 1956 (Apataxis auct.- err.)
PACHYRISSOINA Boettger, 1893
COSSMANNIA Newton, 1891 (fossil)
pro Diasticus Cossmann, 1888 (fossil)
(non Mulsant, 1842)
FOLINIA Crosse, 1868
LAPSIGYRUS Berry, 1958
MICROSTELMA A. Adams, 1863
Amaurella A. Adams, 1867
Crepitacella Guppy, 1867 (fossil)
Dolophanes Gabb, 1873 (fossil)
PSEUDOTAPHRUS Cossmann, 1888 (fossil)
Microtaphrus Cossmann, 1888 ?
Pezantia Cossmann, 1896
SCHWARTZIELLA Nevill, 1881
Atlantorissoina Kosuge, 1965
PANDALOSIA Laseron, 1956
STOSICIA Brusina, 1870
Stossicia, Stossichia auctt. (inv.em.)
ISSELIELLA Weinkauff,1881 (Iselliella, err.)
pro Isellia Semper in Schmeltz,1874 (n.nd.)
ZEBINA H. & A. Adams, 1854
Cibdezebina Woodring, 1928
LEAELLA Cossmann, 1921 ? (18f)
TOMLINELLA Viader, 1938
Dentirissoina Laseron, 1956
Ishimoria Kuroda, 1960
TAKIRISSOINA Oyama, 1962 ?
TIPHYOCERMA Berry, 1958

BARLEEIDAE
BARLEEINAE
BARLEEIA Clark, 1853 (Barleia, Barleeja, err.)
Pseudodiala Ponder, 1967
Ansola Slavoshevskaya, 1975
CAELATURA Conrad, 1865
(non Coelatura Conrad, 1853)

BARLEEINAE (Cont'd.)
FICTONOBA Ponder, 1967
LIROBARLEEIA Ponder, 1983
PROTOBARLEEIA Ponder, 1983
TROPIDORISSOIA Tomlin & Shackleford, 1915

ANABATHRINAE
ANABATHRON Frauenfeld, 1867
SCROBS Watson, 1886
Nannoscrobs Finlay, 1927
Ultiscrobs Iredale, 1955
AFRISCROBS Ponder, 1983
AMPHITHALAMUS Carpenter,1865 (Amphithalmus, err.)
Microfossa Laseron, 1950
Obescrobs Iredale, 1955
NOTOSCROBS Powell, 1927
BADEPIGRUS Iredale, 1955
Laseronula Whitley, 1959
pro Saltatricula Laseron, 1956
(non Burmeister, 1861)
pro Saltatrix Laseron, 1950
(non Klug, 1833)
MICRODRYAS Laseron, 1950
NODULUS Monterosato, 1878
PISINNA Monterosato, 1878
Hagenmulleria Bourguignat, 1881
Estea Iredale, 1915
Feldestea, Nodulestea Iredale, 1955
Microestea Ponder, 1965
PSEUDESTEA Ponder, 1967

EMBLANDIDAE (18g)
EMBLANDA Iredale, 1955

EPIGRIDAE (18f)
EPIGRUS Hedley, 1903

ELACHISINIDAE (18h)
ELACHISINA Dall, 1918
Microdochus Rehder, 1943

ASSIMINEIDAE (Marine, terrestrial)
ASSIMINEINAE
ASSIMINEA Fleming, 1828
Syncera Gray, 1821 (nom.dub.)
Optediceros Leith, 1853
Euassiminea, Pseudomphala Heude, 1882
Assimania Knight, 1900 (err.)
ANGUSTASSIMINEA Habe, 1943
AUSTROPILULA, CROSSILLA, MACRASSIMINEA,
METASSIMINEA, OVASSIMINEA,
SCULPTASSIMINEA all Thiele, 1927
EUSSOIA Preston, 1912
SPHAERASSIMINEA Habe, 1942
TAIWANASSIMINEA Kuroda & Habe, 1950
Assiminopsis Habe, 1942
(non Locard, 1897)
ACMELLA Blanford, 1869
AUSTROASSIMINEA Solem & Girardi, 1982 (20x)
CONACMELLA Thiele, 1927
CYCLOTROPIS Tapparone-Canefri, 1883
ELECTRINA Gray, 1850 ?
PSEUDOGIBBULA Dautzenberg, 1890
SUTERILLA Thiele, 1927
TURBACMELLA Thiele, 1927
VALVATORBIS Bequaert & Clench, 1936

OMPHALOTROPIDINAE
OMPHALOTROPIS Pfeiffer, 1851
Eurytropis Kobelt & Möllendorff, 1898
ATROPIS Pease, 1871
CHALICOPOMA Möllendorff, 1894
ORIELLA Thiele, 1927

OMPHALOTROPIS (Cont'd.)
SCALINELLA Pease, 1867
SPIRATROPIS Kobelt & Möllendorff, 1900
Paratropis Boettger, 1891
STENOTROPIS Kobelt & Möllendorff, 1897
SYCHNOTROPIS Möllendorff, 1898
ALLEPITHEMA Tomlin, 1931 ?
BALAMBANIA Crosse, 1891
CYCLOMORPHA Pease, 1871
GARRETIA Pease, 1873
Diadema Pease, 1868
(non Schumacher, 1817)
GONATORHYPHE Möllendorff, 1898
PSEUDASSIMINEA Thiele, 1927
PSEUDOCYCLOTUS Thiele, 1894
Adelostoma E. Smith, 1885
(non Duponchel, 1827)
Adelomorpha Tapparone-Canefri, 1886
(non Snellen, 1885)
HETEROPOMA Möllendorff, 1894
QUADRASIELLA Möllendorff, 1894

SKENEOPSIDAE
SKENEOPSIS Iredale, 1915
STARKEYNA Iredale, 1930 ?
Stipator Iredale, 1924

TRACHYSMATIDAE
TRACHYSMA Sars, 1878

TROCHACLIDIDAE
TROCHACLIS Thiele, 1912

CINGULOPSIDAE
EATONINA Thiele, 1912
Saginofusca Ponder, 1965
CAPTITONIA Ponder & Yoo, 1980
CORIANDRIA Tomlin, 1917
pro Microsetia Monterosato, 1884
(non Stephens, 1829)
Minosetia Crosse, 1855 (err.)
Cingulopsis Fretter & Patil, 1958
Globisetia F. Nordsieck, 1972 ?
OTATARA Ponder, 1965
SKENELLA Pfeffer in von Martens
& Pfeffer, 1886 (18f)
Eatoniopsis Thiele, 1912
BOOGINA Thiele, 1913
pro Watsonella Thiele, 1912
(non Berg, 1898)
RUFODARDANULA Ponder, 1965 ?
PILITONIA Ponder, 1965
MISTOSTIGMA Berry, 1947 ?
PICKENIA Ponder, 1982
PSEUDOPISINNA Ponder, 1980
TUBBREVA Ponder, 1965

ORBITESTELLIDAE (Microdisculidae)
ORBITESTELLA Iredale, 1917
MICRODISCULA Thiele, 1912
HELISABIA Laseron, 1954 (?)
MICROCARINA Laseron, 1954 (?)

CAECIDAE
CAECUM Fleming, 1813
Anellum Carpenter, 1857
Brochus T. Brown, 1827
Odontidium Philippi, 1836
Odontina Zborzewsky, 1834
Spirolidium O. G. Costa, 1861
BROCHINA Gray, 1857
Armata de Folin, 1875
ELEPHANTELLUM Bartsch, 1920 ?

CAECIDAE (Cont'd.)
ELEPHANTULUM Carpenter, 1857
Quadrulata de Folin, 1867
FARTULUM Carpenter, 1857
Defolinia Weisbord, 1962
Levia de Folin, 1869
MEIOCERAS Carpenter, 1858
MICRANELLUM Bartsch, 1920
PICTOCAECUM Habe, 1978
PSEUDOPARASTROPHIA Distaso, 1905
STREBLOCERAS Carpenter, 1858
WATSONIA de Folin, 1879

CTILOCERATIDAE
CTILOCERUS Watson, 1885
PARASTROPHIA de Folin, 1869
Moreletia de Folin, 1869
(non Gray, 1855)
Pedumicra Iredale & Laseron, 1957

VITRINELLIDAE (Circulidae)
VITRINELLA C. B. Adams, 1850
Dierama Pfeiffer, 1877
Tomura Pilsbry & Olsson, 1952
DECOMPHALA Bartsch, 1907
PSEUDORBIS Monterosato, 1884
VITRINELLOPS Pilsbry & Olsson, 1952
ANTICLIMAX Pilsbry & McGinty, 1946
Calliglyptum, Climacea Dall, 1903
Climacina Aguayo & Borro, 1946
SUBCLIMAX Pilsbry & Olsson, 1950
AOROTREMA Schwengel & McGinty, 1942
CALLOMPHALA Adams & Angas, 1864 (Callopmphalidae ?)
CALLODIX Laseron, 1954
PARISANDA Laseron, 1954
ROTOSTOMA Laseron, 1954
CIRCULUS Jeffreys, 1865
CITHNA A. Adams, 1854
DIDIANEMA Woodring, 1928
DIAGONAULUS Pilsbry & McGinty, 1945
DISCOPSIS de Folin & Perrier, 1870
ALLEORUS Strong, 1938
DISCREVINIA Laseron, 1956 (18f)
EPISCYNIA Mörch, 1875
LYDIPHNIS Melvill, 1906
CYMATOPTERYX Pilsbry & Olsson, 1946
MICROTHYCA A. Adams, 1863 ?
MORCHIELLA Thiele, 1925
Morchia A. Adams, 1860
(non Albers, 1850)
PARVITURBOIDES Pilsbry & McGinty, 1950
PLEUROMALAXIS Pilsbry & McGinty, 1945
CALODISCUS Rehder, 1934
PSEUDOLIOTIA Tate, 1898
SANSONIA Jousseaume, 1892 (18f)
MECOLIOTIA Hedley, 1899
MICROLIOTIA Boettger, 1901 (fossil)

SANSONIA (Cont'd.)
PICKWORTHIA Iredale, 1917
REYNELLONA Iredale, 1917
SCISSILABRA Bartsch, 1907
SCRUPUS Finlay, 1927 ? (18f)
SOLARIORBIS Conrad, 1865
EULEREMA Pilsbry & Olsson, 1952
HAPALORBIS Woodring, 1957
SYSTELLOMPHALIS Pilsbry & Olsson, 1941
TEINOSTOMA H. & A. Adams, 1854
CALCIOLATA Iredale, 1864
Calceolina A. Adams, 1864
ESMERALDA Pilsbry & Olsson, 1952
IDIORAPHE Pilsbry, 1922
MEGATYLOMA Cossmann, 1915
PSEUDOROTELLA P. Fischer, 1857
Parkeria Gabb, 1881
VITRIDOMUS Pilsbry & Olsson, 1955
VITRINORBIS Pilsbry & Olsson, 1951
WOODRINGILLA Pilsbry & Olsson, 1951

TORNIDAE
TORNUS Turton & Kingston, 1830
Adeorbis Wood, 1842
SIGARETORNUS Iredale, 1936
CHAMLONGIA Brandt, 1968
COCHLIOLEPIS Stimpson, 1958
Discopsis de Folin, 1870
LACINIORBIS von Martens, 1897
TYLAXIS Pilsbry, 1953
NARICAVA Hedley, 1913
Tropidorbis Iredale, 1936
MACROMPHALINA Cossmann, 1888 ?
Gyrodisca Dall, 1896
MUNDITIELLA Kuroda & Habe, 1952
PHANETA H. Adams, 1874
PYGMAEOROTA Kuroda & Habe, 1961
SOYOROTA Habe, 1961
UZUMAKIELLA Habe, 1958

RASTODENTIDAE
RASTODENS Ponder, 1966
TRIDENTIFERA Ponder, 1966

Placement ?
CYTHNIA Carpenter, 1864
(ex Eulimoidea: Warén, 1983)
Cythnoa Dall, 1924
pro Cythna Carpenter, 1864
(non Cithna A. Adams, 1863)
RETROTORTINA Chaster, 1896 (sinistral)
ZEROTULA Finlay, 1927 (ex Architectonicoidea)

LOXONEMATOIDEA (24)
ABYSSOCHRYSIDAE
ABYSSOCHRYSOS Tomlin, 1927

References:
Backhuys, W.& H.Boeters "Kenntnis Marokkanscher Binnenmolusken" Arch.Moll.104(4-6):107.1974(1)

Bodin, M. & F. Giusti "A new valvatoid hydrobiid: Liguria" Arch.Moll.117(1-3):61-71 -1986

Boeters, H. "Horatia, Plagigeyeria, Litthabitella" Arch.Moll.104(1-3):85-92 - 1974 (2)
—— "Unbekannte Westeuropaische Prosobranchia" Arch.Moll.111(1-3):55-61 - 1981 (2a)
—— "Proposed (systematic revis.) Belgrandiella" Arch.Moll.114(1-3):17-24 - 1983 (2b)

Brandt, R.A.M. "The non-marine aquatic mollusca: Thailand" Arch.Moll.105(1-4):1-423.1974 (4)

Burch, J.B. "Freshwater snails of North America" U.S.Envir.Prot.Agcy. - 1982 (5)

—— "Outline of classification of the recent freshwater Gasropoda: N.Am.(No. of Mexico)" Journ.Conch.115(1-2):3-9 - 1978 (5a)

Clarke, A. N. "Canadian Freshwater Molluscs" Malacologia 13 (1-2):1-510 - 1973 (6)

References: (Rissoidea, Cont'd.)
Climo, F. M. "Notes: New Zealand Hydrobiid Fauna" J.Roy.Soc.N.Z.7(1):67-77 - 1977 (8)

Davis, G. M. "Origin and evolution of Pomatiopsidae and Triculinae" Monog.Acad.Nat.Sci.Philad.20:1-120 -1979 (9)
---- "Evolution, adaptive radiation:Pomatiopsidae" Malacologia 21(1-2):209-262 - 1981 (9a)
---- Mazurkiewicz & Mandriacchia "Spurwinkia: morphology, systematics, ecology: new genus" Proc.Acad.Nat.Sci.Phila.134:143-179-1982 (9b)
---- G. M. & M. C. Pons da Silva "Potamolithus -" Malacologia 25(1):73-108 - 1984 (9c)
---- Y. Kuo, K.E. Hoagland, P. Chen, H. Yang & D. Chen "Systematics of Fenoulia ---" Proc.Nat.Sci.Phila.135:177-199 - 1983
---- Ibid. 136:165-193 - 1984
---- Ibid. 137:48-78 - 1985 (9d)

Golikov, A. N. & O. G. Kusakin "Supplement to the testaceous Gastropoda of the Kuril I." (10) S.I.D.1:289-299 (ES) (Veliger Review) 1974

Giusti, F. & E.Pezzoli "Hydrobioidea nuevo o poco conosciure dell' Italia appenninica" Arch.Moll.111(4-6):207-222 - 1981 (11)
---- "Il genre Avenionia in Italy" Arch.Moll.111(4-6:223-242 - 1981 (11a)
---- "Gasteropodi, 2.(Gastropoda, Prosobranchia, Hydrobioidea, Pyrguloidea) Collana Progetto finalezzato".Consiglio Nazionale delle Richerche. AQ 1(47):1-66 - 1982b (11b)
---- "Notes on the small Hydrobioidea in Italian subterranean waters: Catalogue, etc." Malacologia 22(1-2):463-468 - 1982 (11c)

---- & M.Bodin"Nuevo Hydrobiidae;Italy Nord-Occ." Arch.Moll.114(4-6):157-181 - 1984 (11d)

Hadzisce, S., J. B. Burch & C. M. Patterson "The systematic postition of Gocea ohridana" Malacologia 16(1):139 - 1977 (abstract) Malacological Review 9:1-14 - 1976 (12)

Haszprunar, G. "The Heterobranchia- a new concept in the phylogeny of the higher Gastropoda" Zeitschrift fur zoologische Systematik und Evolutionsforschung 23(1):15-37 - 1985 (13)

Hershler, R. "Systematic revision of Hydrobiidae of Coahuila, Mex." Malacologia 26(1-2):31-123 - 1985 (14)
---- "North Amerocan Hydrobiidae" Nautilus 101(1):25-32.1987 (14c)
---- & R. & G. Longley "Phreatic Hydrobiids of Edwards Aquifer region, S.Central Florida" Malacologia 27(1):127-172 - 1986 (14a)
---- "On Phreatoceras" Proc.Biol.Soc.Wash.100(2):402 - 1987 (14b)

Hubricht, L. "New species of Hydrobiidae" Nautilus 76(4):138-140 - 1963 (15)
---- "New Hydrobiidae from Ozark caves" Nautilus 84(3):93-96 - 1971 (15a)

Izzatulaev, Z.I., T.Ya.Sitnikova & Ya.A. Starobogatov "Taxonomic position: mid. Asean pseudamnicols" Bull.Mosk.Obsch.Ispyt.Prir (Otd.Biol.) 90(50):52-60 - 1985 (15x)

Malek, E.A. "The South American hydrobioid genus Idiopyrgus" Nautilus 97(1):16-20 -1983 (16)

Moore, D. R. "Systematic position: Caecidae" 1962 Bull.Mar.Sci.Gulf.Carib. 12(4):695-701 (17)

Ponder, W.F."New Family of Rissoidae: N.Zealand" Rec.Dom.Mus.N.Z. 5(18):177-184 - 1966 (18)
---- "Classification: Rissoidae, Orbitestellidae" Trans.Roy.Soc.N.Z. 9(17):193-224 -1967 (18a)
---- "Review of the Genera of Cingulopsinae" Rec.Aust.Mus. 33(1):1-88 - 1980 (18b)
---- "Review of the Genera of Barleeinae" Rec.Aust.Mus.35(5-6): - 1983 (18c)
---- "Rissoaform gastropods from the Antarctic and Sub-Antarctic" Br.Ant.Survey Rep.108 - 1983 (18d)
---- "Review of the genera of Iravadiidae" Malacologia 25(1):21-71 - 1984 (18e)
---- "A review of the genera of Rissoidae" Rec.Aust.Mus.Supp.4:1-221 - 1985 (18f)
---- "The Anatomy and Relationships of Emblanda emblematica: Emblandidae, new family" Rec.Aust.Mus.37:343-351 - 1985 (18g)
---- "Anatomy and relationships: Elachisina Dall" Jour.Moll.Stud.51(1):23 - 1986 (18h)

Radoman, P. "New Classification: fresh & brackish water Prosobranchia from Balkans, Asia Minor" Mus.D'Hist.Nat.de Belgrad #3231-30.1973 (19x)
---- "Species formation in the genus Belgrandiella, related genera: Balkans" Glasnik Prir.Mozeya Beogradu 30:29-69 (Veliger 20(4):394.1975(19)
---- Hydrobiidae: der balkan halbisel;Kleinasien" Arch.Moll.(4-6):203-223 - 1977 (19a)
---- "Neue Vertreter der gruppe Hydrobioidea von der Balkanhabinsel" Arch.Moll.109(1-3):27-44 - 1978 (19b)
---- "Hydrobioidea, superfamily of Prosobranchia" Monogr.Serbian Acad.Sci.Arts 547 (Dept.Sci.57) 1983 (19c)

Schütt, H."Zur kenntnis griechischer Hydrobiiden" Arch.Moll.110(4-6):115-149 - 1980 (20)
---- "Die mollusken fauna der Sufswasser in Einzusgebiet des Orontes" Arch.Moll.106(1-6):17-91-225-228 1983 (20a)

Solem, A., E.L. Girardi, S. Slack-Smith & G. W. Kendrick "Austroassiminiea, Gen.nov., Southwest Australia" Jour.Roy.Soc.W.Aust. 65(4):119-129 - 1982 (20x)

Taylor,D.W."Remarkable snail fauna from Coahuila, Mexico" Veliger 9(2):152-228 - 1966 (21)
---- "Tertiary gastropod Orygoceras found living" Arch.Moll.104(1-3):93-96 - 1974 (21a)
---- Freshwater molluscs: New Mexico and vicinity" N.M.Bur.Mines,Min.Res.Bull.116:5-49-1987(21b)

Thompson, F. G. "The systematic relationships; status of the subfamily Nymphophiliinae" Malacological Rev.12:41-49 - 1979 (22)

---- "North American freshwater snails of the hydrobiid subfamily Lithoglyphinae" Malacologia 25(1):109-141 - 1984 (22a)

Thiele, J. "Handbuch der Systematischen Weichtierkunde" - 1931 (Rep. 1963) (23)

LOXONEMATOIDEA Reference:
Houbrick, R. S. "Classification and systematic relationships of the Abyssochrysidae" Smith.Contr.Zool. 290:1-20 - 1979 (24)

CERITHIOIDEA
THIARIDAE (Freshwater)
THIARINAE
THIARA Röding, 1798
Melania Lamarck, 1799
Melas Montfort, 1810
Amarula Sowerby, 1842
Spirilla Humphrey in Gray, 1847
Hydrognoma, Lithoparches Gistl, 1848
Tiaropsis Brot, 1871 (non Agassiz, 1850)
BALANOCOCHLIS Fischer, 1885
PLOTIA Röding, 1798
PLOTIOPSIS Brot, 1874
SERMYLA H. & A. Adams, 1853
FAUNUS Montfort, 1810
Ebena Schumacher, 1817
Melanamona Bowdich, 1822
Pirena Lamarck, 1822
FIJIDOMA Morrison, 1952
HEMISINUS Swainson, 1840
Aulacostoma Spix in Wagner, 1827 ?
Aulacostoma Agassiz, 1846
Bassistoma Lea, 1852
Semisinus P. Fischer, 1885
CUBAEDOMUS Thiele, 1928
VERENA H. & A. Adams, 1854
MELANOIDES Olivier, 1804
Striatella Brot, 1870
Eumelania Rovereto, 1899
Pallarya Hesse, 1916
HOREA Bourguignat, 1888
Nyassella Bourguignat, 1889
MICRONYASSIA Bourguignat, 1888
NYASSIA Bourguignat, 1889
NYASSOMELANIA Bourguignat, 1889
PIRENOPSIS Brot, 1879
PACHYMELANIA E. Smith, 1893
Claviger Haldeman, 1842
(non Preyssler, 1790)
Vibex Gray, 1847 (non Oken, 1815)
Hemipirena Rovereto, 1899
Clavigerina von Martens, 1903
Itameta Ihering, 1909
STENOMELANIA Fischer, 1885
Radina Preston, 1915
TAREBIA H. & A. Adams, 1854

PLANAXIDAE (2h)
PLANAXIS Lamarck, 1822 (non Risso, 1826)
? Tritia Risso, 1826
Proplanaxis Thiele, 1929
ANGIOLA Dall, 1926
FISSILABIA Macgillivray, 1836
Quoyia Gray, 1839 (inv.) (Quoya - err.)
Leucostoma Swainson, 1840
(non Meigen, 1803)
Fissilabria Brown in Gray, 1847 (err.)
Fissilabra Brown in Paetel,1875 (err.)
HINEA Gray, 1847
HOLCOSTOMA H. & A. Adams, 1854
SUPPLANAXIS Thiele, 1929
? Nucleus Lamarck

MELANOPSIDAE (Freshwater)
MELANOPSIS Férussac, 1807
CANTHIDOMUS Swainson, 1840
LYRCAEA H. & A. Adams, 1854
SISTANIANA Izzatulaev & Starobogatov, 1984
ESPERIANA Bourguignat, 1877 ?
FAGOTIA Bourguignat, 1884
Pseudhemisinus Nevill, 1884
MICROCOLPIA Bourguignat, 1884
ZEMELANOPSIS Finlay, 1927

MODULIDAE
MODULUS Gray, 1842
Haplodon Agassiz, 1846
(non Wagler, 1830)
Turbinopsis Conrad, 1860
Pseudotrochus Heilprin, 1887
(non Herrmannsen, 1847)

CERITHIIDAE (2)
CERITHIINAE
CERITHIUM Bruguière, 1789
Contumax Hedley,1899
CONOCERITHIUM Sacco, 1895
ISCHNOCERITHIUM Thiele, 1929
SEMIVERTAGUS Cossmann, 1889
THERICIUM Monterosato, 1890
Drillocerithium, Gladiocerithium,
Hirtocerithium, Lithocerithium,
Pliocerithium Monterosato, 1910
Gourmierium Jousseaume, 1894
Ptychocerithium, Tiaracerithium,
Pithocerithium, Procerithium Sacco,1895
Vulgocerithium Cossmann, 1906
Stomocerithium F. Nordsieck, 1975 ?
LITTERACERITHIUM Gründel, 1983
ALUCO Von Martens, 1880
ARGYROPEZA Melvill & Standen, 1901
BITTIUM Leach in Gray, 1847 (2a, 2b)
Cerithiolum Tiberi, 1869
Semibittium Cossmann, 1896
Bittiolum Cossmann, 1906
Stylidium Dall, 1907
Lirobittium Bartsch, 1911
Inobittium, Manobittium Monterosato,1917
Bittinella Dall, 1924 ?
Zebittium Finlay, 1927
Cacozeliana Strand, 1928
pro Cacozelia Iredale, 1924 (non Grote, 1878)
Bittiscalia Finlay & Marwick, 1937
Paracerithium Cotton, 1932
(non Cossmann, 1902)
Eubittium Cotton, 1937
Brachybittium Weisbord, 1962
Dahlakia Biggs, 1971 (2b)
Rasbittium Gründel, 1976
CERITHIOCLAVA Olssen & Harbison,1953 (recent)(2g)
CLAVOCERITHIUM Cossmann, 1889 (fossil)
INDOCERITHIUM Chavan, 1952 (recent)
CLYPEOMORUS Jousseaume, 1888
COLINA H. & A. Adams, 1854
Colinia Cossmann, 1889
FASTIGIELLA Reeve, 1848 (2f)
GLYPTOZARIA Iredale, 1924 (1)
GOURMYA Fischer, 1884
LIOCERITHIUM Tryon, 1887
PLESIOTROCHUS Fischer, 1878
PSEUDOVERTAGUS Vignal, 1904
RHINOCLAVIS Swainson, 1840
Clava Martyn, 1784
Vertagus Schumacher, 1817
(non Link, 1807)
LONGICERITHIUM Houbrick, 1978
OCHETOCLAVA Woodring, 1928
PROCLAVA Thiele, 1929
ROYELLA Iredale, 1912
TROCHOCERITHIUM Cossmann & Sacco, 1896
VARICOPEZA Gründel, 1976
Placement ?
HEMICERITHIUM Cossmann, 1893
SEILOPSIS Tomlin, 1931 (ex Cerithiopsidae)

ORECTOSPIRINAE
ORECTOSPIRA Dall, 1925

DIALIDAE ? (5)
DIALA A. Adams, 1861
DIALOPSIS Cossmann, 1889
LAEVITESTEA Laseron, 1950
MELLITESTEA Laseron, 1956
PARADIALA Laseron, 1956
RISSOALABA Oyama, 1954

LITIOPIDAE (5; 2e)
LITIOPA Rang, 1829
Bombyxinus Lesson, 1834 (Bombycinus, em.)
ALABA H. & A. Adams, 1853
Styliferina A. Adams, 1860
Gibborissoa Cossmann, 1895
Diffalaba, Obstopala Iredale, 1936
Dialessa Iredale, 1955
Australaba Laseron, 1956

CAMPANILIDAE (2c)
CAMPANILE Bayle in Fischer, 1884
Ceratoptilus Bouvier, 1887
HYPOTROCHUS Cotton, 1832

DIASTOMATIDAE (Diastomidae)
DIASTOMA Deshayes, 1850
Neodiastoma Cotton, 1932
ALABINA Dall, 1902 (6)
Elachista Dall & Simpson, 1901
(non F. Treitschke, 1833)
CASSIELLA Gofas, 1988 (?)
CERITHIDIUM Monterosato, 1884
FESANDELLA Gründel, 1976
TRUIDELLA Gründel, 1976
OBTORTIO Hedley, 1899 ? (5)
Finella A. Adams, 1860 (error) ?
pro Fenella A. Adams, 1864
(non Westwood, 1840)
Eufenella Kuroda & Habe, 1954
CALOOSALABA Olsson & Harbison, 1953
CLATHROFENELLA Kuroda & Habe, 1952
SCALIOLA A. Adams, 1860

POTAMIDIDAE
POTAMIDINAE (2d)
CERITHIDEA Swainson, 1840
Phaenommia Mörch, 1860
Aphanistylus P. Fischer, 1884
CERITHIDEOPSIS, CERITHIDEOPSILLA Thiele, 1929
PIRENELLA Gray, 1847
PYRAZUS Montfort, 1810
TELESCOPIUM Montfort, 1810
TEREBRALIA Swainson, 1840
TYMPANOTONUS Klein in Schumacher, 1817
Tympanotomus Gray, 1842

BATILLARIINAE
BATILLARIA Benson, 1842
Lampania Gray, 1847
BATILLARIELLA Thiele, 1929
LAMPANELLA Mörch, 1876
VELACUMANTUS Iredale, 1936
Velacumanthus Wenz, 1938 (err.)
ZEACUMANTUS Finlay, 1927
RHINOCORYNE von Martens, 1900

PLEUROCERIDAE (Freshwater)
PLEUROCERINAE
PLEUROCERA Rafinesque, 1818
Ceriphasia Swainson, 1840
Telescopella Gray, 1847
Trypanostoma Lea, 1862
OXYTREMA Rafinesque, 1819 ?
STREPHOBASIS Lea, 1861

PLEUROCERINAE (Cont'd.)
ANCULOSA Say, 1821
Ellipsoma Rafinesque, 1818 ?
Leptoxis Rafinesque, 1819 ?
Anculotus Say, 1825
Ancylotus Herrmannsen, 1846
GONIOBASIS Lea, 1862
Elimia, Melasma H.& A. Adams, 1854
Megara H. & A. Adams, 1854 (part)
GYROTOMA Shuttleworth, 1845
Schizostoma Lea, 1842 (non Bronn, 1835)
Schizocheilus Lea, 1852
Apella Mighels in Tryon, 1883
HUA Chen, 1943
Wanga Chen, 1943 (?)
NAMRUTUA Abbott, 1948
IO Lea, 1831 ?
Melafusus Swainson, 1840
JUGA H. & A. Adams, 1854
CALIBASIS, OREOBASIS Taylor, 1966
LITHASIOPSIS Pilsbry, 1910
LITHASIA Haldeman, 1840
Athearnia Morrison, 1971 ?
pro Eurycaelon Lea, 1865
ANGITREMA Haldeman, 1841
Glottela Gray, 1847
MUDALIA Haldeman, 1840
Nitrocris H. & A. Adams, 1854
SEMISULCOSPIRA Boettger, 1886

AMPHIMELANIINAE
AMPHIMELANIA Fischer, 1885
Melanella Swainson,1840 (non Bowdich,1822)
Holandriana Bourguignat, May, 1884

LAVIGERIINAE
LAVIGERIA Bourguignat, 1888
EDGARIA Bourguignat, 1888
Nassopsis E. Smith, 1890
Nassopsidia von Martens, 1897
Hirtha Ancey, 1898
JOUBERTIA, RANDABELIA Bourguignat, 1888

MELANATRIINAE
MELANATRIA Bowdich, 1822
ADAMIETTA Brandt, 1974
AMNIPHILA Pilsbry, 1956 (Potadominae)
BROTIA H. Adams, 1866
ANTIMELANIA Fischer & Crosse, 1892
PSEUDOPOTOMUS Brot, 1894
SENCKENBERGIA Yen, 1939
NEORADINA Brandt, 1974 ?
PACHYCHILUS Lea, 1850
Sphaeromelania Rovereto, 1899
CERCIMELANIA, GLYPTOMELANIA, OXYMELANIA
all Fischer & Crosse, 1892
DORYSSA H. & A. Adams, 1854

POTAMANAX Pilsbry, 1892
PARACROSTOMA Cossmann, 1900
POTADOMA Swainson, 1840 (Potadominae ?)
Nigritella Brot,1871 (non v.Martens,1860)
Nigriculina Rovereto, 1899
RHINOMELANIA von Martens, 1901
SULCOSPIRA Troschel, 1857
Acrostoma Brot,1871 (non Le Sauvage,1826)
Brotella Rovereto, 1899
TYLOMELANIA P. & F. Sarasin, 1898

PALUDOMINAE
PALUDOMUS Swainson, 1840
Hemimitra Swainson, 1840
Rivulina Lea, 1850

PALUDOMUS (Cont'd.)
PHILOPOTAMIS Layard, 1855
Heteropoma Benson, 1856
TANALIA Gray, 1847
Ganga Layard, 1855
Serenia Benson, 1856
STOMATODON Benson, 1862
BAIZEA Bourguignat, 1885 (8)
Ponsonbya Ancey, 1890
BATHANALIA Moore, 1898
BRIDOUXIA Bourguignat, 1885
CHLOROSTRACIA Mabille, 1889
Claucostracia Ancey, 1898
CHYTRA Moore, 1897
CLEOPATRA Troschel, 1856 (8)
ZANGEBARIA Fischer, 1881
GIRAUDIA Bourguignat, 1885
Reymondia Bourguignat, 1885
LARINA A. Adams, 1854
Robinsonia H. Nevill, 1869
LECHAPTOISIA Ancey, 1894
Horea E. Smith, 1889
(non Bourguignat, 1888)
LIMNOTROCHUS E. Smith, 1880
PARAMELANIA E. Smith, 1880
Bourguignatia Giraud, 1885
BYTHOCERAS Moore, 1898
PSEUDOCLEOPATRA Thiele, 1928 (8)
RUMELA Bourguignat, 1885
SPEKIA Bourguignat, 1879
STANLEYA Bourguignat, 1885
Coulboisia Bourguignat, 1888
TANGANYICIA Crosse, 1881
Cambieria, Tanganikia Bourguignat,1885
Hauttecoeuria Bourguignat, 1888
Tanganyikia Moore, 1899 (em.)

TIPHOBIINAE
TIPHOBIA E. Smith, 1880
Hilacantha Ancey, 1886
Typhobia Moore, 1898 (non Pascoe,1869)

FOSSARIDAE
FOSSARUS Philippi, 1841
Maravigna Aradas & Maggiore, 1841
ANAFOSSARUS, HALOTAPADA Iredale, 1936
CONRADIA A. Adams, 1863 ?
FOSSARELLA Thiele, 1925 ?
GOTTOINA A. Adams, 1863
PENDROMA Dall, 1927 ?
CHILKAIA Preston, 1915 (brackish water)
CONSTANTIA A. Adams, 1860 ? (1)
COUTHOUYIA A. Adams, 1860
Escharella, Micreschara Cossmann, 1888
RADINISTA, ZERADINA Finlay, 1927 ?
MEGALOMPHALUS Brusina, 1871 ?
Discopsis Auctt. (non de Folin, 1870)
Chonebasis Pilsbry & Olsson, 1945
Korovina Iredale, 1918
Larinopsis Gatliff & Gabriel, 1916
NILSIA Finlay, 1927 (1)

SYRNOLOPSIDAE
SYRNOLOPSIS E. Smith, 1880
ANCEYA Bourguignat, 1884
BURTONILLA E. Smith, 1904
MARTELIA Dautzenberg, 1908

TURRITELLIDAE
TURRITELLINAE
TURRITELLA Lamarck, 1799
TORCULA Gray, 1847
TORCULOIDELLA Sacco, 1895

TURRITELLINAE (Cont'd.)
ARCHIMEDIELLA Sacco, 1895
ARMATUS Golikov, 1986 (1x)
COLPOSPIRA Donald, 1900 (Colospira)
Colpospirella Powell, 1951 (fossil ?)
Leptocolpus Finlay & Marwick, 1937 (?)
ACUTOSPIRA Kotaka, 1959
CTENOCOLPUS Iredale, 1925
PLATYCOLPUS Donald, 1900
GAZEMEDA Iredale, 1924 ?
HAUSTATOR Montfort, 1810
KUROSIOIA Ida, 1952
MAORICOLPUS Finlay, 1927
MESALIA Gray, 1847
CALLOSTRACA E. Smith, 1909
Smithia Maltzan, 1883
MESALIOPSIS Thiele, 1929
NEOHAUSTATOR Ida, 1950
PEYROTIA Cossmann, 1912
TACHYRHYNCHUS Mörch, 1860
ZARIA Gray, 1847
ZEACOLPUS Finlay, 1927
STIRACOLPUS Finlay, 1927

PROTOMINAE
PROTOMA Baird, 1870
PROTOMELLA Thiele, 1929

PAREORINAE
PAREORA Marwick, 1981 (fossil)

TURRITELLOPSINAE
TURRITELLOPSIS G. O. Sars, 1878
Kimberia Cotton & Wood, 1935

VERMICULARIINAE
VERMICULARIA Lamarck, 1789
STEPHOPOMA Mörch, 1860
Vermatoma Kuroda, 1928
LILAX Finlay, 1927 ?

SILIQUARIIDAE
SILIQUARIA Bruguière, 1789
Tenagodus Guettard, 1774 ?
AGATHIRSES Montfort, 1810
PYXIPOMA Mörch, 1860

VERMETIDAE (3)
VERMETUS Daudin, 1800
Bivona, Siphonium Gray, 1847 (?)
Bivoniopsis Sacco, 1896
Dofania Mörch, 1860
THYLAEODUS Mörch, 1860
DENDROPOMA Mörch, 1861 (nom.cons.ICZN 1425)
Spiroglyphus Daudin, 1800 (Suppr.)
Stoa De Sevres, 1855 (Suppr.)
Siphonium Mörch, 1859 (Suppr.?)
Veritoma Kuroda, 1928
Veristoa Iredale, 1937
EUALETES Keen, 1971
NOVASTOA Finlay, 1927
PETALOCONCHUS H. C. Lea, 1843
MACROPHRAGMA Carpenter, 1857
SEGMENTELLA Thiele, 1925 ?
SERPULORBIS Sacco, 1827
Aletes Carpenter, 1857
Anguinella Conrad, 1846
Hatina Gray, 1847
Lemintina Risso, 1826 ?
Serpuloides Gray, 1850
Tetranemia Mörch, 1859
Thylacodes Mörch, 1862
TRYPSYCHA Keen, 1961

References: (Cerithioidea)
Garrard, T.A. "Genus Haustator, subgenus Kurosioia: Australia "J.Mal.Soc.Aust.3(1):15-17-1974 (1a)
---- "Supplement: Australian recent Turritellidae" J.Mal.Soc.Aust.5(3-4):195-200 - 1981 (1b)

Golikov, A.N."Sytematics, Evolution: Turritellidae, N.Hemisphere" Zool.Zh.65(8):1140-1150.1986

Houbrick, R. S. "Family Cerithiidae; Indo-Pacific" Mono.Mar.Mollusca.Part I.1-130 - 1978 (2)
---- "Review of the deep sea genus Argyropeza" Smiths.Cont.Zool.321:30 - 1979 (2a)
---- "Reappraisal: gastropod genus Varicopeza" Proc.Biol.Soc.Wash.93(3):525-535 - 1980 (2b)
---- "Anatomy: systematics; Campanile symbolicum" Malacologia 21(1-2):263-289 -1981 (2c)
---- "Revision of the higher taxa in Cerithidea (Potamididae)" AMU Bull.2:1-20 - 1984 (2d)
---- "Anatomy of Alaba, Litiopa: systematic implications" Nautilus 101(1):9-18-1987(2e)

---- R. Robertson & R. T. Abbott "--Fastigiella" Nautilus 101(3):101-110 - 1987 (2f)
---- "Discovery of living Cerithioclava: Caribbean" Proc.Biol.Soc.Wash.99(2):257-260 - 1986 (2g)
---- "Anatomy, Reproductive Biology and Philogeny: Planaxidae" Smith.Cont.Biol.445:1-57-1987 (2h)

Keen, A. M. "Reclassification: Vermetidae" - 1961 Bull.Brit.Mus.Nat.Hist.Zool.7(3):183-213 (3)

Ponder, W. F."Review of the genera of Rissoidae" Rec.Aust.Mus.Supp.4:1-221 - 1985 (5)

Rehder, H. A. "Marine Mollusks of Easter Island" Smith.Contr.Zool.289:1-166 - 1980 (6)

STROMBOIDEA

STRUTHIOLARIIDAE

STRUTHIOLARIA Lamarck, 1816
- PELICARIA Gray, 1857
- TYLOSPIRA G. Harris, 1897

PERRISODONTA von Martens, 1883

APORRHAIDAE

APORRHAIS Da Costa, 1778

DREPANOCHEILUS Meek, 1864 (fossil)
- ARRHOGES Gabb, 1868 (recent)

STROMBIDAE

STROMBUS Linnaeus, 1758
  - Strombella Schlüter, 1838
- CANARIUM Schumacher, 1817
  - Strombidea Swainson, 1840
  - Canarium Jousseaume, 1888
  - Oostrombus Sacco, 1893 (fossil ?)
- CONOMUREX Fischer, 1884
- DOLOMENA Iredale, 1931
- DOXANDER Iredale, 1931
- EUPROTOMUS Gill, 1870
  - Lobatus Swainson, 1837
  - Monodactylus Herrmannsen, 1847
- GIBBERULUS Jousseaume, 1888
- LABIOSTROMBUS Oostingh, 1925
  - Gallinula Mörch, 1852
- LAEVISTROMBUS Kira, 1955

STROMBUS (Cont'd.)
- LENTIGO Jousseaume, 1886
- TRICORNIS Jousseaume, 1886
  - Lobatus Iredale, 1921 (non Swainson, 1837)
  - Aliger Thiele, 1929
  - Eustrombus Wenz, 1940

LAMBIS Röding, 1798
  - Pterocera Lamarck, 1799
  - Pteroceras Link, 1807 (err.)
  - Pteroceres Montfort, 1810 (err.)
  - Digitator Fabricius, 1823 (nom. nud.)
  - Pterocerus Brongniart, 1829 (err.)
  - Heptadactylus Mörch, 1852
- HARPAGO Mörch, 1852
  - Pterocera Anton,1839 (non Lamarck, 1799)
- MILLEPES Mörch, 1852

RIMELLA Agassiz, 1840 (fossil)
- RIMELLOPSIS Lambiotte, 1979 ?
- VARICOSPIRA Eames, 1952 (recent)

STROMBICONUS Marks, 1951

TEREBELLUM Röding, 1798
  - Artopoia, Lucis Gistl, 1848

TIBIA Röding, 1798
  - Rostellaria Lamarck, 1799
  - Rostellum Montfort, 1810
  - Gladius H. & A. Adams, 1854
  - Platyoptera Conrad, 1855
- ROSTELLARIELLA Thiele, 1929

References:
Abbott, R.T. "Genus Strombus in the Indo-Pacific" Indo-Pacific Mollusca 1(2):33-146 - 1960
---- "The genus Lambis in the Indo-Pacific" Indo-Pacific Mollusca 1(3):147-174 - 1961

HIPPONICOIDEA

HIPPONICIDAE (2)

HIPPONIX Defrance, 1819 (Hipponyx)
  - Amalthea Schumacher, 1817 (non Rafinesque, 1815)
  - Sabia Gray, 1839 ? (non Reeve, 1842)
  - Cochlolepas H. & A. Adams, 1854
  - Capulonix Iredale, 1929
  - Capulonyx Martyn (non-binomial)
- MALLUVIUM Melvill, 1906

ANTISABIA Iredale, 1937

COCHLEAR Mörch, 1877 ?

PILOSABIA Iredale, 1929

SABIA Reeve, 1842 ?

HIPPONICIDAE (Cont'd.)

CHEILEA Modeer, 1793
  - Mitrularia Schumacher, 1817
  - Lithedaphus Owen, 1842
  - Calyptra H. & A. Adams, 1854

VANIKORIDAE

VANIKORO Quoy & Gaimard, 1832
  - Merria Gray, 1839
  - Leucotis Swainson, 1840
  - Nioma Gray, 1842
  - Narica Récluz in Orbigny, 1842

CALEDONIELLIDAE (3)

CALEDONIELLA Souverbie, 1869
  - Epistethe Preston, 1912

References: (Hipponicoidea)
Cowan, I. M. "West American Hipponicidae"
Veliger 16(4):377-380 - 1974 (1)

Thiele, J. "Weichtierkunde" - 1963 (4)

Ponder, W. F. "Review of the Genera of Rissoidae"
Rec.Aust.Mus.Supp.4:1-221 - 1984 (2)

Rosewater, J. "Gross anatomy and classification of
Caledoniella" Veliger 11(14):345-50 - 1969 (3)

CREPIDULOIDEA

CREPIDULIDAE (Calyptraeidae)

CREPIDULA Lamarck, 1799 (4)
Crypta, Galerus Humphrey, 1797
Dispotaea Say, 1826
Ianacus Mörch, 1852 (Janacus)
Ergaea H. & A. Adams, 1854
Garnotia, Noicia Gray, 1867
Maoricrypta, Zeacrypta Finlay, 1927
Proscenula Perry in Férussac, 1820 ?
Proxenula auctt. (err.)
Bostrycapulus Olsson & Harbison, 1953
SIPHOPATELLA Lesson, 1830
CALYPTRAEA Lamarck, 1799
Mitrula Gray, 1821
Mitella Leach in Gray, 1847 ?
Mitrella Gray, 1847
(non Risso, 1826)
Trochilina Gray, 1847
Poculina Gray, 1867
TROCHITA Schumacher, 1817
Infundibulum Sowerby, 1815
(non Montfort, 1810)
Trochatella Lesson, 1830
CREPIPATELLA Lesson, 1830
VERTICUMBO S. S. Berry, 1940
CRUCIBULUM Schumacher, 1817
Calypeopsis Lesson, 1850
Neleta Gray, 1867 ?
Catillina Gray, 1867
SANDALIUM Schumacher, 1817 ? (4)
(non Retzius, 1788) (nom.nud.)
Sandalinum Récluz, 1841 (err.)
Elara H. & A. Adams, 1856
Elana Gray, 1867
SIGAPATELLA Lesson, 1830
Haliotidea Swainson, 1840
Trochella Gray, 1867
CLYPEOLA Gray, 1867
Zegalerus Finlay, 1927 ?
BICATILLUS Swainson, 1840
Trelania Gray, 1867

CAPULIDAE

CAPULUS Montfort, 1810
Pileopsis Lamarck, 1812
HYALORISIA Dall, 1889
KREBSIA Mörch, 1877
CAPULELLA Dall, 1924
TENPETASUS Iredale, 1929
NEOJANACUS Suter, 1907 ?

TRICHOTROPIDAE (7)

TRICHOTROPIS Broderip & Sowerby, 1829
Cerithioderma Conrad, 1860
Mesostoma Deshayes, 1861
ARIADNARIA Habe, 1961
Pro Ariadna Fischer, 1864
Verona Gray, 1857
(non H. & A. Adams, 1854
Alora H. Adams, 1861 ?
(non H. & A. Adams, 1853)
Trichosirius Finlay, 1927
IPHINOE H. & A. Adams, 1854
Neoiphinoe Habe 1976 (unnec. em.)
IPHINOELLA Habe, 1958
IPHINOPSIS Dall, 1924
PROVANNA Dall, 1918
TURRITROPIS Habe, 1961 (3)
AMAMICONCHA Habe, 1961
AKIBUMIA Kuroda & Habe in Kuroda, 1959
ALVINICONCHA Ohta, 1988 (?) (5)
ICUNCULA Iredale, 1924 ?
LAMINILABRUM Kuroda & Habe in Habe, 1961
Labrolabrum Kuroda & Habe, 1961 (n.nud.)
LIPPISTES Montfort, 1810 ?
SEPARATISTA Gray, 1847
LYOCYCLUS Thiele, 1925
SIRIUS Hedley, 1900
DOLICHOSIRIUS Iredale, 1931
OPPOSIRIUS Iredale, 1931
TRICHAMATHINA Habe, 1962
TORELLIA Lovén in Jeffreys, 1867 (6)
Trachyoma Seguenza, 1879
Trichoconcha E. A. Smith, 1907
NEOCONCHA E. A. Smith, 1907
Antitrichotropis Powell, 1951 ?
DISCOTRICHOCONCHA Powell, 1951
HALOCERAS Dall, 1889
ZELIPPISTES Finlay, 1927 ?

XENOPHOROIDEA

XENOPHORIDAE (6)

XENOPHORA Fischer, 1807 (1)
Phorus Montfort, 1810
Xenophorus Philippi, 1849 (err.)
Endoptygme Gabb, 1877
STELLARIA Schmidt in Möller, 1832
(non Nardo, 1834)
Haliphoebus Fischer in Kiener, 1880
ONUSTUS Gray, 1847 (Onustis auctt. - error)
Tugurium Fischer in Kiener, 1876
Trochotugurium Sacco, 1896

References:
Clench, W. J. & C. G. Aguayo "Xenophora, Tugurium;
West Atlantic" Johnsonia 8:1:1-16 - 1943 (1)

Graham, A. "Trichotropis: position of Capulidae"
J.Mar.Biol.Ass.U.K. 33:129-144 - 1954 (2)

Habe, T. "Trichotropidae in Japan" 1962
Bull.Nat.Sci.Mus.(Tokyo) 6(2):67-77 (3)

Hoagland, K. E.- "Systematic review: Crepidula"
Malacologia 16(2):353-420 - 1977 (4)

Ohta, S. "New Gastropod Mollusk from the
hydrothermal vents in the West Pacific"
Venus 47(1):1-9 - 1988 (5)

Ponder, W. F. "Xenophoridae of the World"
Aust.Mus.Memoir 17:1-126 - 1983 (6)

Warén, A., P. M. Arnaud & J. R. Cantera "Two new
Gastropods of the Trichotropidae: Indian O."
Veliger 29(2):157-165 - 1986 (7)

CYPRAEOIDEA

CYPRAEIDAE

CYPRAEA Linnaeus, 1758
Porcellana Humphrey, 1797
Cypriarius Duméril, 1806
Tigris Troschel, 1863
Erythraea Mörch, 1884
Vulpicella Cossmann, 1884
Vulgusella Jousseaume, 1884
ADUSTA Jousseaume, 1884
Ficadusta Habe & Kosuge, 1966
ARABICA Jousseaume, 1884
ALBACYPRAEA Cotton & Steadman, 1936
ANNEPONA Iredale, 1935
ARESTORIDES Iredale, 1930
Porcellana Roberts, 1870
(non Humphrey, 1797)
BARYCYPRAEA Schilder, 1927
BERNAYA Jousseaume, 1884
BISTOLIDA Cossmann, 1920
Stolida Jousseaume, 1884
Derstolida Iredale, 1935
BLASICRURA Iredale, 1930
Eclogavena Iredale, 1930
CALLISTOCYPRAEA, CHELYCYPRAEA Schilder, 1927
CRIBRARULA Strand, 1929
pro Cribraria Jousseaume, 1884 (invalid)
Ocellaria Weinkauff, 1881
CYPRAEOVULA Gray, 1824
Cypreovula Fleming, 1828 (err.)
Cypraeova Swainson, 1840
Cypraeovulum Sowerby, 1842
Cyprovula Gray, 1847
Cypraeovum Schaufuss, 1869
Crossia Shikama, 1974 ?
Hiraseadusta Shikama, 1976 ?
Kurodadusta, Nakayasuia Shikama, 1976
EROSARIA Troschel, 1863
Perisserosa Iredale, 1930
ERRONEA Troschel, 1863
EUSTAPHYLAEA Steadman & Cotton, 1946
EVENARIA, GRATIADUSTA Iredale, 1930
GUTTACYPRAEA Iredale, 1935
IPSA Jousseaume, 1884
LEPORICYPRAEA Iredale, 1930
LUPONIA Broderip, 1837
Lupina Bronn, 1853
Gaskoinia Roberts, 1870
Luponaria, Luponica Weinkauff, 1881
Luperia Oppenheim, 1901
LURIA Jousseaume, 1884
Basilitrona Iredale, 1930
LYNCINA Troschel, 1863
Prolyncina Schilder, 1927
MACROCYPRAEA Schilder, 1930
Erythraea Mörch, 1877 (non 1884)
MAURITIA Troschel, 1863
Maurina, Mauxenia Jousseaume, 1884
MONETARIA Troschel, 1863
Aricia Broderip, 1837
MURACYPRAEA Woodring, 1957
MELICERONA, MYSTAPONDA Iredale, 1930
NARIA Broderip, 1837
NESIOCYPRAEA Azuma & Kurohara, 1967
NIVIGENA, PALMADUSTA Iredale, 1930
NOTADUSTA Schilder, 1935
NEOBERNAYA, NOTOCYPRAEA Schilder, 1927
NUCLEOLARIA Oyama, 1959
pro Nuclearia Jousseaume, 1884
ORNAMENTARIA Schilder & Schilder, 1936
OVATIPSA, PALANGEROSA Iredale, 1931
PANTHERINARIA Sacco, 1890
PONDA Jousseaume, 1884

CYPRAEA (Cont'd.)
PAULONARIA Iredale, 1930
Cupinota, Opponaria Iredale, 1939
PROPUSTULARIA, PROTOCYPRAEA Schilder, 1927
PSEUDOZONARIA Schilder, 1927
PUPEROSA Iredale, 1935
PURPURADUSTA Schilder, 1939
PUSTULARIA Swainson, 1840
Epona H. & A. Adams, 1854
RAVITRONA Iredale, 1930
SCHILDERIA Tomlin, 1930
Globulina Cerulli-Irelli, 1911
SIPHOCYPRAEA Heilprin, 1887
Cypraeactaeon White, 1887
SOLVADUSTA Iredale, 1935
STAPHYLAEA, THELIXINOVUM Jousseaume, 1884
TALOSTOLIDA Iredale, 1930
TALPARIA Troschel, 1863
TESSELLATA Jousseaume, 1886
TRONA Jousseaume, 1884
Basterotia, Etrona Jousseaume, 1884
Cavicypraea Cossmann, 1896
UMBILIA, ZOILA, ZONARIA Jousseaume, 1884

OVULIDAE (3)

OVULINAE

Tribe Eocypraeini

PSEUDOCYPRAEA Schilder, 1932
Sulcocypraea Conrad, 1865
Spaerocypraea Schilder, 1927
Marginocypraea Ingram, 1947

Tribe Jenneriini

JENNERIA Jousseaume, 1884
Pustularia Melvill, 1888 (nom.van.)
Transovula de Gregorio, 1880 (nom.nud.)

Tribe Ovulini

OVULA Bruguière, 1789
Ovulus Montfort, 1810 (em.)
Ovulum Sowerby, (em.)
Ovularia Link, 1830 (em.)
Anula Griffith & Pidgeon, 1834
Semiporcellana Da Costa, 1776
Licium Humphrey, 1797 (nom.nud.)
Amphiperas Gronovius, 1781 (nom.nud.)
Parlicium Iredale, 1935
APERIOVULA Cate, 1973
CALPURNUS Montfort, 1810
Cypraella Swainson, 1840
Cypraela Sowerby, 1842 (err.)
PROCALPURNUS Thiele, 1939
CARPISCULA Cate, 1973
CRENAVOLVA Cate, 1973
SERRATOVOLVA Cate, 1973 (1)
CUSPIVOLVA Cate, 1973
CYPHOMA Röding, 1798
Ultimus Montfort, 1810
Dinovoluta Schlüter, 1838
Carinea Swainson, 1840
Cyphonia Gray, 1847 (err.)
DELONOVOLVA Cate, 1973
DENTIOVULA Habe, 1961
GALERAVOLVA Cate, 1974
pro Galera Cate, 1973
(non Gray, 1842)
GLOBOVULA Cate, 1973
HABUPRIONOVOLVA Azuma, 1970
LACRIMA Cate, 1973
MARGOVULA Iredale, 1935
NEOSIMNIA Fischer, 1884
PELLASIMNIA Iredale, 1931
PRIONOVOLVA Iredale, 1930
PRIMOVULA Thiele, 1925
ADAMANTIA Cate, 1973

OVULINAE (Cont'd.)
PSEUDOCYPHOMA Cate, 1973
PSEUDOSIMNIA Schilder, 1927
DIMINOVULA Iredale, 1930
INFLATOVULA Cate, 1973
LABIOVOLVA Cate, 1973
ROTAOVULA Cate & Azuma, 1973
SANDALIA Cate, 1973
SIMNIA Risso, 1826
Calpurna Fleming, 1828
SIMNIALENA Cate, 1973
SPICULATA Cate, 1973 ?
STOHLEROMA Cate, 1973
SUBSIMNIA Cate, 1973
TESTUDOVOLVA Cate, 1973
XANDAROVULA Cate, 1973
Tribe Volvini (3a, 3b)
VOLVA Röding, 1798
Radius Montfort, 1810
Birostra Swainson, 1840
ACLYVOLVA Cate, 1973
CYMBOVULA Cate, 1974
pro Cymbula Cate, 1973
(non H. & A. Adams, 1854)
DISSONA Cate, 1973
HIATAVOLVA Cate, 1974
pro Hiata Cate, 1973
(non Zetek & McLean, 1936)
KUROSHIOVOLVA Azuma & Cate, 1971
PHENACOVOLVA Iredale, 1930
Radius Schumacher, 1817
(non Montfort, 1810)
CALCAROVULA Cate, 1974 (1)
pro Calcaria Cate, 1973 (non Porat, 1878)
TAKASAGOVOLVA Azuma, 1974 ?
KURODAOVULA Azuma, 1987 (?)
TURBOVULA Cate, 1973 (1)
PROSIMNIA Schilder, 1927

TRIVIIDAE
TRIVIINAE
TRIVIA Broderip, 1837
Trivea Swainson, 1840
Coccinella Herrmannsen, 1847
(non Linnaeus, 1758)
DECORIATRIVIA Cate, 1979
SULCOTRIVIA Schilder, 1933
FOSSATRIVIA Iredale, 1931
GALEATRIVIA Cate, 1979
NIVERIA Jousseaume, 1884
CLEOTRIVIA Iredale, 1930
ELLATRIVIA Iredale, 1931
PSEUDOTRIVIA Schilder, 1936
ROBERTOTRIVIA Cate, 1979
PUSULA Jousseaume, 1884
Pustula Roberts, 1885
Vusula Cossmann, 1896
DOLICHUPIS Iredale, 1930
Trivellona Iredale, 1931
CIRCUMSCAPULA Cate, 1979

TRIVIIA (Cont'd.)
DISCOTRIVIA Cate, 1979
TRIVIELLA Jousseaume, 1884
TRIVIROSTRA Jousseaume, 1884

ERATOINAE (4)
ERATO Risso, 1826
ERATOPSIS Hörnes & Auinger, 1880
ALAERATO Cate, 1977
Lachryma Iredale, 1931 (invalid)
HESPERERATO Schilder, 1932
NOTOFICULA Thiele, 1917 (7)
Ficulina Thiele, 1912 (non Gray, 1867)
PROTERATO Schilder, 1927 (fossil)
Lachryma Sowerby (1832) ? (5)
CYPRAEERATO Schilder, 1932
ERATOENA Iredale, 1935
SULCERATO Finlay, 1930

PEDICULARIIDAE
PEDICULARIA Swainson, 1840
Thyreus Philippi, 1844
Dentiora Pease, 1862 ? (Thiele)
Pediculariella Thiele in Kuckensthal,1925
Pedicularionа Iredale, 1935

LAMELLARIIDAE (2)
LAMELLARIINAE
LAMELLARIA Montagu, 1815
Marsenia Leach in Oken, 1823
Cryptothyra Menke, 1830
Cryptocella H. & A. Adams, 1853
Ermea Gray, 1857
CORIOCELLA Blainville, 1824
Chelinotus Swainson, 1840
Chelynotus Bergh, 1853
DJIBOUTIA Vayssière, 1912
MARSENIELLA Bergh, 1886
LAMELLARIOPSIS Vayssière, 1906
LAROCHEA Finlay, 1927 ?
MARSENINA Gray, 1850 (2)
Oithonella Mörch, 1856
MARSENIOPSIS Bergh, 1886
MYSTICONCHA Allen, 1936 ?

VELUTININAE
VELUTINA Fleming, 1822
LIMNERIA H. & A. Adams, 1853
Morvillia Gray, 1857
VELUTELLA Gray, 1847
CAPULACMAEA Sars, 1859
Pilidium Middendorff, 1851
(non Forbes & Hanley, 1849)
Piliscus Lovén, 1859
ONCHIDIOPSIS Bergh, 1853
ATLANTOLIMAX Dall, 1917

PSEUDOSACCULIDAE (6)
PSEUDOSACCULUS Hirase, 1928
Sacculus Hirase, 1927

References:

Azuma, M. "Systematic studies on recent Japanese Ovulidae" Venus 45(4):226-230 - 1987 (1)

Behrens, D. W. "Lamellariidae of the Northeastern Pacific" Veliger 22(4):323-344 - 1980 (2)

Cate, C. N. "Ovulidae" Veliger 15.Supp. - 1973 (3) 16(2):238-239 (3a): 16(3):307-313 (3b)

---- "Eratoidae" Veliger 19(3):341-366-1977 (4)

---- "Triviacea"San D.Soc.Nat.Hist.Memoir.10(4a)

Cernohorsky, W.O. "Fossil & Recent Cypraeacea of N.Z." Rec.Auck.Inst.Mus.8:103-129 -1971 (5)

Hirase, S. "Sacculus okai etc." Annot.Zool.Jap.11:115-127 - 1927 (6)

Oliver, P. G. "Notoficula Thiele; Antarctica" Br.Ant.Survey Bull.61:1-6 - 1983 (7)

HETEROPODA (Atlantacea)

ATLANTIDAE
PROTATLANTA Tesch, 1908
Protoatlanta Wenz, 1941
ATLANTA Lesueur, 1817
Steira Eschscholtz, 1825
Atlanta Orbigny, 1834 (1)
Atlantidea Pilsbry, 1922
OXYGYRUS Benson, 1835
Helicophlegma Orbigny, 1835 (1)
Brownia Orbigny, 1841
Ladas Cantraine, 1841

CARINARIIDAE (Pelagic)
CARINARIA Lamarck, 1801
Monophora Quoy & Gaimard, 1824 ?
Timoriena Quoy & Gaimard, 1824
Tithyonia Cavolini in Tiberi, 1880

CARINARIIDAE (Cont'd.)
CARDIAPODA Orbigny, 1835
Carinaroida Souleyet, 1852
PTEROSOMA Lesson, 1827
Aloysia Issel, 1908 ? (2)

PTEROTRACHEIDAE
PTEROTRACHEA Forskål, 1775
Firola Bruguière, 1791
Hypterus Rafinesque, 1814
Anops Orbigny, 1835
Euryops Tesch, 1906
Eupterotrachea Bonnevie, 1920
Heterodens Bonnevie, 1920
FIRLOIDA Lesueur, 1817
Demarestia Griffith & Pidgeon, 1834
Cerophora Orbigny, 1841
Averna, Ceratophora Gistl, 1848
Firolella Troschel, 1855

References:
Abbott, R. T. "American Seashells" - 1974 (1)
Thiele, J. "Weichtierkunde" - 1963 (2)

NATICOIDEA
NATICIDAE
AMPULLOSPIRINAE
GLOBULARIA Swainson, 1840
Cernina Gray, 1842
Anomphala Jonas in Herrmannsen, 1846
AMAUROPSIS Mörch, 1857
PSEUDOPOLINICES Golikov & Sirenko, 1983

POLINICINAE
POLINICES Montfort, 1810
Albula Röding, 1798
(non Osbeck, 1762)
Polynices Menke, 1830
Naticina Guilding, 1834
Naticella Guilding in Swainson, 1840
Mamma Klein in Mörch, 1852
Pollinices Mörch, 1852 (nom. null.)
Uber Humphrey in Dunker, 1882
CONUBER Finlay & Marwick, 1937
Uber Humphrey, 1797
EUSPIRA Agassiz, 1838
GLOSSAULAX Pilsbry, 1929
GYRODES Conrad, 1860
HYPTERITA Woodring, 1957
LUNATIA Gray, 1847
MAMMILLA Schumacher, 1817
Naticaria Swainson, 1840
Ruma Chemnitz in Gray, 1847
Reuma Chemnitz in Récluz, 1851
(nom.null.)
NEVERITA Risso, 1826
Mammillaria Swainson, 1840
PAYRAUDEAUTIA Bucquoy, Dautzenberg
& Dollfus, 1883
PROPESINUM Iredale, 1924
SIGARETOTREMA Sacco, 1831
SIGATICA Meyer & Aldrich, 1886
BULBUS Brown, 1839
Bulbus Humphrey, 1797 (non-binom.)
Acrybia H. & A. Adams, 1853
CALINATICINA Burch & Campbell, 1963
CHORISTES Carpenter, 1872 (3)
FALSILUNATIA Powell, 1951 (5)
FRIGINATICA Hedley, 1916 (5)
KERGUELENATICA Powell, 1951 (4)
UBERELLA Finlay, 1928 (5)

GLOBISININAE ? (8)
GLOBISINUM Marwick, 1924

NATICINAE
NATICA Scopoli, 1777
Neritoides Meuschen, 1779
Cochlis Röding, 1798
Nacca Risso, 1826
Notocochlis Powell, 1933
ALCONATICA Shikama, 1971 ?
CRYPTONATICA Dall, 1892
GLYPHEPITHEMA Rehder, 1943
LUNAIA Berry, 1964
NATICARIUS Duméril, 1806
Naticus Montfort, 1810
Naticarius Duméril, Macpherson
& Gabriel, 1962 (n.null.)
PROXIUBER Powell, 1933
STIGMAULAX Mörch, 1852
TECTONATICA Sacco, 1890
BOREONATICA Golikov & Kusakin, 1974 ? (3)
SULCONATICA Golikov, 1975
SCARLATIA Shileyko, 1977 ? (10)
TANEA Marwick, 1931 (8)

SININAE
SINUM Röding, 1798
Sigaretus Lamarck, 1799
Cryptostoma Blainville, 1818
? Catinus Klein in H. & A.
Adams, 1853 (11)
Ectosinum Iredale, 1931 (4)
HELICONATICA Dall, 1924 ?
EUNATICINA Fischer, 1885
Naticina Gray, 1840 (nom. nud.)
Naticina Gray, 1847
(non Guilding, 1834)
Pervisinum Iredale, 1931
GENNAEOSINUM Iredale, 1929
HALIOTINELLA Souverbie, 1875

Placement ?
FROVINA Thiele, 1912 (11)
PROLACUNA Thiele, 1913
Sublacuna Thiele, 1912
(non Pilsbry, 1895)
Frigidilacuna Tomlin, 1930

References: (Naticoidea)

Bouchet, P. & A. Warén, A. "The abyssal molluscan fauna of the Norwegian Sea" Sarsia 64:211-243 - 1979 (1)

Golikov, A. N. & O. G. Kusakin "Supplement to the testaceous gastropoda fauna: Kuril I." S.I.D.1:289-299 (ES) - 1974 (3)

Kilburn, R. A. "Revision of the Naticidae of South Africa and Mozambique" Ann.Natal Mus.22(3):829-884 - 1976 (4)

Marincovich, L. "New Tertiary & Recent Naticidae; Eastern Pac." Veliger 18(2):168-173-1975 (5)

---- "Cenozoic Naticidae of the Northeastern Pacific" Bull.Am.Pal.70(294):169-494-1977 (6)

Powell, A. W. B. "New Zealand Mollusca" - 1974 (8)

Oyama, K. "Sytematic revision of the Japanese Naticidae; Preliminary report" Venus 28(2):69-88 - 1969 (9)

Shileyko, A. A. "Material on the morphology of the Naticoida and problems of the taxonomy" Veliger 22(4):395 (Review) - 1977 (10)

Thiele, J. "Weichtierkunde" - 1963 (11)

Wenz, W. "Gastropoda" Tiel I, in Schindewolf (Ed.) "Handbuch der Palaozoologie" 1941 (12)

TONNOIDEA

TONNIDAE

TONNA Brünnich, 1772
- Cadus Röding, 1798
- Dolium Lamarck, 1801
- Cadium Link, 1807
- Perdix Montfort, 1810 (non Brisson, 1760)
- Macgillivrayia Forbes, 1852 (larval form)
- Galea Klein in Mörch, 1852
- Foratidolium Rovereto, 1899
- Parvitonna Iredale, 1931

EUDOLIUM Dall, 1889
- Doliopsis Monterosato, 1872 (non Voigt, 1852)
- Endolium Dall in Preston, 1924 (error)

MALEA Valenciennes, 1883
- QUIMALEA Iredale, 1929

FICIDAE

FICUS Röding, 1798
- Pyrula Lamarck, 1798
- Pirula Montfort, 1810 (non Wood, 1835)
- Otus Risso, 1826 (non Pennant, 1769)
- Ficula Swainson, 1835
- Sycotypus Gray, 1847
- Syctopus Conrad, 1856 (error)
- Syctyphus Gabb, 1861 (error)

THALASSOCYON Barnard, 1960 (2)

CASSIDAE

CASSINAE

CASSIS Scopoli, 1777
- Cassida Brünnich, 1772 (non Linnaeus, 1758)
- Cassidea Bruguière, 1789
- Cassis Röding, 1798
- Cassinia Rafinesque, 1815 (error)
- Goniogalea Mörch, 1857
- Fimbriola Megerle in Scudder, 1882 (nom. nud.)
- Cassisoma Rovereto, 1899
- Pro Cassis Klein, 1753 (non-binom.)
- Galeodocassis Sacco, 1890 (fossil)
- HYPOCASSIS Iredale, 1927
  - Nannocassis Iredale, 1927

CASSINAE (Cont'd.)

CYPRAECASSIS Stutchbury, 1837
- Cypraeicassis Cossmann, 1903 (error)
- LEVENIA Gray, 1847

GALEODEA Link, 1807 (4)
- Morio Montfort, 1810 (non Latreille, 1810)
- Cassidaria Lamarck, 1812
- Echinora Schumacher, 1817
- Echinophora Latreille, 1825
- Doliopsis Conrad, 1865 (non Voigt, 1852)
- Galeodaria Conrad, 1865
- GALEOOCORYS Kuroda & Habe, 1957

SCONSIA Gray, 1847

HERCULEA Hanley in H. & A. Adams, 1858

(MORUM Röding, 1798 - removed to Harpidae, p.52)

OOCORYTHINAE (5)

OOCORYS Fischer, 1883
- BENTHODOLIUM Verrill & Smith, 1884

DALIUM Dall, 1889

HADROOCORYS Quinn, 1980

PHALIINAE

PHALIUM Link, 1807
- Bezoardica Schumacher, 1817
- Cassis Children, 1823
- Cassidea Swainson, 1840 (non Bruguière, 1789)
- Bezoardicella Habe, 1961

CASMARIA H. & A. Adams, 1853
- Casmeria Jousseaume, 1888

ECHINOPHORIA Sacco, 1890 (fossil)
- BATHYGALEA Woodring & Olsson, 1957 (recent)

SEMICASSIS Mörch, 1852
- Faurotis Jousseaume, 1888
- Bezoardica Dall, 1909 (non Schumacher, 1817)
- Tylocassis Woodring, 1928
- Xenophalium Iredale, 1927
- Xenogalea Iredale, 1927
- Kahua Marwick, 1928
- ANTEPHALIUM Iredale, 1927

RANELLIDAE (Cymatiidae) (2e)

RANELLINAE

RANELLA Lamarck, 1816

RANELLA (Cont'd.)
Gyrina Schumacher, 1817
Eugyrina Dall, 1904
Mayena Iredale, 1917
Gyrinopsis Dall, 1927
ARGOBUCCINUM Herrmannsen, 1846
Gondwanula Finlay, 1927
Mediargo Terry, 1968
PRIENE H. & A. Adams, 1858
Liohindsia Coen, 1947
FUSITRITON Cossmann, 1903
Cryotritonium von Martens, 1904
GYRINEUM Link, 1807
Apollon Montfort, 1810
Gyrinella Dall, 1924
BIPLEX Perry, 1811

CYMATIINAE
CYMATIUM Röding, 1798
Lotorium Montfort, 1810
(non Pusch, 1837)
Tritocurrus Lesson, 1842
Luterium Herrmannsen, 1847 (err.)
Nyctilochus Gistl, 1848
GUTTURNIUM Mörch, 1852
LOTORIA Emerson & Old, 1963
MONOPLEX Perry, 1811
Lampusia Schumacher, 1817
Cymatriton Clench & Turner, 1957
RANULARIA Schumacher, 1817 (2d)
Ranula Schumacher, 1817 (err.)
Retusum Jousseaume, 1892
Tritonocauda Dall, 1904
pro Ranularia Schumacher in Fischer
(non Schumacher, 1817)
RETICUTRITON Habe & Kosuge, 1966
SEPTA Perry, 1810
Simpulum Mörch, 1852
(non Fabricius, 1823)
Simplum Stoliczka, 1867 (err.)
TURRITRITON Dall, 1904
Tritoniscus Dall, 1904
Particymatium Iredale, 1936
Cabestanimorpha Iredale, 1936
Parlycymatium Iredale
in Wenz, 1941 (err.)
CABESTANA Röding, 1798
Aquillus Montfort, 1810
Dolarium Schlüter, 1838
Aquilus Mörch, 1852 (err.)
Neptunella Gray, 1854
Cymatilesta Iredale, 1936
CHARONIA Gistl, 1848
Tritonium Röding, 1798
(non Müller, 1776)
Triton Montfort, 1810
(non Linnaeus, 1758)
Tritonia Bowdich, 1822
(non Cuvier, 1798)
Tritonium Montfort in
Bowdich, 1822 (em.)
Charonis Gistl in Mörch, 1877 (err.)
Buccinatorium Mörch, 1877
Tritonellium Mörch, 1877
(non Valenciennes, 1858)

CHARONIA (Cont'd.)
Septa Perry in Dall & Simpson, 1901
(non Perry, 1810)
Eutritonium Cossmann, 1904
Eutriton Couturier, 1907 (err.)
LINATELLA Gray, 1857
Zinatella Cossmann, 1903 (err.)
GELAGNA Schaufuss, 1869
Paralagena Dall, 1904
pro Lagena Schumacher, 1817
(non Walker & Boys, 1784)
SASSIA Bellardi, 1872
Semiranella de Gregorio, 1880
Monocirsus Cossmann, 1889
Cymatona Iredale, 1929
Charoniella Powell &
Bartrum, Nov. 1929
(non Thiele, Oct. 1929)
Austrosassia Finlay, 1931
Phanozesta Iredale, 1936
Proxicharonia Powell, 1938
AUSTROTRITON Cossmann, 1903
Negyrina Iredale, 1929
Charoniella Thiele, 1929
CYMATIELLA Iredale, 1924
Vernotriton Iredale, 1936

PERSONINAE (Distorsioninae) (2e,f)
PERSONELLA Conrad, 1865
DISTORSIO Röding, 1798
Distortrix Link, 1807
Persona Montfort, 1810
Distorta Perry, 1811 (err.)
Distortio Gray, 1847 (em.)
Distorsus auctt. (err.)
Distorsia auctt. (err.)
Distrortix auctt. (err.)
Rhysema Clench & Turner, 1957
DISTORSIONELLA Beu, 1978

BURSIDAE (2e,f)
BURSA Röding, 1798
Rana Humphrey, 1797 (non-binom.)
(non Linnaeus, 1758)
Lampadopsis Jousseaume, 1881
Pseudobursa Rovereto, 1899
COLUBRELLINA Fischer, 1884
Bufonariella Thiele, 1929
Dulcerana Iredale, 1931
(unavailable)
Annaperenna Iredale, 1936
Dulcerana Iredale in Oyama, 1964
Tritonoranella Oyama, 1964
BUFONARIA Schumacher, 1817
Chasmotheca Dall, 1904
Bursina Oyama, 1964
MARSUPINA Dall, 1904
pro Buffo Montfort, 1810
(non Lacepede, 1788)
ASPA H. & A. Adams, 1853
CROSSATA Jousseaume, 1881
TUTUFA Jousseaume, 1881
TUTUFELLA Beu, 1981
Pro Lampas Schumacher, 1817
(non Montfort, 1808)

References:
Abbott, R. T. "Helmet Shells of the World" Part I. Cassididae (1) Indo-Pacific Mollusca 2(9):1-200 -1978

Beu, A. G. "The gastropod genus Thalassocyon" New Zealand Journal.Marine, Freshwater Res. 3(3):445-452 - 1969 (2)

References: (Tonnoidea, Cont'd.)

Beu, A. G. "Arrival of Semicassis pyrum and other Tonnacean gastropods in the Southern Oceans during Pleistocene time" J.Roy.Soc.N.Z.6(4):413-432 - 1976 (2a)

---- "New species of Bursa from deep water in the Philippines" J.Malac.Soc.Aust.4(1):23-27 - 1978 (2b)

---- "Marine Fauna of New Zealand; Cymatona" N.Z.Ocean.Inst.Mem.65:44 pp. - 1978 (2c)

---- "Australian gastropods of family Bursidae" I. Tonnacea, Bursidae; Tutufa revision. Rec.Aust.Mus.33(5):248-324 - 1980 (2d)

---- & W.O. Cernohorsky "Taxonomy: Ranellidae (Cymatiidae) and Bursidae" 1986
I. Ranellidae: N.Z.Journ.Zool.13:241-266 (2e)
II. New sp., Revisions: Ibid.13:273-355 (2f)

Cernohorsky, W. O. "Bursidae, Cymatiidae and Colubrariidae of Fiji" Veliger 9(3):310-329 -1966 (3)

Ponder, W. F. "New species of Galeodea: Australia" J.Malac.Soc.Aust.6(1-2):91-97 - 1983 (4)

Quinn, J. F. Jr. "New genus: Oocorythidae from the Westerm Atlantic" Nautilus 94(4):149-158 - 1980 (5)

Suborder Heteroglossa (1)

CERITHIOPSOIDEA (3)
CERITHIOPSIDAE
ALIPTINAE
ALIPTA Finlay, 1927
MENDAX Finlay, 1927
EUSEILA Cotton, 1951
Binda Laseron, 1951
ALIPTINA Marshall, 1978
CERITHIOPSIDELLA Bartsch, 1911
VATOPSIS Gründel, 1980
STILUS Jeffreys, 1885
PILAFLEXIS Laseron, 1951

EUMETULINAE
EUMETULA Thiele, 1912
CERITHIOPSILLA Thiele, 1912
CERITHIOPSIDA Bartsch, 1911
CHASTERIA Iredale, 1915 ?
CLATHROPSIS Laseron, 1956
FURUKUWAIA Kuroda & Habe, 1961
LASKEYA Iredale, 1918
pro Eumeta Mörch, 1868
(non Walker, 1855)
ALTISPECULA Powell, 1930
ONCHODIA Dall, 1924
RETILASKEYA Marshall, 1978
MARSHALLASKEYA Gründel, 1980
SOCIENNA Finlay, 1927
SPECULA Finlay, 1927
ZACLYS Finlay, 1927
Zaclys Allan, 1926 (nom.nud.)
Miopila Allan, 1926 (nom.nud.)

CERITHIOPSINAE
Tribe Cerithiopsini
CERITHIOPSIS Forbes & Hanley, 1850 (3)
Pyramis Brown, 1827
(non Schumacher, 1817)
Conciliopsis Laseron, 1956

CERITHIOPSINAE (Cont'd.)
CYRBASIA Harris & Burrows, 1891
Tiarella Cossmann, 1889
(non Swainson, 1840)
DIZONIOPSIS Sacco, 1895 (3)
HOROLOGICA Laseron, 1956 ?
JOCULATOR Hedley, 1909 (3)
POTENATOMUS Laseron, 1956
PROLIXODENS Marshall, 1978
SUNDAYA Oliver, 1915
SYNTHOPSIS Laseron, 1956
TUBERCLIOPSIS Laseron, 1956
Tribe Seilini
SEILA A.Adams, 1861
HEBESEILA Finlay, 1927
LYROSEILA Finlay, 1927
NOTOSEILA Finlay, 1927
Notoseila Allan, 1926 (nom.nud.)
CINCTELLA Monterosato, 1884
PARASEILA Laseron, 1951
PROSEILA Thiele, 1929 (?)

TRIFORIDAE (Cerithiellidae) (4)
TRIFORINAE
TRIFORIS Deshayes, 1834 (Fossil)
GRANULOTRIFORIS Kosuge, 1967
PARAMENDAX Powell, 1937
TRITUBA Jousseaume, 1884

SHERBORNIINAE
SHERBORNIA Iredale, 1917

CERITHIELLINAE
CERITHIELLA Verrill, 1882
Lovenella G. O. Sars, 1878
(non Hicks, 1869)
Newtonia Cossmann, 1893
(non Schlegel, 1866)
Newtoniella Cossmann, 1893
Cerithiolinum Locard, 1903
GLIBERTIELLA Gründel, 1980
ATAXOCERITHIUM Tate, 1893
LAIOCOCHLIS Dunker & Metzger, 1874

References:
Haszprunar, G. "Heterobranchia; a new concept of the phylogeny of the higher Gastropoda" Z.zool.Syst.Evolut.forsch.23:15-37 - 1985 (1)

Marshall, B. A. "Recent New Zealand Triforis" N.Z.Jour.Zool.(4):101-110 - 1977 (2)

---- "The Cerithiopsidae of New Zealand" N.Z.Jour.Zool.(5):47-120 - 1978 (3)

---- "The systematic position of Triforis " N.Z.Jour.Zool.(7):85-88 - 1980 (4)

TRIPHOROIDEA
TRIPHORIDAE
TRIPHORINAE (10)
TRIPHORA Blainville, 1828
Tristoma Deshayes in Menke, 1830
Triphoris Deshayes, 1832
(non Triforis Deshayes, 1834)
ACLOPHORA Laseron, 1958
ACLOPHOROPSIS Marshall, 1983
BOUCHETRIPHORA Marshall, 1983
CAUTOR Finlay, 1927
CHEIRODONTA Marshall, 1983
COSMOTRIPHORA Olsson & Harbison, 1953
EUTHYMELLA Thiele, 1929
pro Euthymia Jousseaume, 1884
(non Stål, 1876)
Torresophora Laseron, 1958
EUTRIPHORA Cotton & Godfrey, 1931
HEDLEYTRIPHORA Marshall, 1983
HYPOTRIPHORA Cotton & Godfrey, 1931
INELLA Bayle, 1879
pro Ino Hinds, 1843
(non Samsuelle, 1817)
INIFORIS Jousseaume, 1884
ISOTRIPHORA Cotton & Godfrey, 1931
LATITRIPHORA Marshall, 1983
LINIPHORA Laseron, 1958
LITHARIUM Dall, 1924
MAGNOSINISTER Laseron, 1954
Macrosinister Laseron, 1954 (err.)
MARSHALLORA Bouchet, 1985
MASTONIA Hinds, 1843
MASTONIAEFORIS Jousseaume, 1884
Contraforis Laseron, 1958
Epiforis Laseron, 1958

TRIPHORINAE (Cont'd.)
MESOPHORA Laseron, 1958
Coriophora Laseron, 1958
MONOPHORUS Grillo, 1877
Biforina Bucquoy, Dautzenberg & Dollfus, 1884
Notosinister Finlay, 1927
NANAPHORA Laseron, 1958
NOTOTRIPHORA Marshall, 1983
OBESULA Jousseaume, 1898
OPIMAPHORA Laseron, 1958
RISBECIA Kosuge, 1966
SAGENOTRIPHORA Marshall, 1983
SIMILIPHORA Bouchet, 1985 (1)
STROBILIGERA Dall, 1924
SUBULOPHORA Laseron, 1958
SYCHAR Hinds, 1843
TALOPHORA Gründel, 1975
TERETRIPHORA Finlay, 1927
Distophora Laseron, 1958
TETRAPHORA Laseron, 1958
VIRIOLA Jousseaume, 1884
Sinistroseila Oliver, 1915
Solosinister Laseron, 1954
Orbitriphora Laseron, 1958
VIRIOLOPSIS Marshall, 1983

METAXIINAE (3)
METAXIA Monterosato, 1884
SEILAREX Iredale, 1924

ADELACERITHINAE (5)
ADELACERITHIUM Ludbrook, 1941

References:
Bouchet, P. "Les Triphoridae de Mediterranee et du proche Atlantique" Lavori della Soc. Malacologica Italiano 21:5-58 - 1985 (1)

Haszprunar, G. "The Heterobranchia" (2) Z.zool.Syst.Evolut.forsch.23:15-37 - 1985

Marshall, B.A. "Dextral triforid genus Metaxia,SW. Pacific" N.Z.Journ.Zool.4:111-117 - 1977b (3)

---- "Additional dextral triphorids" N.Z.Journ.Zool.6:397-404 - 1979 (3a)

---- "A revision of the Triphoridae of Southern Australia" Rec.Aust.Mus.Supp.2:1-119 - 1983 (4)

---- "Adelacerithiinae: new subfamily of the Triphoridae" J.Moll.Stud.50:78-84.1984 (5)

EPITONIOIDEA
EPITONIIDAE (2, 4)
EPITONIINAE
EPITONIUM Röding, 1798
Scala Bruguière, 1792
Cyclostoma Lamarck, 1799
(non Lamarck, 1801)
Scalaria Lamarck, 1801
Scatlatarius Duméril, 1806
Scalarius, Scalarus Montfort, 1810
Aciona Leach, 1815
Parviscala de Boury, 1887
Linctoscala Monterosato, 1890
Sphaeroscala Monterosato, 1890
Boreoscala Kobelt, 1902
Arctoscala Dall, 1909
Acutiscala de Boury, 1909
Anguliscala de Boury, 1909
Evolutiscala de Boury, 1909
Graciliscala de Boury, 1909
Lepidiscala de Boury, 1909
Liriscala de Boury, 1909

EPITONIUM (Cont'd.)
Minutiscala de Boury, 1909
Pyramiscala de Boury, 1909
Solvaclathrus Iredale, 1936
ASPERISCALA de Boury, 1909
Cinctiscala,
Decussiscala de Boury, 1909
CONNEXISCALA de Boury, 1909
FOLIACEISCALA de Boury, 1912
GLOBISCALA de Boury, 1909
GYROSCALA de Boury, 1887
Turbona Brown in Mörch, 1874
(non Turbona Leach, 1847)
Pistoscala Dall, 1917
HIRTOSCALA Monterosato, 1890
Foveoscala de Boury, 1909
Pudentiscala Iredale, 1936
HYALOSCALA de Boury, 1909
Glabriscala de Boury, 1909
LAEVISCALA de Boury, 1909
LAMELLISCALA de Boury, 1909
LIBRARISCALA Kilburn, 1985 (3)

EPITONIUM (Cont'd.)
LIMISCALA de Boury, 1909 (Limniscala - err.)
Dulciscala de Boury, 1909
Crenuliscala Iredale, 1936
NIPPONISCALA Masahito & Habe, 1973
PUPISCALA Masahito, Kuroda & Habe, 1971
SODALISCALA de Boury, 1921 (3)
Firmiscala de Boury, 1909
VICINISCALA de Boury, 1909 ?
ACIRSA Mörch, 1857
Hemiacirsa de Boury, 1890
Pseudacirsa Kobelt, 1903
Plesioacirsa de Boury, 1909
ACRILLOSCALA Sacco, 1891
BIFIDOSCALA Cossmann, 1888
ALEXANIA Strand, 1928
Alexandria Tomlin, 1926
(non Pfeiffer, 1881)
Tomlinula Strand, 1932 ?
Habea Kuroda, 1943
Stenacme Pilsbry, 1945
ALORA H. & A. Adams, 1853
Teramachiacirsa Kuroda & Ito, 1961
AMAEA H. & A. Adams, 1853
ACRILLA H. Adams, 1860
Adiscoacrilla Sacco, 1890
Texiscala de Boury, 1911
Fragilopalia Azuma, 1972
? FILISCALA de Boury, 1911
SCALINA Conrad, 1865
Ferminoscala Dall, 1908
Elegantiscala de Boury, 1911
CIRSOTREMA Mörch, 1852
Caloscala Tate, 1885
Cirsostrema Conrad, 1865 (error)
Cirsotremopsis Thiele, 1928
PLASTISCALA Iredale, 1936 (?)
? RECTACIRSA Iredale, 1936
CLAVISCALA de Boury, 1909
CYCLOSCALA Dall, 1889
CYLINDRISCALA de Boury, 1912
Cylindriscala de Boury, 1909 (nom.nud.)
Torquatiscala de Boury, 1909 ?
DEPRESSISCALA de Boury, 1909
FORATISCALA de Boury, 1887
GREGORIOSCALA Cossmann, 1912
Bria de Gregorio, 1890
Compressiscala Masahito & Habe, 1976
NITIDISCALA de Boury, 1909 (3)
OPALIA H. & A. Adams, 1853
Psychrosoma Tapparone-Canefri, 1876
DENTISCALA de Boury, 1886
Granuliscala de Boury, 1909
NODISCALA de Boury, 1889
PLICISCALA de Boury, 1887
PERIAPTA Bouchet & Warén, 1986
PUNCTISCALA de Boury, 1890
RUTELLISCALA Kilburn, 1985
SAGAMISCALA Masahito,
Kuroda & Habe, 1971
STHENORHYTIS Conrad, 1862
Tenuiscala de Boury, 1887
Pseudostenorytis Sacco, 1891
Stenorhyscala de Boury, 1912 (err.)
Stenorhytiscala de Boury, 1912 (err.)
Stenorhytis Cossmann, 1912 (err.)
Pseudostenorhytis Cossmann, 1912 (err.)

EPITONIINAE
Unassigned Genera/Subgenera: (1)
AMICISCALA Jousseaume, 1912
AVALITISCALA Jousseaume, 1912
BERTHAIS Melvill, 1904

EPITONIINAE (Cont'd.)
CERITHISCALA de Boury, 1887
CHUNISCALA Thiele, 1928
CIRRATISCALA de Boury, 1909
CLATHROSCALA de Boury, 1890
CLATHRUS Oken, 1815
CORONISCALA de Boury, 1909
COUTHOUYELLA Bartsch, 1909
CREBRISCALA de Boury, 1909
CRISPOSCALA de Boury, 1909
DANNEVIGENA Iredale, 1936
DELICATISCALA de Boury, 1909
DISCOSCALA Sacco, 1890
Mammiscala de Boury, 1909
EGLISIA Gray, 1847
FORACEISCALA de Boury, 1911
FRAGILISCALA Azuma, 1962
FUNISCALA de Boury, 1891
FUSICOSCALA Monterosato, 1890
GRADATISCALA de Boury, 1909
INNESISCALA Jousseaume, 1912
KURODACIRSA Masahito & Habe, 1975
LAMPROPALIA Kuroda & Ito, 1961
LANISCALA Lan, 1976
LITTORINISCALA de Boury, 1887
MAZESCALA Iredale, 1936
NARVALISCALA Iredale, 1936
PAPYRISCALA de Boury, 1909
PERLUCIDISCALA de Boury, 1912
POMISCALA Iredale, 1936
PROPESCALA Cotton & Godfrey, 1931
RETICULISCALA de Boury, 1909
Resticuliscala Cossmann, 1912 (err.)
SPINISCALA de Boury, 1909
Eburniscala de Boury, 1909
SUBULISCALA de Boury, 1909
Longiscala de Boury, 1910
TASMALIRA Dall, 1956
TURBINISCALA de Boury, 1909
TURRISCALA de Boury, 1890
Rudiscala de Boury, 1909
VARICISCALA de Boury, 1909

NYSTIELLINAE (2)
OPALIOPSIS Thiele, 1928 (2)
Nystiella Clench & Turner, 1952
ECCLESIOGYRA Dall, 1892
Solutiscala de Boury, 1909
Abyssicala de Boury, 1911
IPHITUS Jeffreys, 1883 (2)
Iphitella Thiele, 1925
(non Rafinesque, 1815)
(unnec.nom.nov.pro Iphitus)
Sculptifer Beu & Climo, 1974 (6)
MURDOCHELLA Finlay, 1926
NARRIMANIA Taviani, 1984
PAPULISCALA de Boury, 1911
Pustuliscala de Boury in
Cossmann & Peyrot, 1921

JANTHINIDAE
JANTHINA Röding, 1798
Ianthina auctt.
Fragilis Lamarck, 1816
Communis Lamarck, 1822
ACHATES Mörch, 1860
AMETHISTINA Mörch, 1860
JODES Mörch, 1860
Iodes auctt.
JODINA Mörch, 1860
Iodina auctt.
VIOLETTA Iredale, 1929
RECLUZIA Petit, 1853

References: (Epitonioidea)
Abbott, R.T. "The Family Epitoniidae" Standard Catalog,1978 (1)

Bouchet, P. & A. Warén "Revision: NE. Atlantic bathyl and abyssal Aclididae, Eulimidae, and Epitoniidae" Soc.Ital.di Malac. Boll.Malac.Supp.2:299-576 -1986 (2)

Clench, W. L. & R. D. Turner "Sthenorhytis etc. in the Western Atlantic" Johnsonia 2(29) - 1950
---- "The genus Epitonium in the Western Atlantic" Ibid.2(30) - 1951
---- II. "The genera Depressescala etc. in the Western Atlantic" Ibid.2(31) - 1952

DuShane, H. "The family Epitoniidae in the Northeastern Pacific" Veliger 22(2):91-134 - 1979 (3)

Kilburn, R. N. "Epitoniidae in South Africa" Ann.Natal Mus.27(1):239-337 - 1985 (4)

Warén, A. "On the systematic postition of Couthouyella (Epitoniidae)" Nautilus 94(3):105-107 - 1980 (5)

---- "Generic revision of the family Eulimidae" Journ.Moll.Stud., Supp.13 - 1983 (6)

EULIMOIDEA (1)
EULIMIDAE (Melanellidae)
EULIMA Risso, 1826
Leiostraca H. & A. Adams, 1853
Liostraca Cossmann, 1921 (err.)
Subularia Monterosato, 1884
Hordeulima Sacco, 1892
Sulcosubularia Sacco, 1892
Cuspeulima Laseron, 1955
ACROCHALIX Bouchet & Warén, 1986
AMAMIBALCIS Kuroda & Habe, 1950
ANNULOBALCIS Habe, 1965
ARCUELLA Nevill, 1874 ?
ASTEROLAMIA Warén, 1980
AURICULIGERINA Dautzenberg, 1925
BACULA H. & A. Adams, 1863
Subeulima Souverbie, 1875
BALCIS Leach, 1847
BATHEULIMA F. Nordsieck, 1968
BATHYCRINICOLA Bouchet & Warén, 1986
BULIMEULIMA Bouchet & Warén, 1986
CAMPYLORHAPHION Bouchet & Warén, 1986
CHILEUTOMIA Tate & Cossmann, 1898
CONCAVIBALCIS Warén, 1980
COSTACLIS Bartsch, 1947 (? = Hemiaclis)
CRINOLAMIA Bouchet & Warén, 1979
CRINOPHTHEIROS Bouchet & Warén, 1986
ECHINEULIMA Lützen & Nielsen, 1975 (3c)
Luetzenia Warén, 1980
(non Rehder, 1980)
ECHIUROIDICOLA Warén, 1980
ERSILIA Monterosato, 1872
Hersilia Monterosato, 1872
EULIMITRA Laseron, 1955 (? doubtful)
EULIMOSTRACA Bartsch, 1917
EULITOMA Laseron, 1955
FUSCAPEX Warén, 1981
FUSCEULIMA Laseron, 1955
GOUBINIA Dautzenberg, 1923 ?
HALLIELLA Monterosato, 1878
HALLIELLOIDES Bouchet & Warén, 1986
HEBEULIMA Laseron, 1955
HEMIACLIS Sars, 1878
HEMILIOSTRACA Pilsbry, 1917
HOPLOPTERON Fischer, 1876 ?
MELANELLA Bowdich, 1822
Eulimaustra Laseron, 1955
MENON Hedley, 1900 (4)
MOLPADICOLA Grusov, 1957
MUCRONALIA A. Adams, 1860
Mucronella Ivanov in Fishelson, 1973 (nom.nud.)
NISO Risso, 1826
Bonellia Deshayes, 1838
(non Rolando, 1821)

EULIMIDAE
NISO (Cont'd.)
Janella Grateloup, 1838
Neovolusia Emerson, 1965
pro Volusia A. Adams, 1861
(non Robineau-Desvoidy, 1830)
OCEANIDA De Folin, 1870 ?
Spiroclimax Mörch, 1875
Athleenia Bartsch, 1946
Hoplopteropsis Morgan, 1915
OPHIEULIMA Warén & Sibuet, 1981
OPHIOARACHNICOLA Warén, 1980 ?
OPHIOLAMIA Warén & Carney, 1981
PALISADIA Laseron, 1956
PARASTILBE Cossmann, 1900
Stilbe Jeffreys, 1884
(non De Kay, 1842)
PARVIORIS Warén, 1981
PELSENEERIA Koehler & Vaney, 1908
Stylina Fleming, 1828
(non Lamarck, 1816)
Rosenia Schepman, 1914
pro Turtonia Rosen, 1910
(non Alder, 1848)
Venustilifer Powell, 1938
Parastilifer Ivanov, 1954
PICTOBALCIS Laseron, 1955
PISOLAMIA Bouchet & Lützen, 1976 (?)
PLAGYOSTILA Fischer, 1872 ? (4)
pro Plagiostyla Fischer in De Folin & Perier, 1871 (nom.nud.)
POLYGIREULIMA Sacco, 1892
Acicularia Monterosato, 1884
PROSTILIFER Warén, 1980
PULICOCHLEA Ponder & Goodring, 1978
PSEUDORETUSA Ponder & Goodring, 1978
PUNCTIFERA Warén, 1981
PYRAMIDELLOIDES Nevill, 1885 (4)
Isselia auctt. (nom.nud.)
(non Bourguignat, 1877)
CHRYSTELLA Laseron, 1956 ?
TERETIANAX Iredale, 1919 ?
RECTILABRUM Bouchet & Warén, 1986
SABINELLA Monterosato, 1890
Eulimoda Laseron, 1955
SCALARIBALCIS Warén, 1980
SCALENOSTOMA Deshayes, 1863
Hyperlia Pilsbry, 1918
Stilimella Laseron, 1955
STICTEULIMA Laseron, 1955
Lentigobalcis Habe, 1961
STROMBIFORMIS da Costa, 1778 ?
TROCHOSTILIFER Warén, 1980 (5a)
TRIPIOMETRICOLA Warén, 1981
TURVIERIA Berry, 1956

EULIMIDAE (Cont'd.)
UMBILIBALCIS Bouchet & Warén, 1986
VITREOBALCIS Warén, 1980
VITREOLINA Monterosato, 1884
CURVEULIMA Laseron, 1955

STILIFERIDAE
STILIFER Broderip & Sowerby,1832
Stylifer auctt. (error)
APICALIA A. Adams, 1862
Chryseulima Laseron, 1955
GOODINGIA Lützen, 1972
HYPERMASTUS Pilsbry, 1899
LEUTZENIA Rehder, 1980 ?
pro Robillardia E. A. Smith in Gooding & Lützen, 1973
(non E. A. Smith, 1889)
MICROSTILIFER Warén, 1980
MEGADENUS Rosen, 1910
MONOGAMUS Lützen, 1976
PARAMEGADENUS Humphreys & Lützen, 1972
PEASISTILIFER Warén, 1980
ROBILLARDIA E. A. Smith, 1889
STILAPEX Iredale, 1925
Stylapex auctt. (error)
Lambertia Souverbie, 1869 ?
(non Robineau-Desvoidy, 1853)

ASTEROPHILIDAE ? (3)
ASTEROPHILA Randall & Heath, 1965

PAEDOPHOROPODIDAE
PAEDOPHOROPUS Ivanov, 1933
MOLPADICOLA Grusov, 1957

ENTOCONCHIDAE (3b)
ENTOCONCHA Müller, 1852
Helicosyrinx Baur, 1864
ENTOCOLAX W. Voigt, 1888
ENTEROXENOS Bonnevie, 1902 (3a)
Comenteroxenos Tikasingh, 1961
GASTEROSIPHON Koehler & Vaney, 1905
Entosiphon Koehler & Vaney, 1903
(non Stein, 1878)

ENTOCONCHIDAE (Cont'd.)
THYONICOLA Mandahl-Barth, 1941
Parenteroxenos Ivanov, 1945
DIACOLAX Mandahl-Barth, 1946

THYCIDAE (5)
THYCA H. & A. Adams, 1854
Granulithyca Habe, 1976
BESSOMIA Berry, 1959
KIRAMODULUS Kuroda, 1949

ACLIDIDAE (1)
ACLIS Lovén, 1846
Cyclonidea Laseron, 1956
Menippe Jeffreys, 1867
(non de Haan, 1833)
Pherusa Jeffreys, 1869 (pro Menippe)
(non Oken, 1807)
Pherusina Norman, 1888 (pro Pherusa)
Martelliella Dautzenberg & Durouchoux, 1914 (pro Pherusa)
Eusetia Cotton, 1944
Schwengelia Bartsch, 1947

Placement ? (ex Aclididae - 1)

AUSTRORISSOPSIS Ponder, 1965
AWANUIA Powell, 1927
BERMUDACLIS Bartsch, 1947
BOURYISCALA Cossmann, 1902
CIMA Jeffreys, 1898
COENACULUM Iredale, 1924
Parascala Cotton & Godfrey, 1931
DISSOPALIA Iredale, 1936
GRAPHIS Jeffreys, 1867
Cioniscus Jeffreys, 1869
Pseudochemnitzia Boettger, 1901
Miraclis Boettger, 1906
MURCHISONELLA Mörch, 1875
Murchisonieaella Fischer, 1884 (err.)
HENRYA Bartsch, 1927
LAROCHELLA Powell, 1927
RISSOPSETIA Dell, 1956
RUAPUKEA Dell, 1952
SCALARONOBA Powell, 1927 (5)
SELMA A. Adams, 1864
THALEIA Warén, 1979

References:
Bouchet, P. & A. Warén "Revisions: NE. Atlantic bathyl and abyssal Aclididae, Eulimidae, and Epitoniidae" Soc.Ital. di Malac. Boll.Malac.Supp.2:299-576 - 1986 (1)

Grusov E. N. "The Endoparasitic mollusk Asterophila japonica, and its relation to the parasitic gastropods" Malacologia 3(1):111-181 - 1965 (2)

Lützen J. "On a new genus of Prosobranchia parasitic on the tropical sea-urchin Echinometra mathei" Jour.Zool.Israel 25:38-51 - 1976 (3)

---- "Studies on the life history of Enteroxenus bonnevie" Ophelia 18(1):1-51 - 1979 (3a)

---- "Unisexuality in the parasitic family Entoconchidae" Malacologia 7(1):7-15 - 1968 (3b)

---- & K. Nielsen "Contributions to the anatomy and biology of Echineulima" Vidensk.Medd.Dansk.Naturhist.forening 138:171-199 (Veliger review) - 1975 (3c)

Ponder, W. F. "Review of the genera of Rissoidae" Rec.Aust.Mus.Supp.4:1-221 - 1985 (4)

Warén, A. "Revisions: Thyca, Stilifer etc." Zool. Scr. 9:187-210 - 1980 (5)

---- "Descriptions of New Taxa of Eulimidae" Zool.Scr.9:283-306 - 1980b (5a)
---- "Revision of Genera Apicalia & Stilapex" Zool.Scr.10:133-154 - 1981 (5b)
---- "Generic revision of Family Eulimidae" Journ.Moll.Stud., Suppl.13 - 1984 (5c)

---- , B. L. Burch & T. A. Burch "Description of Five new Species of Hawaiian Eulimidae" Veliger 26(3):17-178 - 1984 (5d)

Suborder Neogastropoda (Stenoglossa)

MURICOIDEA
MURICIDAE
MURICINAE
MUREX Linnaeus, 1758
Purpura Martini, 1777 (non-bin.)
Aranea Perry, 1810
(non Linnaeus, 1758)
Tribulus Kobelt, 1877
(non H. & A. Adams in Klein, 1853)
Acupurpura Jousseaume, 1880
Tubicauda Jousseaume, 1880
PROMUREX Ponder & Vokes, 1988 (16b)
ASPELLA Mörch, 1877
ATTILIOSA Emerson, 1968 (26)
BOLINUS Pusch, 1837 (16b)
Rhinocantha H. & A. Adams, 1853
Rhynocantha Bellardi, 1872 (err.)
Purpura Jousseaume, 1880
(non Bruguière, 1789)
Brandaria Monterosato, 1917
CALOTROPHON Hertlein & Strong, 1951
Pseudosalpinx Olsson & Harbison, 1953
Pseudosyrinx auctt. (error)
Hertleinella Berry, 1958
CHICOREUS Montfort, 1810
Triplex Perry, 1810
Chicoracea Griffith &
Pidgeon, 1834 (err.)
Cichorium Voigt, 1834 (err.)
Chicorax Pusch, 1837
Frondosaria Schlüter, 1838
Cichoreus Agassiz, 1846 (nom.nud.)
Cichoraceus Herrmannsen, 1847 (em.)
Cichoreum Paetel, 1875 (err.)
Euphyllon Jousseaume, 1880
Pirtus de Gregorio, 1885
Torvamurex Iredale, 1936
Torramurex Salisbury, 1937 (err.)
Foveomurex Wenz, 1941
CHICOMUREX Arakawa, 1964
NAQUETIA Jousseaume, 1880 (22)
Triplex Humphrey in Harris, 1897
(non Perry, 1810)
Rhizophorimurex Oyama, 1950
PHYLLONOTUS Swainson, 1833
Phyllonota Conrad, 1847 (err.)
Phyconotus Simroth, 1907 (err.)
DERMOMUREX Monterosato, 1890
pro Poweria Monterosato, 1884
(non Bonaparte, 1840)
Hexachorda Cossmann, 1903 (fossil)
GRACILIMUREX Thiele, 1929
TAKIA Kuroda, 1953
TRIALATELLA Berry, 1964
VIATOR E. H. Vokes, 1974
HAUSTELLUM Schumacher, 1817 (16b)
Haustellum Deshayes, 1830 ?
(pro Klein in Bruguière, 1792)
Brontes Montfort, 1810
(non Fabricius, 1801)
Brontesia Reichenbach, 1828 err.)
Brontis Griffith & Pidgeon, 1834 (err.)
Bronta Pusch, 1837 (error pro Brontes)
Haustellaria Swainson, 1833
? SIRATUS Jousseaume, 1880 (16b)
HEXAPLEX Perry, 1810
Purpura Röding, 1798
(non Bruguière, 1789)
Exaplex Férussac, 1820 (err.)
Muricanthus Swainson, 1840
(non Swainson, 1833)

HEXAPLEX (Cont'd.)
pro Centronotus Swainson, 1833
(non Schneider, 1801)
Bassiella Wenz, 1941
pro Bassia Jousseaume, 1880
(non Quoy & Gaimard, 1830)
MURICANTHUS Swainson, 1833 (25)
Muricantha Fischer, 1884
Muricantha Swainson in Suter, 1913
Aaronia Verrill, 1950
TRUNCULARIOPSIS Cossmann, 1921
pro Truncularia Monterosato, 1917
(non Wiegmann, 1832)
Murithais Grant & Gale, 1931
NIPPONOTROPHON Kuroda & Habe, 1971
POIRIERIA Jousseaume, 1880
Poirrieria Fischer, 1884 (err.)
FLEXOPTERON Shuto, 1969 (living fossil)
PANAMUREX Woodring, 1959
PAZIELLA Jousseaume, 1880
Bathymurex Clench & Farfante, 1945
Dallimurex Rehder, 1946
PAZINOTUS E. H. Vokes, 1970
PTERYNOTUS Swainson, 1833
Pterymurex Rovereto, 1899
pro Pteronotus Swainson, 1833
(non Rafinesque, 1815)
Marchia Jousseaume, 1880 (17)
Morchia Baker, 1891 (err.)
Triplex Newton, 1891 ?
(non Perry, 1810)
PTEROCHELUS Jousseaume, 1880
Alipurpura P. Fischer, 1884
PURPURELLUS Jousseaume, 1880
Triremus Bayle in Fischer, 1884
Triremia Korobokov, 1955
TIMBELLUS de Gregorio, 1885

MURICINAE s.l. (ex Typhinae; 7, 13a)
CINCLIDOTYPHIS DuShane, 1969
PONDERIA Houart, 1986 (13a)
PROTOTYPHIS Ponder, 1972
PTEROTYPHIS Jousseaume, 1881
pro Perotyphis Jousseaume, 1880
Trigonotyphis Jousseaume, 1881 ?
TRIPTEROTYPHIS Pilsbry & Lowe, 1932
Nothotyphis Fleming, 1962 ?

MURICOPSINAE (27)
MURICOPSIS Bucquoy, Dautzenberg & Dollfus, 1882
Muricidea auctt.
(non Swainson, 1840)
Jania Bellardi in Cossmann, 1882
(non Bellardi)
RISOMUREX Olsson & McGinty, 1958 (27)
ACANTHOTROPHON Hertlein & Strong, 1951
BIZETIELLA Radwin & D'Attilio, 1972
FAVARTIA Jousseaume, 1880
Minnimurex Woolacott, 1957
CARIBIELLA Perrilliat, 1972
HOMALOCANTHA Mörch, 1852
Homalacantha Kobelt, 1877 (err.)
Homolocantha Ludbrook, 1958 (err.)
LATAXIENA Jousseaume, (1833)
Bedevina Habe, 1946
LINDAPTERYS Petuch, 1987 (?)
MAXWELLIA Baily, 1950
MUREXIELLA Clench & Farfante, 1945
(= Favartia Subgenus ?)
SUBPTERYNOTUS Olsson & Harbison, 1953
MUREXSUL Iredale, 1915
Murexul Habe, 1961 (err.)
PYGMAEPTERYS E. H. Vokes, 1978 (24)

OCENEBRINAE

OCENEBRA Gray, 1847
Tritonalia Fleming in Gray, 1847
(non Fleming, 1828)
Inermicosta Jousseaume, 1880
Heteropurpura Jousseaume, 1880
Occenebra Hörnes & Auinger, 1885 (err.)
OCINEBRINA Jousseaume, 1880
Corallinia Bucquoy & Dollfus in Bucquoy, Dautzenberg & Dollfus, 1882)
Ocenebrina Cossmann, 1903 (err.)
Dentocenebra Monterosato, 1917
HADRIANIA Bucquoy, Dautzenberg & Dollfus, 1882
Hadrina Tryon, 1883 (err.)
Adriania Cossmann, 1903 (em.)
ROPERIA Dall, 1898
CERATOSTOMA Herrmannsen, 1846
pro Cerostoma Conrad, 1837
(non Latreille, 1802)
Cerastoma Herrmannsen, 1846
(non Koch, 1839)
Spinostoma Coen, 1943
Microrhytis Emerson, 1959 (fossil)
CRASSILABRUM Jousseaume, 1880
Antimurex Cossmann, 1903
EUPLEURA H. & A. Adams, 1853
GENKAIMUREX Kuroda, 1953
JATON Pusch, 1837
Iaton Gray, 1847 (em.)
Jatova Jousseaume, 1880
NAMAMUREX Carrington & Kensley, 1969
PTEROPURPURA Jousseaume, 1880
Centrifuga Grant & Gale, 1931
CALCITRAPESSA Berry, 1959
OCINEBRELLUS Jousseaume, 1880
Ocenebrellus Cossmann, 1903 (err.)
Ternaria Coen, 1943
POROPTERON Jousseaume, 1880 (23)
SHASKYUS Burch & Campbell, 1963
PTERORYTIS Conrad, 1863
Pterorhytis auctt. (err.)
Pterorhythis auctt. (err.)
Pterophytis auctt. (err.)
Pterohytus auctt. (err.)
Neurarhytis Olsson & Harbison, 1953 (foss.)
TRACHYPOLLIA Woodring, 1928
Morunella Emerson & Hertlein, 1964
UROSALPINX Stimpson, 1865
Hanetia Jousseaume, 1880

ERGALATAXINAE (25)

ERGALATAX Iredale, 1931
BEDEVA Iredale, 1924
CRONIA H. & A. Adams, 1853
Semiricinula von Martens, 1903
CYTHAROMORULA Kuroda, 1953
MACULOTRITON Dall, 1904
MURICODRUPA Iredale, 1918
ORANIA Pallary, 1900 (27)
Nemofusus Cossmann, 1903
PASCULA Dall, 1908
Evokesia Radwin & D'Attilio, 1972
PHRYGIOMUREX Dall, 1904
PHYLLOCOMA Tapparone-Canefri, 1881
Craspedotriton Dall, 1904
GALFRIDUS Iredale, 1924
UTTLEYA Marwick, 1934
USILLA H. Adams, 1860
XANTHOCHORUS P. Fischer, 1884

TYPHINAE

TYPHIS Montfort, 1810
HAUSTELLOTYPHIS Jousseaume, 1880
RUGOTYPHIS Vella, 1961
TALITYPHIS Jousseaume, 1882
TYPHINELLUS Jousseaume, 1880
TYPHINA Jousseaume, 1880
TYPHISOPSIS Jousseaume, 1880
DISTICHOTYPHIS Keen & Campbell, 1964
MONSTROTYPHIS Habe, 1961
SIPHONOCHELUS Jousseaume, 1880
Cyphonochelus Jousseaume, 1882
Choreotyphis Iredale, 1936
LAEVITYPHIS Cossmann, 1903
TRUBATSA Dall, 1899
TYPHISALA Jousseaume, 1881

TROPHONINAE

TROPHON Montfort, 1810
Polyplex Perry, 1810 (Suppr.)
Muricidea Swainson, 1840
ACTINOTROPHON Dall, 1902
AFRITROPHON Tomlin, 1947
ANATROPHON Iredale, 1929
APIXYSTUS Iredale, 1929
BENTHOXYSTUS Iredale, 1929
BOREOTROPHON P. Fischer, 1884
Nodulotrophon Habe & Ito, 1965
COMPTELLA Finlay, 1927
FUEGOTROPHON Powell, 1951
GEMIXYSTUS Iredale, 1929
LITOZAMIA Iredale, 1929
MINORTROPHON Finlay, 1927 ?
PAGODULA Monterosato, 1884
Pinon de Gregorio, 1885
Enixotrophon Iredale, 1929
PARATROPHON Finlay, 1927
TROPHONOPSIS Bucquoy & Dautzenberg, 1882
Chalmon de Gregorio, 1885
Enatimene Iredale, 1929
XENOTROPHON Iredale, 1929
XYMENE Iredale, 1915
Kalydon Hutton, 1884 (suppressed)
(non Calydon Thompson, 1864)
Lenitrophon Finlay, 1927
Terefundus Finlay, 1927
Axymene Finlay, 1927
Xymenella Finlay, 1927
Zeatrophon Finlay, 1927 (16)
XYMENOPSIS Powell, 1951

THAIDINAE

THAIS Röding, 1798
Trochia Swainson, 1840
Dicathais Iredale, 1936
Menathais Iredale, 1937
AGNEWIA Tenison-Woods, 1878
pro Adamsia Dunker, 1857
(non Forbes, 1840)
CYMIA Mörch, 1860
Cumopsis Rovereto, 1899
pro Cuma Swainson, 1840
LEPSIELLA Iredale, 1912
LEPSITHAIS Finlay, 1928
MANCINELLA Link, 1807
Thalessa H. & A. Adams, 1853
Tribulus Klein in Bruguiere, 1792
Planithais Bayle in Fischer, 1884
PINAXIA H. & A. Adams, 1853
Conothais Kuroda, 1930 ?
STRAMONITA Schumacher, 1817

THAIS (Cont'd.)
THAISELLA Clench, 1947
VASULA Mörch, 1860
ACANTHINA Fischer de Waldheim, 1807
Unicornis Montfort, 1810
Rudolpha Schumacher, 1817
Monoceros Sowerby, 1827
(non Lacepede, 1798)
Monocerus Lamarck, 1822
(non Bloch & Schneider, 1801)
ACANTHINUCELLA Cooke, 1918
CONCHOLEPAS Klein in Bruguière, 1792
Concholepas Lamarck, 1801
(non Bruguière)
Conchulus Rafinesque, 1815
Conchopatella Herrmannsen, 1847
DAPHNELLOPSIS Schepman, 1913 (ex Turridae - 13)
DRUPA Röding, 1798 (11)
Canrena Link, 1807
Sistrum Montfort, 1810
Ricinula Lamarck, 1816
Ricinella Lamarck in Bowdich, 1822 (err.)
(non Schumacher, 1817)
Pentadactylus Mörch, 1852
(non Schulze, 1760)
Ricimula Gould, 1855 (err.)
Ricinulus Desmarest in Chenu, 1859 (err.)
RICINELLA Schumacher, 1817
DRUPINA Dall, 1923
DRUPELLA Thiele, 1925
LINIAXIS Laseron, 1955 ?
MORULA Schumacher, 1817
Oppomorus Iredale, 1937 ?
Tenguella Arakawa, 1965
AZUMAMORULA Emerson, 1968
Pro Morulina Dall, 1923
(non Borner, 1906)
NASSA Röding, 1798
Iopas H. & A. Adams, 1853
Jopas H. & A. Adams, 1853
NEOTHAIS Iredale, 1912
NUCELLA Röding, 1798
Polytropa Swainson, 1840
Polytropalicus Rovereto, 1899
PURPURA Bruguière, 1789 (nom.cons.)
Lepsia Hutton, 1884
Plicopurpura Cossmann, 1903
pro Purpurella Dall, 1871
(non Bellardi, 1882)
Patellipurpura Dall, 1909
Haustrum Perry, 1911
REISHIA Kuroda & Habe, 1971
SPINIDRUPA Habe & Kosuge, 1966

THAIDINAE (Cont'd.)
VEXILLA Swainson, 1840
Provexillum Hedley, 1918
VITULARIA Swainson, 1840 (25a)
Vitulina Swainson, 1840
Transafer Iredale, 1929

RAPANINAE
RAPANA Schumacher, 1817
AUSTROTROPHON Dall, 1902
Straminotrophon Powell, 1951
CHORUS Gray, 1847
FORRERIA Jousseaume, 1880
NEORAPANA Cooke, 1918
ZACATROPHON Hertlein & Strong, 1951

Placement ? - Muricidae:
MINIBRARIA Sarasua, 1984 (18)
(Ex Colubrariinae - 1)

CORALLIOPHILIDAE (Magilidae) (14)
CORALLIOPHILA H. & A. Adams, 1853
Quoyola Iredale, 1912
Latimurex, Lepadomurex Coen, 1922
Latiromurex auctt. (3, 14)
Aradomurex Coen, 1947
Aradasia Settepassi, 1970 ?
(non Gray, 1850)
CORALLIOBIA H. & A. Adams, 1853
CORALLIOFUSUS Kuroda, 1953
PSEUDOMUREX Monterosato, 1872 ?
BABELOMUREX Coen, 1922
Langfordia Dall, 1924
Tolema Iredale, 1929
ECHINOLATIAXIS Kosuge, 1979
LAEVILATIAXIS Kosuge, 1979
LAMELLATIAXIS Habe & Kosuge, 1970
TARANTELLAXIS Habe, 1970
EMOZAMIA Iredale, 1929
HIRTOMUREX Coen, 1922
Fusomurex Coen, 1922
LATIAXIS Swainson, 1840
MAGILUS Montfort, 1810
Magilopsis Sowerby, 1919
Leptoconchus Rüppell, 1834
MIPUS de Gregorio, 1885
Rhombothais Wollacott, 1954
RAPA Röding, 1798
Bulbus Herrmannsen, 1848
(non Brown, 1939)
Rapanus Schumacher in Sowerby, 1839
Rapella Swainson, 1840
RELIQUIAECAVA Massin, 1987 (14x)
RHIZOCHILUS Steenstrup, 1850

References:
Beu, A.G. & P.A. Marshall"Revis.:Plesiotritoninae, Appendix: Pisaniinae related to Colubraria" N.Z.Geol.Surv.Paleo.Bull.54:17-58 - 1987 (1)

Cernohorsky, W. O. "Neothais, Morula etc." Veliger 11(4) - 1969 (2)

Clover, P. "Latiaxis Catalog and Illustrated Check List of Coralliophilidae" - 1982 (3)

D'Attilio, A. "Remarks on Coralliophilidae" Festivus Vol.3(8):2-5 - 1972 (5)

---- "A Catalogue of Coralliophilidae" Festivus Vol.X(10) - 1978 (6)

---- "Convergence in the Typhine form" Festivus XIV(8):94-94 - 1982 (7)

---- & Myers "Illustration; recent Pygmaepterys" Veliger XVII (6):59-63 - 1985 (8)

DuShane, H. "New genus: of Typhinae from the Panamic Province" Veliger 11(4):343-348 - 1969 (9)

Emerson, W. K. "Taxonomic placement of Coralliophila incompta, with the proposal of a new genus, Attiliosa" Veliger 10(4):379-381 - 1968 (10)

References: (Muricidae, Cont'd.)
Emerson, W. K. & W. O. Cernohorsky
"The Genus Drupa in the Indo-Pacific"
Indo-Pacific Mollusca 3(13):1-40 -1973 (11)

Espinosa, J. "La subfamilia Typhinae en Cuba"
Poeyana 300:1-13 - 1985 (12)

Houart, R. "Noteworthy Muricidae from the Pacific Ocean, with description of new species"
Resultats des Campagnes Musorstom I & II; P.I.
Mem.Mus.Nat.Hist.nat.Ser.A.Zool.133 -1986 (13)

---- "Ponderia, Gen. nov. with discussion of related genera, and new species"
Apex.1(3):88-93 - 1986 (13a)

---- "Revision of the Subfamily Trophoninae in Southern Africa —" 1987 (13b)
Apex - Soc.Belge de Malacologie 2(2):25-54

Kosuge, S. & M. Suzuki "Latiaxis and its related groups in the family Coralliophilidae"
Inst.Mal.Tokyo Sp.pub.1 - 1985 (14)

Massin, C. "New Genus of Coralliophilidae" - 1987
Bull.Inst.Roy.Sci.Nat.Belgique.57:79-90 (14x)

McLean, J. H. & W. K. Emerson "Calotrophon, a New World genus" Veliger 13(1):57-62 - 1968 (15)

Ponder, W.F."Notes on some Australian - Muricidae"
J.Malac.Soc.Aust.2(3)215-248 - 1972 (16)
---- "The origen & evolution of the Neogastropoda"
Malacologia 12(2):295-338 - 1973 (16a)

---- & E.H. Vokes "Revision: Indo-W. Pacific Murex, Haustellum"Rec.Aust.Mus.Supp.8:1-160.1988(16b)

Radwin, G. E. & A. D'Attilio "Muricacean Supraspecific groups"
Echo 4:55-67 - 1971 (17)
---- "Murex Shells of the World" (17a)
Stanford Univ.Press, Stanford, Calif. -1976

---- "Comments on superfamilial relationships in Acanthotrophon & Attiliosa"
Tulane St.Geol.& Paleo.(131-134) 1978 (17b)

Sarasua, H. "Genera nuevo, Buccinidae; Subgenus, Colubrariidae" Poeyana 272:1-4 - 1984 (18)

Vokes, E. H. "Supraspecific groups in Muricinae and Tritonaliinae"
Malacologia 2(1):1-41 - 1965 (19)

---- "Catalog: Murex; Muricinae, Ocenebrinae"
Bull.Am.Paleo.Vol.61(268) -1971 (20)

---- "New species and subgenus of Australian Dermomurex"
Jour.Malac.Soc.Aust.3(1):1-14 - 1974 (21)

---- "On the identity of Murex triqueter, Born"
Veliger 16:258-264 - 1974 (22)

---- "Cenozoic Muricidae of the western Atlantic region" Part VI: Aspella; Dermomurex" 1975
Tulane Stud.Geol.Paleont.11:121-162 (22a)

---- "Muricidae from the eastern coast of Africa"
Ann.Natal Mus.23(2):375-418 - 1978 (23)

---- "Pygmaepterys: newly described taxon"
Tul.stud.Geo.Paleo.16(2):45-54 - 1980 (24)

---- "Comparison of Muricidae of Eastern Pacific with West Atlantic, with cognate species"
Shells & Sea Life 16(11):210-215 - 1985 (25)

---- "-Short happy life of Vitularia linguabison"
COA Bull.14(2):19-22 - 1986 (25a)

---- & A. D'Attilio "Review of Genus Attiliosa"
Veliger 25(1):67-71 - 1982 (26)

---- & R. Houart "Evaluation: genera Muricopsis, Risomurex" Tul.Stud.19(2):63-88 -1986 (27)

BUCCINIDAE
BUCCININAE
BUCCINUM Linnaeus, 1758
Tritonium Müller, 1776
Mada Jeffreys, 1867
MADIELLA Wenz, 1944
AENATOR Finlay, 1927
Ellicea Finlay, 1928
AMERICOMINELLA Klappenbach & Ureta, 1972
Echinosipho Kaiser, 1977
ANCISTROLEPIS Dall, 1895
CLINOPEGMA Grant & Gale, 1931
ANTISTREPTUS Dall, 1894
ANOMACME Strebel, 1905
AUSTROFUSUS Kobelt, 1879
Aethocola Iredale, 1915
BABYLONIA Schlüter, 1838
Dipsacus Klein, 1753 (pars)
Eburna Lamarck, 1822 (non 1801)
Latrunculus Gray, 1847
Galanthis Gistl, 1848
PERIDIPSACUS Rovereto, 1900
BAILYA M. Smith, 1944
BATHYANCISTROLEPIS Habe & Ito, 1968
BATHYDOMUS Thiele, 1912
BAYERIUS Olsson, 1971

BUCCININAE (Cont'd.)
BELOMITRA Fischer, 1882 (3)
Pleurobela Monterosato in Locard, 1897
Bathyclionella Kobelt, 1905
Cryptomitra Dall, 1924
Morrisonella Bartsch, 1945
Dellina Beu, 1970 (3)
BERINGION Habe & Ito, 1965
BERINGIUS Dall, 1886
Jumala Friele, 1882
Ukko Friele, 1893
BUCCINULUM Deshayes, 1830 (nom.cons.)
Evarne H. & A. Adams, 1858
Euthrena Iredale, 1918
Evarnula Finlay, 1927
Chathamina Finlay, 1928
EUTHRENOPSIS Powell, 1929 ?
EUTHRIA Gray, 1850 ?
GLYPTEUTHRIA Strebel, 1905
Probuccinum Thiele, 1912
JAPEUTHRIA Iredale, 1918
NODOPELAGIA Hedley, 1915
TASMEUTHRIA Iredale, 1925 (Syn. ?)
BURNUPENA Iredale, 1918
CALLILOCONCHA Lus, 1978

BUCCININAE (Cont'd.)
CHLANIFICULA Powell, 1958
CLEA H. A. Adams, 1855 (freshwater)
Canidia H. Adams, 1862
(non Thomson, 1857)
Quadrasia Crosse, 1886
(ex Planaxidae - 7x)
Anentome Cossmann, 1901
Afrocanidia Connolly, 1929
COLUS Röding, 1798
Sipho Bruguière, 1792
(non Fabricius, 1823)
Tritonofusus Beck, 1847
ANOMALOSIPHO Dautzenberg
& Fischer, 1912
AULACOFUSUS Dall, 1918
LATISIPHO Dall, 1916
LIMATOFUSUS Dell, 1918
PARVISIPHO Cossmann, 1889 (fossil)
SIPHONELLONA Wenz, 1944
pro Siphonella Verrill, 1879
(non Hagenow, 1851)
Neptunella Verrill, 1873
(non Meek, 1864)
SIPHONORBIS Mörch, 1869
Colycriptus Iredale, 1918
COSTARIA Golikov, 1977
EOSIPHO Thiele, 1929
EUTHRIOSTOMA Marche-Marchaud
& Brebion, 1977
EXILIOIDEA Grant & Gale, 1931 (fossil)
EXILIA Conrad, 1860 (recent)
FUSIPAGODA Habe & Ito, 1965
GOLIKOVIA Habe & Sato, 1973 ("1972")
JAPELION Dall, 1916 (1918 ?)
METAJAPELION Tiba & Kosuge, 1980
KAPALA Ponder, 1982
KELLETIA Fischer, 1884
KRYPTOS Jeffreys in Dautzenberg
& Fischer, 1896
LIOMESUS H. & A. Adams, 1853
Buccinopsis Jeffreys, 1867
(non Conrad, 1857,
nor Deshayes, 1865)
MACRON H. & A. Adams, 1853
Tomlinia Peile, 1937
MANARIA E. Smith, 1906
MOHNIA Friele, 1879
NEANCISTROLEPIS Habe & Ito, 1972
NEOBERINGIUS Habe & Ito, 1965
NEOBUCCINUM E. Smith, 1879
NEPTUNEA Röding, 1798
Chrysodomus Swainson, 1840
Atractus Agassiz, 1840
BARBITONIA Dall, 1916
SULCOSINUS Dall, 1916 ?
SULCOSIPHO Dall, 1916
PARANCISTROLEPIS Azuma, 1965
ANTARCTODOMUS Dell, 1972
PENION Fischer, 1884
Austrosipho Cossmann, 1906
(fossil)
Verconella Iredale, 1914
Largisipho Iredale, 1929
PLICIFUSUS Dall, 1902
Parasipho Dautzenberg
& Fischer, 1912
HELICOFUSUS Dall, 1916
LATIFUSUS Dall, 1916
PROSIPHO Thiele, 1912
PSEUDOLIOMESUS Habe &
Sato, 1973 ("1972")
PSEUDONEPTUNEA Kobelt, 1882

BUCCININAE (Cont'd.)
PTYCHOSALPINX Gill, 1867
RETICUBUCCINUM Habe & Ito, 1980
RETIFUSUS Dall, 1916
PARARETIFUSUS Kosuge, 1967
SAVATIERIA Rochebrune &
Mabille, 1885
SIPHONALIA A. Adams, 1863
SIPHONOFUSUS Kuroda
& Habe, 1952
TACITA Lus, 1971
THALASSOPLANES Dall, 1908
Brevisiphonia Lus, 1973
TROMINA Dall, 1918
TROSCHELIA Mörch, 1876
TRUNCARIA Adams & Reeve, 1848 ?
BUCCINOPSIS Deshayes, 1865 ?
COPTAXIS Cossmann, 1901
TURRISIPHO Dautzenberg
& Fischer, 1912
Undacolus F. Nordsieck, 1968
VOLUTHARPA Fischer, 1856

Placement ? (3)
AGASSITULA Olsson & Bayer, 1972
ANTARCTONEPTUNEA Dell, 1972
CAVINEPTUNEA Powell, 1951
CORNEOBUCCINUM Golikov &
Gulbin, 1977
CRASSICANTHARUS Ponder, 1972
FALSIMOHNIA Powell, 1951
HARPOFUSUS Habe & Ito, 1965
HYPOJAPELION Okutani, 1968
LOOCHOOIA MacNeil, 1960
(Fossil & recent)
MINITULA Olsson & Bayer, 1972
PLICIBUCCINUM Golikov & Gulbin, 1977
THYSANOBUCCINUM Golikov & Gulbin
in Golikov, 1980

VOLUTOPSIINAE (8)
VOLUTOPSIUS Mörch, 1857
Strombella Gray, 1857
(non Schlüter, 1838)
Volutopsion Habe & Ito, 1965
FUSIVOLUTOPSIUS Habe & Sato, 1972
HABEVOLUTOPSIUS Kantor, 1984 ?
LUSSIVOLUTOPSIUS Kantor, 1983
PYRULOFUSUS Mörch, 1857
Pyrolofusus auctt. (err.)
Heliotropis Dall, 1873

PHOTINAE
PHOS Montfort, 1810
Rhinodomus Swainson, 1840
CYMATOPHOS Pilsbry & Olsson, 1941
(fossil)
ENGONIOPHOS Woodring, 1928
METAPHOS Olsson, 1964
PHILINDOPHOS Shuto, 1969 (fossil)
STROMBINOPHOS Pilsbry & Olsson, 1941
(fossil)
STRONGYLOCERA Mörch, 1852
ANTILLOPHOS Woodring, 1928 (13)
CHLANIDOTA von Martens, 1878
CHLANIDOTELLA Thiele, 1929
METEUTHRIA Thiele, 1912
PFEFFERIA Strebel, 1908
PRONEPTUNEA Thiele, 1912

PHOTINAE (Cont'd.)
COMINELLA Gray, 1850
Acominia Finlay, 1926
AFROCOMINELLA Iredale, 1918
EUCOMINIA Finlay, 1926
Zephos Finlay, 1926
GODFREYENA Iredale, 1934
JOSEPHA Tenison-Wood, 1878 (Iosepha)
Cominula Finlay, 1926
Cominista Finlay, 1927
FAX Iredale, 1925
SCAEOFAX Iredale, 1931
NASSARIA Link, 1807 (5c)
Hindsia A. Adams, 1851
MICROFUSUS Dall, 1916
Benthindsia Iredale, 1936
Nihonophos MacNeil, 1960
NEOTERON Pilsbry & Lowe, 1932 (5c)
NORTHIA Gray, 1847
PAREUTHRIA Strebel, 1905
PARVIPHOS Sarasua, 1984 (13)
SEARLSIA Harmer, 1915
TRAJANA Gardner, 1948 (5c)
Hindsia auctt.
(non A. Adams, 1851)
TRIUMPHIS Gray, 1847
NICEMA Woodring, 1864

PISANIINAE (5b)
PISANIA Bivona, 1832
Ecmanis Gistl, 1848
pro Proboscidea Schmidt in Möller, 1832
(non Bruguière, 1791)
Polliana E. M. Gray, 1850
Appisania Thiele, 1929 (9)
Sukunaia Cernohorsky, 1966 (4a)
JEANNEA Iredale, 1912 (5b) (Syn. ?) (10)
PISANIANURA Rovereto, 1899 ?
Anura Bellardi, 1871
(non Hodgson, 1841)
TAENIOLA Dall, 1904 (10)
CADUCIFER Dall, 1904
MONOSTIOLUM Dall, 1904 (5b)
CANTHARUS Röding, 1798
Anna Risso, 1826
Muricantharus Hertlein & Strong, 1971
CLIVIPOLLIA Iredale, 1929
ENGINELLA Monterosato, 1917
POLLIA Gray in Sowerby, 1834
Pusio Gray in Griffith
& Pidgeon, 1834
Tritonidea Swainson, 1840
Algrus de Gregorio, 1855 ?
Gemophos Olsson & Harbison, 1953 (5b)
PRODOTIA Dall, 1924
ZEAPOLLIA Finlay, 1927 (fossil)
CHAUVETIA Monterosato, 1884
Lachesis Risso, 1826
(non Daudin, 1803)
Nesaea Risso, 1826
(non Lamarck, 1812
Donovania Bucquoy, Dautzenberg
& Dollfus, 1882
(non Leach, 1814)
Syntagma Iredale, 1918
Chauvetiella F. Nordsieck, 1968
Donovaniella F. Nordsieck, 1968 (?)
FOLINAEA Monterosato, 1884
ENGINA Gray, 1839
Enzina Gray, 1847 (nom.null.)
Enzinopsis Iredale, 1940
SOLENOSTEIRA Dall, 1890
Fusinosteira Olsson, ?

COLUMBELLIDAE
COLUMBELLINAE
COLUMBELLA Lamarck, 1799
Pyrene Röding (authors)
(non Röding, 1798)
Columbellarius Duméril, 1806
Columbus Montfort, 1810
Peristera Rafinesque, 1815
Colombella Duclos, 1840 (error)
Pterygia Wenz, 1941
(non Röding, 1798)
ALIA H. & A. Adams, 1853
EUPLICA Dall, 1889
MINIPYRENE Coomans, 1967
NITIDELLA Swainson, 1840
Pardalina Jousseaume, 1888
(non Gray, 1847)
RHOMBINELLA Radwin, 1968
ZAFRONA Iredale, 1916
Metanachis Thiele, 1924

PYRENINAE
PYRENE Röding, 1798
Aidone H. & A. Adams, 1853
Conoidea Swainson, 1840
Cionella Swainson, Gray, 1847 (err.)
Brachylixella Sacco, 1890
Thiarinella Sacco, 1890 ?
AESOPUS Gould, 1860
Parviterebra Pilsbry, 1904
GLYPTAESOPUS Pilsbry & Olsson, 1941 ?
ITHIAESOPUS Olsson & Harbison, 1953
AMPHISSA H. & A. Adams, 1853
ANACHIS H. & A. Adams, 1953
Ecostoanachis Sacco, 1890
ANTIZAFRA Finlay, 1927
CLATHRANACHIS Kuroda & Habe, 1952 ?
COSTOANACHIS Sacco, 1890
GLYPTANACHIS Pilsbry & Lowe, 1932
LITOTREMA Weisbord, 1962
MACROZAFRA Finlay, 1927
MICROCITHARA Fischer, 1844
PARVANACHIS Radwin, 1968
RETIZAFRA Hedley, 1913
ZAFRA A. Adams, 1860
Mitropsis Pease, 1868
ANTIMITRELLA Powell, 1937
AOTEATILIA Powell, 1939
ASTYRIS H. & A. Adams, 1853
FLUELLA Dall, 1924
GRAPHICOMASSA Iredale, 1929
LAVESOPUS Iredale, 1929
PARASAGENA Dall, 1924
PLECTARIA Dall, 1924
BRACHYSTYLOMA Weisbord, 1962
COSMIOCONCHA Dall, 1913
CONELLA Swainson, 1840
COLUMBELLOPSIS Bucquoy, Dautzenberg
& Dollfus, 1882
Tetrastomella Bellardi, 1889
DECIPIFUS Olsson & McGinty, 1958
EXOMILOPSIS Powell, 1964
LIRATILIA Finlay, 1926 ?
MAZATLANIA Dall, 1913
pro Euryta H. & A. Adams, 1853
(non Gistl, 1848)
METULELLA Gabb, 1873
MITRELLA Risso, 1826
Mitsella Risso in
Mörch, 1859 (err.)
Angulatomitrella Sacco, 1890
Arcuatomitrella Sacco, 1890

PYRENINAE
MITRELLA (Cont'd.)
ALCIRA H. Adams, 1860
ATILIA H. & A. Adams, 1853
CILARA Thiele, 1924
Striomitrella Grant & Gale, 1931
DENTIMITRELLA Ludbrook, 1958
LONGITRELLA Olsson, 1942
PARATILIA Thiele, 1924
PAXULA Finlay, 1927
PLEUROTOMITRELLA Habe, 1958
PYRENOLA Iredale, 1918
SULCOMITRELLA Habe, 1958
SUNDAMITRELLA Oostingh, 1952
ZELLA Iredale, 1924
ZEMITRELLA Finlay, 1927
NASSARINA Dall, 1889
Pyramimitra Cossmann, 1910
CIGCLIRINA Woodring, 1928
RADWINIA Shasky, 1970
STEIRONEPION Pilsbry & Lowe, 1932
Psarostola Rehder, 1943
Turrijaumelia Sarasua, 1975
ZANASSARINA Pilsbry & Lowe, 1932
NODOCHILA Rehder, 1980
PARAMETARIA Dall, 1916
Conella H. & A. Adams, 1853
(non Swainson, 1840) ?
Meta Reeve, 1859
(non Koch, 1835)
PICTOCOLUMBELLA Habe, 1962
PSEUDAMYCLA Pace, 1902
PSEUDANACHIS Thiele, 1924
PUSIOSTOMA Swainson, 1840
RUTHIA Shasky, 1970
SALTIRA Marincovich, 1973 ?
SEMINELLA Pease, 1868
STROMBINA Mörch, 1852
Strombocolumbus Cossmann, 1901
BIFURCIUM Fischer, 1884 ?
Bifurcina Cossmann, 1901
COTONOPSIS Olsson, 1942
SUTUROGLYPTA Radwin, 1968
ZETEKIA Dall, 1918

NASSARIIDAE (5d)
NASSARIINAE
NASSARIUS Duméril, 1806
Nassa Lamarck, 1799
(non Röding, 1798)
Arcularia Link, 1807
ACICULINA A. Adams, 1853
Hannonia Pallary, 1914
(non Hoek, 1881)
ADINOPSIS Odhner, 1923
(non Cameron, 1918)
ALECTRION Montfort, 1810
Alectryon Berthold in
Latreille, 1827
ALLANASSA Iredale, 1929
CAESIA H. & A. Adams, 1853
Zaphon H. & A. Adams, 1853
Schizopyga Conrad, 1856
(non Gravenhorst, 1829)
Demondia Addicott, 1956 (fossil)
CATILLON Addicott, 1965
CRYPTONASSARIUS Cernohorsky, 1975
GUSSONEA Monterosato, 1912
Amycla H. & A. Adams, 1853
(non Rafinesque, 1815)
Amyclina Iredale, 1918
Fackia F. Nordsieck, 1972

NASSARIINAE
NASSARIUS (Cont'd.)
HIMA Leach in Gray, 1852
pro Tritonella A. Adams, 1852
(non Swainson, 1839)
Mirus Marwick, 1931 (fossil)
Reticunassa Iredale, 1936
HINIA Gray, 1847
Tritia A. Adams, 1853
(non Risso, 1826)
Miohinia F. Nordsieck, 1972
NASSODONTA H. Adams, 1867
NAYTIA H. & A. Adams, 1853
NAYTIOPSIS Thiele, 1929
NIOTHA H. & A. Adams, 1853
Niothia Brazier, 1877 (em.)
Tavaniotha Iredale, 1936
PALLACERA Woodring, 1964
PHRONTIS H. & A. Adams, 1853
PLICARCULARIA Thiele, 1929
Eione Risso, 1826
Parcanassa Iredale, 1936
Austronassaria C. & J. Laseron, 1956
Chelenassa Shuto, 1969
PROFUNDINASSA Thiele, 1929
PYGMAEONASSA Annandale, 1924
SPHAERONASSA Locard, 1886
Nasseburna de Gregorio, 1890
TELASCO H. & A. Adams, 1853
Tarazeuxis Iredale, 1936 (nom.nud.)
UZITA H. & A. Adams, 1853
VARICINASSA Habe, 1946
ZEUXIS H. & A. Adams, 1853
Venassa von Martens, 1881 (deformed)
Glabrinassa Shuto, 1969
Bathynassa Ladd, 1976
CYCLOPE Risso, 1826
Cyclops Montfort, 1810
(non O. F. Müller, 1776)
Nana Schumacher, 1817
Nanina Risso, 1826
Panormella Da Costa, 1840
Cyclonassa Swainson, 1840
Nannia Philippi, 1844
Cyclocyrtia Agassiz, 1847 (em.)
Cencus Gistl, 1848
Neritula Herrmannsen, 1852
PRONERITULA Thiele, 1929
DEMOULIA Gray, 1838
Desmoulea Gray, 1847
Desmoulinsia Woodward, 1851
Streptorhega Bronn, 1856
Moulinsia Tournouer, 1874
(non Grateloup, 1840)
HEBRA H. & A. Adams, 1853
Scabronassa Peile, 1939
ILYANASSA Stimpson, 1865
Paranassa Conrad, 1867 (fossil)

DORSANIINAE
DORSANUM Gray, 1947
FLUVIODORSUM Boettger, 1885
AMBULLINA Palmer, 1937 (fossil)
BULLIA Gray in Griffith
& Pidgeon, 1834
Leiodomus Swainson, 1840
Bulliana M. E. Gray, 1842
Pseudostrombus H. & A. Adams, 1853
Ancillopsis Conrad, 1865 (fossil)
CEREOBULLIA Melvill & Peile, 1924
BUCCINANOPS Orbigny, 1841
BULLIOPSIS Conrad, 1862 (fossil)
DESORINASSA Nuttall & Cooper, 1973 (fossil)

CYLLENINAE
CYLLENE Gray in Griffith & Pidgeon, 1834
Radulphus Iredale, 1924
CYLLENINA Bellardi, 1882 [fossil]

MELONGENIDAE
MELONGENA Schumacher, 1817
Galeodes Röding, 1798
[non Oliver, 1791]
REXMELLA Olsson & Harbison, 1953 ?
BUSYCON Röding, 1798
Fulgur Lamarck, 1799
Sycopsis Conrad, 1867
BUSYCOARCTICUM Hollister, 1958 ?
BUSYCOTYPUS Wenz, 1943
Sycotypus Gill, 1867
[non Mörch, 1852]
Sycofulgur Marks, 1956
FULGUROPSIS Marks, 1950
Pyrofulgur Hollister, 1958
SINISTROFULGUR Hollister, 1958
PUGILINA Schumacher, 1817
Volegalea Iredale, 1938
HEMIFUSUS Swainson, 1840
Semifusus Agassiz, 1846 [em.]
Cochlidium Gray, 1850
MEGALATRACTUS Fischer, 1884 ?
SYRINX Röding, 1798
TAPHON H. & A. Adams, 1853
VOLEMA Röding, 1798
Cassidula Swainson, 1840
[non Férussac, 1821]
Myristica Swainson, 1840

FASCIOLARIIDAE
FASCIOLARIINAE
FASCIOLARIA Lamarck, 1799
Terebraspina Conrad, 1862
CINCTURA Hollister, 1957
TARANTINAEA Monterosato, 1917
BENIMAKIA Habe, 1958
CYRTULUS Hinds, 1843 ?
EGESTAS Finlay, 1927
FALSILATIRUS Emerson & Moffitt, 1983 [7a]
FASCINUS Hedley, 1903
FUSOLATIRUS Kuroda & Habe, 1971
GLAPHYRINA Finlay, 1927
HARASEWYCHIA Petuch, 1987 [?]
MICROFULGUR Finlay & Marwick, 1937
PLEUROPLOCA Fischer, 1884
PLEIA Finlay, 1930
SAGINAFUSUS Iredale, 1931
TARON Hutton, 1883
TERALATIRUS Coomans, 1965

FUSININAE
FUSINUS Rafinesque, 1815
Fusus auctt.
[non Helbling, 1779]
Colus Humphrey, 1797 [invalid]
Sinistralia H. & A. Adams, 1853

FUSININAE
FUSINUS [Cont'd.]
Pseudofusus Monterosato, 1884
Falsifusus Grabau, 1904 ?
APTYXIS Troschel, 1868
BARBAROFUSUS Grabau & Shimer, 1909
Harfordia Dall, 1921
BERYLSMA Iredale, 1924
FUSINELLA Thiele, 1917
Buccinella Thiele, 1912
[non Perry, 1811]
GRACILPURPURA Jousseaume, 1880
Gracilipurpura Jousseaume, 1882 [err.]
HEILPRINIA Grabau, 1904
MICROCOLUS Cotton & Godfrey, 1932
PROPEFUSUS Iredale, 1924
SIMPLICIFUSUS Kira, 1962
BUCCINOFUSUS Conrad, 1866 [9]
Boreofusus Sars, 1878
GRANULOFUSUS Kuroda & Habe, 1952 ?
TROPHONOFUSUS Kuroda, Habe & Oyama, 1971 ?

PERISTERNIINAE
PERISTERNIA Mörch, 1852
DOLICHOLATIRUS Bellardi, 1884
Latirofusus Cossmann, 1889
Fusilatirus McGinty, 1955
LATIRUS Montfort, 1810
Lathyrus auctt. [err.]
Eolatirus Bellardi, 1883
Plesiolatirus Bellardi, 1883 [13] ?
Hemipolygona Rovereto, 1899 [13]
pro Chascax Watson, 1873
[non Ritgen, 1828]
FRACTOLATIRUS Iredale, 1936
LATIRULUS Cossmann, 1901
Lathyrulus Cossmann, 1889
POLYGONA Schumacher, 1817
Plicatella Swainson, 1840
LEUCOZONIA Gray, 1847
LATIROLAGENA G. F. Harris, 1897
MAZZALINA Conrad, 1860 ?
OPEATOSTOMA Berry, 1958
PSEUDOLATIRUS Bellardi, 1897 ?

COLUBRARIINAE [Fusidae] [2]
COLUBRARIA Schumacher, 1817
Epidromus Mörch, 1852
Obex Iredale, 1929
ESBELTA Sarasua, 1975 [13]
NIVITRITON Iredale, 1929 ?
BARTSCHIA Rehder, 1943
FUSUS Helbling, 1779
IREDALULA Finlay, 1926
KANAMARUA Kuroda, 1951
METULA H. & A. Adams, 1853 [7]
Acamptochetus Cossmann, 1901
Antemetula Rehder, 1943
Colubrarina Kuroda & Habe, 1971
RATIFUSUS Iredale, 1929

References:

Abbott, R. T. "The Family Columbellidae" Standard Catalogue - 1978 [1]

Beu, A. G. & P. A. Maxwell "Revision: Fossil and Living gastropods related to Plesiotriton: Appendix: Genera of Buccinidae [Pisaniinae] related to Colubraria Schumacher, 1817" N.Z.Geo.Surv.Paleo.Bull.54:16-58 - 1987 [2]

Bouchet, P. & A. Warén "Taxonomical notes on tropical deep water Buccinidae " 1986 [Resultats des Campagnes Musorstrom I, II Mem.Mus.Natn.Hist.nat.Ser.A.,Zool.133 [3]

Bullock, R. C. "Contribution to the systematics of some West Indian Latirus" Nautilus 88[3]:69-79 - 1974 [4]

References: (Buccinidae, cont'd.)
Cernohorsky, W.O."New genus, species: Buccinidae from Fiji" Veliger 9(2):229-282 - 1966 (5)

---- "Indo-Pacific Pisaniinae; related buccinid genera"Rec.Auck.Inst.Mus.8:137-167.1971 (5a)
---- "Supplementary notes on the taxonomy of the buccinid species of the subfamily Pisaniinae" Ibid.12:175-211 - 1975 (5b)
---- "The family Buccinidae I. Nassaria, Trajana, Neoteron" Monog.Marine Mollusca 2:1-52 - 1981 (5c)
---- "Systematics of the family Nassariidae" Bull.Auck.Inst.Mus.14:1-356 - 1984 (5d)

Coomans, H. E. "Teralatirus, new genus in the Fasciolariidae" Basteria 29(1-4:10-14 - 1985 (6)

Emerson, W. K. "On the type species of Metula" Nautilus 100(1):27-30 - 1986 (7)

---- & R. B. Moffitt "New genus and species of (? Fasciolariidae): Mariana Islands" Veliger 31(1-2):43-45 - 1988 (7a)

Houbrick, R. S. "Transfer of Quadrasia from Planaxidae to Buccinidae" Proc.Biol.Soc.Wash.99(2):359-362 - 1986 (7x)

Kantor, Yu. I. "Volutopsius complex: Kurile I." Zool.Journ.(USSR)63(11):1616-1628 - 1984 (8)

---- "Compendium of Subfamily Volutopsiinae" La Conchiglia XVIII(204-205):8-15 -1986 (8a)

Okutani, T. "Buccinofusus ----" Venus 41(2):109-120 - 1982 (9)

Ponder, W. F. "Notes on some Australian species and genera of the Buccinidae" J.Malac.Soc.Aust.2(3):249-265 - 1972 (10)

---- "Review of the Australian species of Penion" Jour.Mal.Soc.Aust.2(4):401-434 - 1973 (10a)

Poorman, L. "New species of Fusinus: tropical Pacific" Veliger 23(4):339-344 - 1981 (11)

---- "Comments on two misunderstood fusinids from the tropical Eastern Pacific" Veliger 23(4):345-397 - 1981b (11a)

Radwin, G. E. "Columbellidae of the W. Atlantic" Veliger 19(4):403-417 - 1977 (12)

---- "Columbellidae" II. Ibid. 20(2):328-343 -1978

Sarasua, H. "Nuevos genero, subgenero y especies de moluscas marinos Neogastropodos" Poeyana 140:1-15 - 1975 (13)

Thiele, J. "Handbuch der Systematischen Weichtierkunde" - 1963 (15)

VOLUTIDAE (26)

VOLUTINAE

VOLUTA Linnaeus, 1758
- Plejona Röding, 1798
- Volutarius Duméril, 1806
- Harpula Swainson, 1831
- Musica Gray, 1847
- Chlorosina Gray in H. & A. Adams, 1858
- Volutolyria Crosse, 1877
- Austrovoluta Cotton, 1949 (fossil ?)

ATHLETINAE

VOLUTOCORBIS Dall, 1890
- Volutilithes Dall, 1890

TERNIVOLUTA von Martens, 1897
- Volutocorbis von Martens, 1929 (non Dall, 1890)

LYRIINAE

LYRIA Gray, 1847
- Otocheilus Conrad, 1865
- CORDILYRIA F. M. Bayer, 1971
- ENAETA H. & A. Adams, 1853
- HARPEOLA Dall, 1907
  - Harpella Gray, 1858
- LYRENETA Iredale, 1937

FULGORARIINAE

FULGORARIA Schumacher, 1817
- GUIVILLEA Watson, 1886
  - Wyvillea Watson, 1882
  - Pseudocymbium Cossmann, 1899
- KURODINA Rehder, 1969
- MUSASHIA Hayashi, 1960
- PSEPHAEA Crosse, 1871
  - Miopleione Dall, 1907
- SAOTOMEA Habe, 1943
- VOLUTIPYSMA Rehder, 1969

FULGORARIINAE (Cont'd.)

DALLIVOLUTA Okutani, 1982 (17)

ERICUSA H. & A. Adams, 1858
- MESERICUSA Iredale, 1929

FALSILYRIA Pilsbry & Olsson, 1954

FESTILYRIA Pilsbry & Olsson, 1954

HARPOVOLUTA Thiele, 1912

IREDALINA Finlay, 1926

LIVONIA Gray, 1855
- Mamillana Crosse, 1871
- Pterospira Harris, 1897

CYMBIINAE

CYMBIUM Röding, 1798
- Cymba Broderip in Sowerby, 1826
- Yetus Gray, 1855
- Cymba Marche-Marchaud, 1968

CALLIPARA Gray, 1847

CYMBIOLA Swainson, 1831
- Ausaba H. & A. Adams, 1853
- AULICA Gray, 1847
  - Volutocorona Pilsbry & Olsson, 1954
- AULICINA Rovereto, 1899
  - Scapha Gray, 1847 (non Motschulsky, 1845)
  - Cymbiola Rovereto, 1899 (non Swainson, 1831)
  - Vespertilio Klein in Morch, 1852 (non Linnaeus, 1758)
- CYMBIOLENA Iredale, 1929

CYMBIOLACCA Iredale, 1929
- Pseudocymbiola McMichael, 1961

MELO Broderip & Sowerby, 1826
- Cymbium Marche-Marchaud, 1968 (non Röding, 1798)
- MELOCORONA Pilsbry & Olsson, 1954

SIGATULA Rehder, 1967 (22a)

ZIDONINAE
ZIDONA H. & A. Adams, 1853
Volutella Orbigny, 1841
(non Perry, 1810)
ADOMELON Dall, 1906
Janeithoe Pilsbry & Olsson, 1954
WEAVERIA Clench & Turner, 1964
ALCITHOE H. & A. Adams, 1853
Gilvostia Iredale, 1937
LEPOREMAX Iredale, 1937
Carolluta Iredale, 1937
ARCTOMELON Dall, 1915
Boreomelon Dall, 1918
COTTONIA Iredale, 1934
HARPULINA Dall, 1906
Harpula Swainson, 1853
(non Swainson, 1831)
NEOIMBRICARIA Ihering, 1907 ? (4)
PACHYCYMBIOLA Ihering, 1907
PACHYMELON Marwick, 1926
PALOMELON Finlay, 1926
PROVOCATOR Watson, 1882
TEREMELON Marwick, 1926

ODONTOCYMBIOLINAE
ODONTOCYMBIOLA Clench & Turner, 1964
Adomelon Dall in Pilsbry & Olsson, 1954
(non Dall, 1906)
MIOMELON Dall, 1907
Proscaphella Ihering, 1907
TRACTOLIRA Dall, 1896
VOLUTOCONUS Crosse, 1871

SCAPHELLINAE
SCAPHELLA Swainson, 1832
Maculopeplum Dall, 1906
AURINIA H. & A. Adams, 1853
Rehderia Clench, 1946
Auriniopsis Clench, 1953
CLENCHINA Pilsbry & Olsson, 1953
AMORIA Gray, 1855
AMORENA Iredale, 1929
RELEGAMORIA Iredale, 1936
ZEBRAMORIA Iredale, 1929
Pilidia Valenciennes, 1863 (nom.obl.)
AMPULLA Röding, 1798
Halia Risso, 1829
Priamus Beck in Deshayes, 1838
CYMBIOLISTA Iredale, 1929
NANNAMORIA Iredale, 1929
NOTOPEPLUM Finlay, 1927
NOTOVOLUTA Cotton, 1946
PARAMORIA McMichael, 1960
VOLUTIFUSUS Conrad, 1863
Bathyaurinia Clench & Aguayo, 1940

CALLIOTECTINAE
CALLIOTECTUM Dall, 1890
FUSIVOLUTA von Martens, 1902
NEPTUNEOPSIS Sowerby, 1898
TERAMACHIA Kuroda, 1931 (21)
Prodallia Bartsch, 1942
Howellia Clench & Aguayo, 1941 ?

HARPIDAE (22) (nom.cons.- ICZN 1988)
HARPINAE (15) (nom.cons.)
HARPA Röding, 1798
Cithara Klein in Jousseaume, 1881
Harpalis Link, 1806
Harparia Rafinesque, 1815
Lyra Griffith & Pidgeon, 1934
AUSTROHARPA Finlay, 1931 (fossil)
PALAMHARPA Iredale, 1931 (recent)

HARPIDAE (Cont'd.)
MORUMINAE (15)
MORUM Röding, 1798 (14)
Lambidium Link, 1807
Oniscia Sowerby, 1824
Theliostoma Sowerby, 1824 (nom.nud.)
Oniscidia Swainson, 1840 (err.)
Ersina "1840" Gray, 1847 (unavail.)
Plesioniscia Fischer, 1884
ONISCIDIA Mörch, 1852 (2)
Onischidea Olsson, 1931 (err.)
Onimusiro Kuroda, 1955 (nom.nud.)
Pulchroniscia Garrard, 1961
Cancellomorum Emerson & Old, 1963
Onimusiro Kira in Kuroda,
Habe & Oyama, 1971

VASIDAE (Turbinellidae) (24)
VASINAE
VASUM Röding, 1798 (24)
Volutella Perry, 1810
Cynodonta Schumacher, 1817
Scolymus Swainson, 1835
ALTIVASUM Hedley, 1914
GLOBIVASUM Abbott, 1950
SIPHOVASUM Rehder & Abbott, 1951
TUDICLA Röding, 1798
Pyrenella Swainson, 1835
Spirillus Schlüter, 1838
Pyrenella Gray, 1857
(non Swainson, 1835)
Pyropsis Conrad, 1860
TUDIVASUM Rosenberg & Petit, 1987 (23)
pro Tudicula H. & A. Adams, 1863
(non Ryckholt, 1862,
nor Cossmann, 1901)
AFER Conrad, 1858
Streotosiphon Gill, 1867

TURBINELLINAE
TURBINELLA Lamarck, 1799
Xancus Röding, 1798 (rejected)
Turbinellus Lamarck, 1801
Turbinellarius Dumeril, 1806
Buccinella Perry, 1811
Scolymus Deshayes, 1843
(non Swainson, 1835)
Mazza Klein in H. & A. Adams, 1855
Turbofusula Rovereto, 1900

PTYCHATRACTINAE (21)
PTYCHATRACTUS Stimpson, 1865
PIESTOCHILUS Meek, 1864
BENTHOVOLUTA Kuroda & Habe, 1950
CERATOXANCUS Kuroda, 1952
CYOMESUS Quinn, 1981 (21)
Mesorhytis in Dall, 1889 etc.
(non Meek, 1876- fossil)
Prodallia Bartsch, 1942 (part.)
Teramachia in Weaver &
Dupont,1970 (part)
(non Kuroda, 1931)
FUSIMITRA Conrad, 1855 (fossil) (4)
LATIROMITRA Locard, 1897 (2a)
Latriomitra Ponder, 1968
(nom. null.)
METZGERIA Norman, 1879
pro Meyeria Dunker & Metzger, 1878
(non McCoy, 1849)
SURCULINA Dall, 1908
Phenacoptygma Dall, 1918
Chathamidea Dell, 1956 ?

COLUMBARIINAE
COLUMBARIUM von Martens, 1881
FULGUROFUSUS Grabau, 1904
COLUZEA Finlay, 1927
HISTRICOSCEPTRUM Darragh, 1969
PERISTARIUM F. M. Bayer, 1971

OLIVIDAE (16)
OLIVINAE
OLIVA Bruguière, 1789
Cylindrus Breyn, 1732
Strephopoma Browne, 1756
Porphyria Röding, 1798
Olivaria Rafinesque, 1815
Ispidula Gray, 1847
Dactylus H. & A. Adams, 1853
ACUTOLIVA, ANNULATOLIVA, ARCTOLIVA, CARIBOLIVA
MINIACEOLIVA, MULTIPLICOLIVA, MUSTELOLIVA
PROXOLIVA, RUFOLIVA, VIDUOLIVA all
Petuch & Sargent, 1986 (19a)
CARMIONE Gray, 1858
GALEOLA Gray, 1858
(non Klein, 1753) ?
NEOCYLINDRICUS Fischer, 1883
OMOGYMNA von Martens, 1897
PARVOLIVA Thiele, 1929
STREPHONA Mörch, 1852
STREPHONELLA Dall, 1909
BENTHOBIA Dall, 1889
OLIVANCILLARIA Orbigny, 1840
(non Orbigny, 1839)
Scaphula Swainson, 1840
(non Benson, 1834)
Utriculina Gray, 1847
Lintricula H. & A. Adams, 1853
Claneophila Gray, 1858
TATESHIA Kosuge, 1986 (?)

ANCILLINAE (16a)
ANCILLA Lamarck, 1799
Anaulax Roissy, 1805
Ancillaria Lamarck, 1811
? Ancillaria Risso, 1826
(non Lamarck, 1811)
Dipsaccus H. & A. Adams, 1853
Anaulacia, Cymbancilla Fischer, 1881
Spiracilla Vokes, 1936
ANCILLUS Montfort, 1810
Turrancilla von Martens, 1904
ANOLACEA Gray, 1857
SPARELLA Gray, 1857
CHILOTYGMA H. & A. Adams, 1853
Chiloptygma Fischer, 1883 (em.)
EBURNA Lamarck, 1801
Eburnus Montfort, 1810
HESPERANCILLA Kilburn, 1981
SPARELLINA Fischer, 1883 (Syn.?)
AMALDA H. & A. Adams, 1853
Sandella Gray, 1857
BARYSPIRA Fischer, 1883
Gemaspira Olsson, 1956
GRACILISPIRA Olsson, 1956
ALCOSPIRA Cossmann, 1899
PINGUISPIRA Finlay, 1927
SPINASPIRA Olsson, 1956 ?
ANCILLISTA Iredale, 1936
AUSTRANCILLA Habe, 1959
GRACILANCILLA Thiele, 1925

OLIVELLINAE
OLIVELLA Swainson, 1831
Olivina Orbigny, 1839
Micana Gray, 1858

OLIVELLA (Cont'd.)
CALLIANAX H. & A. Adams, 1853
Scaphula Gray, 1858
(non Benson, 1834)
CUPIDOLIVA Iredale, 1924
DACTYLIDELLA Woodring, 1928
DACTYLIDIA H. & A. Adams, 1853
LAMPRODOMA Swainson, 1840
Ramola Gray, 1858
LAMPRODOMINA Marwick, 1931
MACGINTIELLA Olsson, 1956
MANSFIELDELLA Olsson & Harbison, 1953
MINIOLIVA Olsson, 1956
NITEOLIVA Olsson, 1956
ORBIGNYTESTA Klappenbach, 1962
PACHYOLIVA Olsson, 1956
RAMOLIVA Cotton & Godfrey, 1932
TOROLIVA Olsson & Harbison, 1953
ZANOETELLA Olsson, 1956
BELLOLIVA Peile, 1922
GEMMOLIVA Iredale, 1924
OLIVELLOPSIS Thiele, 1929
JASPIDELLA Olsson, 1956

AGARONIINAE
AGARONIA Gray, 1839
Hiatula Swainson, 1840
(non Modeer, 1793)
ANAZOLA Gray, 1858 ? (7)

PSEUDOLIVINAE ? (Position uncertain)
PSEUDOLIVA Swainson, 1840
Gastridium Sowerby, 1842
(non Modeer, 1793)
Gastridia Gray, 1847
Pseudodactylus Herrmannsen, 1847
Sulcobuccinum Orbigny, 1850
Buccinorbis Conrad, 1865
FULMENTUM Fischer, 1884 ?
SYLVANOCOCHLIS Melvill, 1903 ?
Mariona Sowerby, 1889
(non Vayssière, 1879)
MELAPIUM H. & A. Adams, 1853 ?
ZEMIRA H. & A. Adams, 1853 (20a)
Eburna H. & A. Adams, 1853
(non Lamarck, 1801)
Eburnopsis Tate, 1889
Pseudoliva Cossmann, 1901
(non Swainson, 1840)
Latrunculus Cossmann, 1901
(non Gray, 1847)

MARGINELLIDAE (10, 11)
MARGINELLINAE
MARGINELLA Lamarck, 1799
Marginellarius Duméril, 1806 (em.)
Marginellus Montfort, 1810 (em.)
Cucumis Deshayes, 1830
Marginilla Swainson, 1831 (err.)
Porcellana Gray, 1847
(non Lamarck, 1801)
Marginellana H. & A. Adams, 1853 (err.)
Pseudomarginella Maltzen, 1880
Stazzania Sacco, 1890
ALAGINELLA Laseron, 1957
AUSTROGINELLA Laseron, 1957
BAROGINELLA Laseron, 1957
CARINAGINELLA Laseron, 1957
DENTICULOGLABELLA Sacco, 1890
EURYENTOME Cossmann, 1899
GIBBERULA Swainson, 1840
Giberula Sowerby, 1842 (err.)
Eratoidea Weinkauff, 1879

MARGINELLINAE
MARGINELLA (Cont'd.)
MIOGINELLA Laseron, 1957
PROTOGINELLA Laseron, 1957
SIMPLICOGLABELLA Sacco, 1890
STROMBOGINELLA Laseron, 1957
BALANETTA Jousseaume, 1875
OVAGINELLA Laseron, 1957
BULLATA Jousseaume, 1875
Volutella Swainson, 1829
(non Perry, 1810)
Gibberulina Monterosato, 1884
CLOSIA Gray, 1857
CRYPTOSPIRA Hinds, 1844
Crystospira Cotton, 1949 (err.)
CANALISPIRA Jousseaume, 1875
CASSOGINELLA Laseron, 1957
DENTIMARGO Cossmann, 1899
Ebarnospira Olsson & Harbison, 1953
GLABELLA Swainson, 1840
Faba Fischer, 1883
Phaenospira Hinds, 1844
Phoenospira Petit, 1851 (err.)
HIWIA Marwick, 1931 ?
HYALINA Schumacher, 1817 (12)
Neovolvaria Fischer, 1883
MESOGINELLA Laseron, 1957
PLICAGINELLA Laseron, 1957
SINUGINELLA Laseron, 1957
SPIROGINELLA Laseron, 1957
URNIGINELLA Laseron, 1957
MICROSPIRA Conrad, 1868
PACHYBATHRON Gaskoin, 1853
PERSICULA Schumacher, 1817
Persicola Swainson, 1840 (err.)
Rabicea Gray, 1857
PRUNUM Herrmansen, 1852 ?
Porcellanella Conrad, 1863
(non White in MacGillivray, 1852)
Egouena Jousseaume, 1875
Leptegouana Woodring, 1928
Egociena Cotton, 1949
RIVOMARGINELLA Brandt, 1968 (freshwater)
SERRATA Jousseaume, 1875
CONUGINELLA Laseron, 1957
DENTIGINELLA Laseron, 1957
EXIGINELLA Laseron, 1957
HYDROGINELLA Laseron, 1957
VETAGINELLA Laseron, 1957
VOLVARINA Hinds, 1844
Hyalina of authors,
(non Schumacher, 1817)
Volvaria Jousseaume, 1875
(non Lamarck, 1801)
Volvariella Fischer, 1883 ?
Volvorina Jousseaume, 1875 (err.)
DILICULUM Barnard, 1962
HALOGINELLA Laseron, 1957
Pillarginella Gabriel, 1962
LEDAVOLVARINA Rios, 19 ?
NEPTOGINELLA Laseron, 1957
PHYLOGINELLA Laseron, 1957
VOLVARINELLA Habe, 1951
Eburnospira Olsson & Harbison, 1953
Longinella Laseron, 1957
(non Gros & Lestage, 1927)

CYSTISCINAE
CYSTISCUS Stimpson, 1865 (Cysticus auctt. - err.)
Euliginella Laseron, 1957
CRITHE Gould, 1860
Microvulina Habe, 1951
EPIGINELLA Laseron, 1957

CYSTISCINAE (Cont'd.)
DEVIGINELLA Laseron, 1957
EXTRA Jousseaume, 1894
GRANULA Jousseaume, 1875
GRANULINA Jousseaume, 1888
Cypraeolina Cerulli-Irelli, 1911
Merovia Dall, 1920
Mervia Cotton, 1949 (err.)
Microginella Laseron, 1957
HIANOGINELLA Laseron, 1957
KOGOMEA Habe, 1951
LATIGINELLA Laseron, 1957
MARGINELLOPSIS Bavay, 1911
NUDIFABA Eames, 1952
TOPAGINELLA Laseron, 1957 ?
TRIGONELLA Laseron, 1957 ?
(non Da Costa, 1778 ?)

MARGINELLONINAE
MARGINELLONA von Martens, 1903
AFRIVOLUTA Tomlin, 1947

MITRIDAE (3)
MITRINAE (4)
MITRA Lamarck, 1798
Mitra Martyn, 1784 (non binom.)
Mitra Röding, 1798
Mitraria Rafinesque, 1815
Mitrolithes Krüger, 1823 (not avail.)
Mitraxia Binny & Tryon, 1864 (nom.null.)
Cucorbita Megerle in Scudder, 1882
Eumitra Melvill & Standen, 1895
(non Tate, 1889)
Papalaria Dall, 1915
Tiarella Swainson, 1840
Thiarella auctt. (err)
Isara H. & A. Adams, 1853
Mutyca H. & A. Adams, 1853
Phaeomitra von Martens, 1880
Fuscomitra Pallary, 1900
Episcomitra Monterosato, 1917
Atrimitra Dall, 1918
Vicimitra Iredale, 1929
Cryptomitra Dall in Bryan, 1956 (n.nud.)
Volvariella Coan, 1966
(non Fischer, 1883)
EUMITRA Tate, 1889 (fossil)
Diplomitra Finlay, 1927
NEBULARIA Swainson, 1840
Chrysame H. & A. Adams, 1853
DIBAPHUS Philippi, 1847
Mitroidea Pease, 1865
Mauritia H. Adams, 1869
(non Troschel, 1863)
DIBAPHIMITRA Cernohorsky, 1970
STRIGATELLA Swainson, 1840
SOHLIA Cernohorsky, 1970
CHARITODORON Tomlin, 1932

CYLINDROMITRINAE
PTERYGIA Röding, 1798
Dactylus authors
(non Schumacher, 1817)
Pterigia Bucquoy, Dautzenberg
& Dollfus, 1882 (nom.null.)
Ptriegia Shikama, 1963
Cyclindromitra Fischer, 1884
pro Cylindra Schumacher, 1817
Cylinder Melvill & Standen, 1895 (n.nl.)
(non Montfort, 1810)
Cylindrina Risbec, 1955 (nom.null.)
(non Schlüter, 1838)
Acuticylindra Iredale, 1929

IMBRICARIINAE
IMBRICARIA Schumacher, 1817
Conoelix Swainson, 1821
Conohelix, Conehelix, Conalix, Conelix, Conoehelix, Conalex, Conolix, Concelix auctt. (nom.null.)
Imdricaria Habe & Kosuge, 1966 (n.nl.)
SCABRICOLA Swainson, 1840
Scabricula, Scabrecola, Scaricola, Scabricopa auctt. (nom.null.)
SWAINSONIA H. & A. Adams, 1853
SUBCANCILLA Olsson & Harbison, 1953
Tiara authors (non Swainson, 1831)
Taira Shuto, 1969 (nom.null.)
ZIBA H. & A. Adams, 1853
Zeba Garrett, 1880 (nom.null.)
Zeba von Martens, 1881 (nom.null.)
CANCILLA Swainson, 1840
Tiara auctt. (nom.null.)
Cancella, Cancillia, Canilla auctt.(nom.null.)
DOMIPORTA Cernohorsky, 1970
NEOCANCILLA Cernohorsky, 1966

VOLUTOMITRIDAE
VOLUTOMITRA H. & A. Adams, 1853
Volutimitra Gray, 1857 (nom.null.)
Volutamitra Standen & Roebuck, 1892 (n.nl.)
Volutamitra Coates, 1925
PARADMETE Strebel, 1908
CONOMITRA Conrad, 1865 (fossil)
PECULATOR Iredale, 1924
Peculata Kershaw, 1955 (nom.null.)
Peculata Macpherson in May, 1958 (nom.null.)
MICROVOLUTA Angas, 1877
Mitravoluta Preston, 1928 (nom.null.)
WAIMATEA Finlay, 1927
PROXIMITRA Finlay, 1927
Vexillitra Marwick, 1931
PARVIMITRA Finlay, 1930 (fossil)

COSTELLARIIDAE (Vexillidae) (5, 6)
MESORHYTIS Meek,1876 (foss.?) (Ptychatractinae -21)
Mitrodomus Sohl, 1963
VEXILLUM Röding, 1798
Turricula Fabricius & Authors, 1823 (non Schumacher, 1817)
Turricola,Vexilla,Vexillium auctt.(n.nd.)
Turris Montfort, 1810 (non Röding, 1798)
Vulpecula Defrance in Blainville, 1824 (non Jarocki, 1822)
Turricula Cossmann, 1899 (non Schumacher, 1817)
Tiara Swainson, 1831 (Thiara n.null.)
Harpaeformis Lesson, 1842
Tosapusia Azuma, 1965
COSTELLARIA Swainson, 1840
Callithea Swainson, 1840 (non Feisthamel, 1835)
Cyllithea Melvill & Standen, 1895 (n.n.)
Arenimitra, Pulchritima, Mitropifex Iredale, 1929
PUSIA Swainson, 1840
Pasia von Martens, 1903 (nom.null.)
Pussia Friedberg, 1911 (nom.null.)
Ebenomitra Monterosato, 1917
Pusiolina Cossmann, 1921
Pusiola Monterosato, 1917 (non Wallengren, 1863)
Idiochila Pilsbry, 1921
Pusidina Parenzan, 1970 (nom.nud.)
AUSTROMITRA Finlay, 1927
NODICOSTELLARIA Petuch, 1987 (?)
THALA H. & A. Adams, 1853
Micromitra Bellardi, 1888 (non Meek, 1873)
Mitromica Berry, 1958
TURRICOSTELLARIA Petuch, 1987 (?)
ZIERLIANA Gray, 1847
Zerliana Deshayes, 1865 (n.null.)
Zierlina Garrett, 1880 (n.null.)
Ziervogelia Fischer, 1884 (n.van.)
Ziervoglia Cossmann, 1899 (n.null.)
Ziervogeliana Wenz, 1943 (n.van.)

References:

Beu, A.G. "Revision: Southwest Pacific (Oniscidia)" Jour.Malac.Soc.Aust.3(3-4):223-231 - 1976 (2)

Bouchet, P. & A. Waren "Revision of the Northeast Atlantic bathyl and abyssal Neogastropoda" Bolletina Malacologico Supp.1-296 1985 (2a)

Cernohorsky, W. O. "Systematics: Volutomitridae and Mitridae" Bull.Auck.Inst.Mus.8:1-190 -1970 (3)

---- "The Mitridae of the World I. Mitrinae" Indo-Pacific Mollusca 3(17):273-528 - 1976 (4)

---- "Taxonomy of Caribbean-Atlantic Costellaridae" Rec.Auck.Inst.Mus. 15:87-109 - 1978 (5)

---- "Revision of Australian & N.Z. Costellariidae" Rec.Auck.Inst.Mus. 16:109-169 - 1980 (6)

Chavan, A. "Essai de reclassification des Olividae, Ancillinae" - 1965 Bull.Soc.Geol.de France (7)VII:102-109 (7)

Coan, E. "Revision of the Marginellidae" Veliger 8(4) - 1966 (10)

---- & B. Roth "The West American Marginellidae" Veliger 8(4):276-299 - 1966 (11)

---- & B. Roth "Status of the genus Hyalina" J.Moll.Stud.42:217-222 - 1976 (12)

Harasewych, M.G. "Columbariinae: Western Atlantic" Nemouria 27:1-42 - 1983 (13)

---- "Revision of Benthovuluta; Ptychatractinae" Nautilus 10(4):166-181 - 1987 (13a)

Hughes, R. N. "Anatomy & taxonomic displacement of genus Morum" Veliger 29(1):91-180 - 1986 (14)

---- & W. K. Emerson "Anatomical & Taxonomic Characteristics of Harpa and Morum" Veliger 29(4):349-358 - 1987 (15)

Kilburn, R. N. "Olividae with note on Systematics" Annals Nat.Mus.23(1):13-21 - 1977 (16)

---- "Revision of the genus Ancilla" Annals Nat.Mus.24(2):349-463 - 1981 (16a)

Okutani, T. "A new genus: trawled off Surinam" Venus 41(2):109-120 - 1982 (17)

References: (Volutidae, Mitridae, Cont'd.)
Petuch, E. J. "A volutid species radiation -" Proc.Biol.Soc.Wash.94(4):1110-1130 -1981 (19)

---- & D. M. Sargent "Living Olive Shells of the World" C.E.R.F. Charlottsville, Va.-1986(19a)

Ponder, W.F."Morphology: some mitriform gastropods" Malacologia 11(2):295-342 - 1972 (20)

---- & T. A. Darragh "The genus Zemira" J.Mal.Soc.Aust.3(2):89-105 - 1975 (20a)

Quinn, J. F. Jr. "A new genus of Turbinellidae - Caribbean" Nautilus 95(2):72-77 - 1981 (21)

Rehder, H.A. "The family Harpidae of the World" Indo-Pac.Moll. 3(16):207-274 - 1973 (22)
---- "New genus: Volutidae, Turbinellidae; Western Pacific" Pac.Sci.21(2):182-187 - 1967 (22a)

Rosenberg, G. & R.E. Petit "Ryckholt's Melanges: with New Name for Tudicula - " (23) Proc.Acad.Nat.Sci.Phila. 139:53-64 - 1987

Vokes, E.H. "Genus Turbinella in the New World" Tulane Stud.Geol.5:2:39-68 - 1964 (24)

Weaver, C.S. & J.E. duPont "The Living Volutes" Del.Mus.Nat.Hist. - 1979 (26)

CANCELLARIOIDEA
CANCELLARIIDAE
CANCELLARIINAE
CANCELLARIA Lamarck, 1799
Plicaria Fabricius, 1823
BIVETIA Jousseaume, 1887
BIVETIELLA Wenz, 1943
BIVETOPSIA Jousseaume, 1887
EUCLIA H. & A. Adams, 1854
Heteroeuclia Rovereto, 1899
MASSYLA H. & A. Adams, 1854
MERICA H. & A. Adams, 1854
Contortia Sacco, 1894
Nevia Jousseaume, 1887
PYRUCLIA Olsson, 1932
SYDAPHERA Iredale, 1929
AGATRIX Petit, 1967
ANAPEPTA Finlay, 1930
APHERA H. & A. Adams, 1854
AXELELLA Petit, 1988
pro Olssonella Petit, 1970
(non Glibert & Van de Poel, 1967)
BONELLITIA Jousseaume, 1887
BROCCHINIA Jousseaume, 1887 (4f)
CANCELLAPHERA Iredale, 1930
HABESOLATIA Kuroda, 1965
CRAWFORDINA Dall, 1919
pro Crawfordia Dall, 1919
GERDIELLA Olsson & Bayer, 1972
HERTLEINIA Marks, 1949
INGLISELLA Finlay, 1924
Microsveltia Iredale, 1925
MERICELLA Thiele, 1929
MOMOEBORA Kuroda & Habe, 1971
NARONA H. & A. Adams, 1854 (4c)
Panarona Petit, 1975 (4e)
NIPPONAPHERA Habe, 1961 (5a)

CANCELLARIINAE (Cont'd.)
OAMARUIA Finlay, 1927
PEPTA Iredale, 1925
PERPLICARIA Dall, 1890
PROGABBIA Dall, 1918
SCALPTIA Jousseaume, 1887
Trigonaphera Iredale, 1936
SOLATIA Jousseaume, 1887
SOLUTOSVELTIA Habe, 1961 (4b)
SVELTELLA Cossmann, 1889
SVELTIA Jousseaume, 1887
TRIBIA Jousseaume, 1887
TRIGONOSTOMA Blainville, 1825
Trigona Perry, 1811
(non Jurine, 1807)
ARIZELOSTOMA Iredale, 1936
EXACTRIX Korobkov, 1955
OVILIA Jousseaume, 1887
VENTRILIA Jousseaume, 1887
Emmonsella Olsson & Petit, 1964
WAIPAOA Marwick, 1931

ADMETINAE (4)
ADMETE Kroyer, 1842
ADMETULA Cossmann, 1889
MICROCANCILLA Dall, 1924 ?
NEADMETE Habe, 1961
NOTHOADMETE Oliver, 1982
ZEADMETE Finlay, 1927

PLESIOTRITONINAE (1)
PLESIOTRITON Fischer, 1884
AFRICOTRITON Beu & Maxwell, 1987
TRITONOHARPA Dall, 1908
FUSIAPHERA Habe, 1961
TUREHUA Marwick, 1943 (fossil)
PISANELLA van Koenen, 1865 (recent) ?

References:
Beu, A.G. & P.A. Maxwell "Revision of fossil and living gastropods related to Plesiotriton" N.Z.Geo.Surv.Paleo.Bull.54:17-58 - 1987 (1)

Bouchet, P. & A. Warén "Revision of the Northeast Atlantic Bathyl and Abyssal Neogastropoda, excluding Turridae" Bolletino Malac.Supp.I Societa Italiana di Malacologia. - 1985 (2)

Oliver,P.G."New species of Cancellarid: Antarctica" Br.Antarct.Survey Bull.57:15-20 - 1982 (3)

Petit, R. E. "Notes on Cancellariidae" 1967-1976 Tulane Stud.Geol.Paleo.8(2):83-88 etc. (4)

---- "The Cancellariid genus Nipponaphera" Venus Vol. 31(2) - 1972 (4a)
---- "Notes on Japanese Cancellariidae" Venus Vol. 33(3) - 1974 (4b)
---- "The Cancellariid genera Narona-Panarona" Veliger 17(4):387-388 - 1975 (4c)
---- "The Mozambique Cancellariidae" Ann.Natal Mus.24(1):211-216 - 1980 (4d)
---- "The nominal cancellariid genus Panarona" Veliger 25(3):269 - 1983 (4e)
---- "Notes on species of Brocchinia" Nautilus 100(1):23-27 - 1986 (4f)
---- "Axelella, new name for Olssonella:" Nautilus 102(3):130 - 1988 (4g)

CONOIDEA

CONIDAE
CONUS Linnaeus, 1758
Cucullus Röding, 1798
Cucullites Herrmannsen, 1847
Cylindrus Herrmannsen, 1847
(non Deshayes, 1824)
AFRICONUS Petuch, 1975
ASPRELLA Schaufuss, 1869
CHELYCONUS Mörch, 1852
CLEOBULA Iredale, 1930
CONASPRELLA Thiele, 1929
CONOLITHES Swainson, 1840 (fossil ?)
Conolithus Herrmannsen, 1847
Conospirus de Gregorio, 1890
CORONAXIS Swainson, 1840
CYLINDER Montfort, 1810
Cylindrus Deshayes, 1824;
Fischer, 1833 (err.)
(non Pfeiffer, 1840)
CYLINDRELLA Swainson, 1840
DARIOCONUS Iredale, 1930
DAUCICONUS Cotton, 1945
DENDROCONUS Swainson, 1840
DRYASPIS Iredale, 1949
EMBRIKENA Iredale, 1931
ENDEMOCONUS Iredale, 1931
FLORACONUS Iredale, 1930
GASTRIDIUM Modeer, 1793
Rhombus Montfort, 1810
(non Walbaum, 1792)
Rollus Montfort, 1810 ?
Utriculus Schumacher, 1817
Nubecula Herrmannsen, 1846
HERMES Montfort, 1810
KERMASPRELLA Powell, 1958
KURODACONUS Shikama & Habe, 1968
LAUTOCONUS Monterosato, 1923
LEPORICONUS Iredale, 1930
LEPTOCONUS Swainson, 1840
LITHOCONUS Mörch, 1852
LIVIDOCONUS Wils, 1970
MAMICONUS Cotton & Godfrey, 1932
PARVICONUS Cotton & Godfrey, 1932
PHASMOCONUS Mörch, 1852
PIONOCONUS Mörch, 1852
PROFUNDICONUS Kuroda, 1956
PUNCTICULIS Swainson, 1840
Puncticulus Coates, 1925 (err.)
PYRUCONUS Olsson, 1957
REGICONUS Iredale, 1930
RHIZOCONUS Mörch, 1852
STEPHANOCONUS Mörch, 1852
STRIOCONUS Thiele, 1929
TARANTECONUS Azuma, 1972
Cornutoconus Suzuki, 1972
TEXTILIA Swainson, 1840
THELICONUS Swainson, 1840
TULIPARIA Swainson, 1840
(non Blainville, 1830)
TURRICONUS Shikama & Habe, 1968
VIRGICONUS Cotton, 1945
VIRROCONUS Iredale, 1930
XIMENICONUS Emerson & Old, 1962

TURRIDAE
PSEUDOMELATOMINAE
PSEUDOMELATOMA Dall, 1918
Laevitectum Dall, 1919
HORMOSPIRA Berry, 1958
TIARATURRIS Berry, 1958

DRILLIINAE (3b)
DRILLIA Gray, 1838
Douglassia Bartsch, 1934
(non Chambers, 1881)
CLATHRODRILLIA Dall, 1918
AGLADRILLIA Woodring, 1928
EUMETADRILLIA Woodring, 1928
ALICEIA Dautzenberg & Fischer, 1897 (1)
AOTEADRILLIA Powell, 1942
AUSTRODRILLIA Hedley, 1918
REGIDRILLIA Powell, 1942
BELLASPIRA Conrad, 1868
BRACHYTOMA Swainson, 1840
CALLICLAVA McLean, 1971
CARINAPEX Dall, 1924
CERITOTURRIS Dall, 1924
CERODRILLIA Bartsch & Rehder, 1939
CYMATOSYRINX Dall, 1889 ?
LISSODRILLIA Bartsch & Rehder, 1939
VIRIDRILLIA Bartsch, 1943
Viridrillina Bartsch, 1943
CLAVUS Montfort, 1810
Clavicantha Swainson, 1840
Tylotia Melvill, 1917
Eldridgea Bartsch, 1934
PLAGIOSTROPHA Melvill, 1927
CONTICOSTA Laseron, 1954
CRASSOPLEURA Monterosato, 1884
Crossopleura Monterosato, 1889 (err.)
DRILLIOLA Cossmann, 1903 (1)
Microdrillia Casey, 1903
Acropota F. Nordsieck, 1977
pro Acrobela Thiele, 1925
(non Förster, 1862)
ELAEOCYMA Dall, 1918
GLOBIDRILLIA Woodring, 1928
HAEDROPLEURA Bucquoy, Dautzenberg & Dollfus, 1883
IMACLAVA Bartsch, 1944
INQUISITOR Hedley, 1918
Pseudoinquisitor Powell, 1942
IREDALEA Oliver, 1915
Brephodrillia Pilsbry & Lowe, 1932
KYLIX Dall, 1919
Kylix Dall, 1918 (nom.nud.)
LEPTADRILLIA Woodring, 1928
MAORITOMELLA Powell, 1942
Narraweena Laseron, 1954
NEODRILLIA Bartsch, 1943
PARACUNEUS Laseron, 1954
PHENATOMA Finlay, 1924
PULSARELLA Laseron, 1954 (?)
SPIROTROPIS Sars, 1878
SPLENDRILLIA Hedley, 1922
FENIMOREA Bartsch, 1934
HAURTURUA Powell, 1942
SYNTOMODRILLIA Woodring, 1928
TOMOPLEURA Casey, 1904
Cryptomella Finlay, 1924
TYLOTIELLA Habe, 1958

CONORBINAE (10)
CONORBIS Swainson, 1840
BENTHOFASCIS Iredale, 1936
GENOTA H. & A. Adams, 1853
Genotia Fischer, 1883 (em.)

TURRINAE
TURRIS Röding, 1798
Turris Müller, 1766 (non-binominal)
Pleurotoma Lamarck, 1799
Pleurotomus Montfort, 1810
ANNULATURRIS Powell, 1966
AUSTROTURRIS Laseron, 1954

TURRINAE (Cont'd.)
CRYPTOGEMMA Dall, 1918
EPIDIRELLA Iredale, 1931
Austrogemmula Laseron, 1954
EPIDIRONA Iredale, 1931
Epideira Hedley, 1918 ?
FUSITURRIS Thiele, 1929
Tyrrhenoturris Coen, 1929
GEMMULA Weinkauff, 1875
Eugemmula Iredale, 1931
PARAGEMMULA Shuto, 1984 (fossil ?)
PINGUIGEMMULA MacNeil, 1960
PTYCHOSYRINX Thiele, 1925
Bathybermudia Haas, 1949
Subulata von Martens, 1901 (invalid)
UNEDOGEMMULA MacNeil, 1960
LOPHIOTOMA Casey, 1904
Lophioturris Powell, 1964
HETEROTURRIS Powell, 1964 ?
XENOTURRIS Iredale, 1929
Clamturris Iredale, 1931
LUCERAPEX Iredale, 1936
NASKIA Syslov & Ivanov, 1985 (11)
POLYSTIRA Woodring, 1928
Pleuroliria de Gregorio, 1890 (auctt.)
Oxytropa Glibert, 1955
TURRIDRUPA Hedley, 1922
VIRIDOTURRIS Powell, 1964

Placement ?
CORINNAETURRIS Bouchet & Warén, 1980 (1)
FAMELICA Bouchet & Warén, 1980 (1)
HEMICYTHARA Kuroda, Habe & Oyama, 1971
HORAICLAVUS Oyama, 1954 (turrid ?) (10)
MAMILLAEDRILLIA Kuroda, Habe & Oyama, 1971
PARASPIROTROPIS Sysoev, 1984
RETICULOTURRIS Shuto, 1984 (fossil ?)

TURRICULINAE (Cochlespirinae)
TURRICULA Schumacher, 1817
Surcula H. & A. Adams, 1853
Surgula Weinkauff, 1875 (err.)
Corbulospira Vincent,1913: Gardner (n.nd.)
ABYSSOCOMITAS Sysoev & Kantor, 1986
AFORIA Dall, 1889 (11a)
ABYSSAFORIA Sysoev & Kantor, 1987
DALLAFORIA Sysoev & Kantor, 1987
PALAEOAFORIA Sysoev & Kantor, 1987
STEIRAXIS Dall, 1896
ANCISTROSYRINX Dall, 1881 (1) (?)
ANTICLINURA Thiele, 1934
Clinuropsis Thiele, 1931
(non Vincent, 1913)
Clinuromella Beets, 1943
ANTICOMITAS Powell, 1942
ANTIMELATOMA Powell, 1942
ANTIPLANES Dall, 1902
RECTIPLANES Bartsch, 1944
RECTISULCUS Habe, 1958
APIOTOMA Cossmann, 1889
AUSTROCARINA Laseron, 1954
BATHYBELA Kobelt, 1905
Bathypota F. Nordsieck, 1977
BELATURRICULA Powell, 1951
CARINOTURRIS Bartsch, 1944
CLAVOSURCULA Schepman, 1913
COCHLESPIRA Conrad, 1865
Candelabrum Dall, 1878
(non Blainville, 1830)
Rouaultia Bellardi, 1878
Coronasyrinx Powell, 1944
Pagodasyrinx Shuto, 1969 (fossil ?)
THATCHERIASYRINX Powell, 1969 ?

TURRICULINAE (Cont'd.)
COMITAS Finlay, 1926
Carinacomitas Powell, 1942
DECOLIDRILLIA Habe & Ito, 1965
FUSITURRICULA Woodring, 1928
FUSISYRINX Bartsch, 1934
IRENOSYRINX Dall, 1908 (1)
IWAOA Kuroda, 1953
KNEFASTIA Dall, 1919
KURILOHADALIA Sysoev & Kantor, 1986
LEUCOSYRINX Dall, 1889
MAKIYAMAIA Kuroda in MacNeil, 1960
MARSHALLARIA Finlay & Marwick, 1937
MARSHALLENA Allan, 1927
Marshallena Finlay, 1927
Sugitanitoma Kuroda, 1959,
pro Sugitania Kuroda, 1958
(non Matsumura, 1926)
MEGASURCULA Casey, 1904
MICROPLEUROTOMA Thiele, 1931
NIHONIA MacNeil, 1960
Fusosurcula Kuroda in
Taki, 1951 (nom.nud.)
PARACOMITAS Powell, 1942
PARADRILLIA Makiyama, 1940
Alticlavatula MacNeil, 1960
CORONACOMITAS Shuto, 1983 (fossil ?)
PHYMORHYNCHUS Dall, 1908 (1)
PLICISYRINX Sysoev & Kantor, 1986
PYRGOSPIRA McLean, 1971
RHODOPETOMA Bartsch, 1944
TYPHLOSYRINX Thiele, 1925
VEXITOMINA Powell, 1942

CRASSISPIRINAE
CRASSISPIRA Swainson, 1840
BIRMADRILLIA Shuto, 1984 (fossil ?)
BURCHIA Bartsch, 1944
CRASSICLAVA McLean, 1971
CRASSISPIRELLA Bartsch & Rehder, 1939
DALLSPIRA Bartsch, 1950
GIBBASPIRA McLean, 1971
GLOSSISPIRA McLean, 1971
MONILISPIRA Bartsch & Rehder, 1939
STRIOSPIRA Bartsch, 1950
Adanaclava Bartsch, 1950
BUCHEMA Corea, 1934
CARINODRILLIA Dall, 1918 (8)
DOXOSPIRA McLean, 1971
HINDSICLAVA Hertlein & Strong, 1955
Turrigemma Berry, 1958
LIOGLYPHOSTOMA Woodring, 1928
Glyphostomops Bartsch, 1934
MAESIELLA McLean, 1971
MIRACLATHURELLA Woodring, 1928

STRICTISPIRINAE (8)
STRICTISPIRA McLean, 1971
CLEOSPIRA McLean, 1971

ZONULISPIRINAE (8)
ZONULISPIRA Bartsch, 1950
COMPOSODRILLIA Woodring, 1928
PILSBRYSPIRA Bartsch, 1950
NYMPHISPIRA McLean, 1971
PTYCHOBELA Thiele, 1925

CLAVATULINAE
CLAVATULA Lamarck, 1801
BENTHOCLIONELLA Kilburn, 1974
CLIONELLA Gray, 1847
TOXICLIONELLA Powell, 1966

CLAVATULINAE (Cont'd.)
PERRONA Schumacher, 1817
TOMELLANA Wenz, 1943
pro Tomella Swainson, 1840
(non Robineau-Desvoidy, 1830)
PUSIONELLA Gray, 1847
Netrum Philippi, 1850

BORSONIINAE
BORSONIA Bellardi, 1839
BATHYTOMA Harris & Burrows, 1891
pro Dolichotoma Bellardi, 1875
(non Hope, 1839)
MICANTAPEX Iredale, 1936
PARABATHYTOMA Shuto, 1961
RIUGUHDRILLIA Oyama, 1951
BORSONELLA Dall, 1908
BORSONELLOPSIS McLean, 1971
CORDIERIA Rouault, 1848
Phlyctis Harris & Burrows, 1891
pro Phlyctaenia Cossmann, 1889
CRUZITURRICULA Marks, 1951
DARBYA Bartsch, 1934
INODRILLIA Bartsch, 1943
Inodrillara Bartsch, 1943
Inodrillina Bartsch, 1943
OPHIODERMELLA Bartsch, 1944
PARABORSONIA Pilsbry, 1922 (living fossil)
SCRINIUM Hedley, 1922
SUAVODRILLIA Dall, 1918 (? Drillinae)
TYPHLOMANGELIA G. O. Sars, 1878
TARANIS Jeffreys, 1870
Fenestrosyrinx Finlay, 1926
Allo Jousseaume in Lamy, 1934
Feliciella Lamy, 1934
TROPIDOTURRIS Kilburn, 1986

MITROMORPHINAE (2b)
MITROMORPHA Carpenter, 1865 ?
(non A. Adams, 1865 =
Antimitra Iredale, 1917)
LOVELLONA Iredale, 1917 (2)
APATURRIS Iredale, 1917
DIPTYCHOPHLIA Berry, 1964
HELENELLA Casey, 1904
MAORIMORPHA Powell, 1939
MITROLUMNA Bucquoy, Dautzenberg & Dollfus, 1883
Clinomitra, Diptychomitra Bellardi,1889
Mitrithara Hedley, 1922 (2)
Itia Marwick, 1931
Arielia Shasky, 1961
Vexiariella Shuto, 1983 (?)
CYMAKRA Gardner, 1937

CLATHURELLINAE
CLATHURELLA Carpenter, 1857
CROCKERELLA Hertlein & Strong, 1951
ETREMA Hedley, 1918 (2a)
ETREMOPA Oyama, 1953
Iraqetrema Dance & Eames, 1966
EUCITHARA Fischer, 1883 (2a)
Cythara (authors)
PLEUROTOMOIDES Bronn ?
LEIOCITHARA Hedley, 1922
CYRTOCYTHARA F. Nordsieck, 1977 ?
RUGOCYTHARA F. Nordsieck, 1977 ?
GLYPHOSTOMA Gabb, 1872
Rhiglyphostoma Woodring, 1970
EUGLYPHOSTOMA Woodring, 1970
LIENARDIA Jousseaume, 1884 (2a)
ACRISTA Hedley, 1922
HEMILIENARDIA Boettger, 1895
THETIDOS Hedley, 1899

CLATHURELLINAE (Cont'd.)
MACTEOLA Hedley, 1918 (2a)
NANNODIELLA Dall, 1919
STROMBINOTURRIS Hertlein & Strong, 1951

MANGELIINAE
MANGELIA Risso, 1826
Mangilia auctt. (err.)
Villiersia Locard, 1886
Villiersiella Monterosato, 1890,
pro Vielliersia Monterosato, 1884 (err.)
(non Orbigny, 1837)
BRACHYCITHARA Woodring, 1928
GLYPHOTURRIS Woodring, 1928
SACCAROTURRIS Woodring, 1928
STELLATOMA Bartsch & Rehder, 1939
ACMATURRIS Woodring, 1928
AGATHOTOMA Cossmann, 1899
pro Ditoma Bellardi, 1877
(non Illinger, 1807)
VITRICYTHARA Fargo, 1953
ANACITHARA Hedley, 1922
ANTIGURALEUS Powell, 1939
Paraguraleus Powell, 1944 ?
APISPIRALIA Laseron, 1954
APITUA Laseron, 1954
AUSTROPUSILLA Laseron, 1954
METACLATHURELLA Shuto, 1983 (fossil ?)
BACTROCYTHARA Woodring, 1928
BELA Gray, 1847
Smithiella Monterosato, 1890
pro Smithia Monterosato, 1884
(non Milne-Edwards & Haime, 1851)
Ginniana Monterosato, 1884
Ishnula Clark in Gray, 1847 (nom.nud.)
Ichnusa Jeffreys, 1847 (nom.obl.)
BELLACYTHARA McLean, 1971
BENTHOMANGELIA Thiele, 1925
Metuonella Sorgenfrei, 1958
CACODAPHNELLA Pilsbry & Lowe, 1932
CLATHROMANGELIA Monterosato, 1884
CONOPLEURA Hinds, 1884
Kenyonia Brazier, 1896
CONORBELA Powell, 1951
CRYOTURRIS Woodring, 1928
CYTHARELLA Monterosato, 1875
Rissomangilia Monterosato, 1917
Citharella Thiele, 1929 (err.)
ETREMOPSIS Powell, 1942
EUCLATHURELLA Woodring, 1928
FILODRILLIA Hedley, 1922
GLYPTAESOPIS Pilsbry & Olsson, 1941
GURALEUS Hedley, 1918
EUGURALEUS Cotton, 1947
MITRAGURALEUS Laseron, 1954
HETEROCITHARA Hedley, 1922
ITHYCYTHARA Woodring, 1928
KURTZIA Bartsch, 1944
KURTZIELLA Dall, 1918
GRANOTURRIS Fargo, 1953
KURTZINA Bartsch, 1944
RUBELLATOMA Bartsch & Rehder, 1939
LIRACRAEA Odhner, 1924
MANGILIELLA Bucquoy, Dautzenberg
& Dollfus, 1883
LYROMANGELIA Monterosato, 1917
MARITA Hedley, 1922
NEOGURALEUS Powell, 1939
NODOTOMA Bartsch, 1941 (10)
Canetoma, Curtitoma, Nematoma,
Venustoma all Bartsch, 1941
NOTOCYTHARELLA Hertlein & Strong, 1955
OBESOTOMA Bartsch, 1941

MANGELIINAE (Cont'd.)
OENOPOTA Mörch, 1852
Propebela Iredale, 1918
Bela, Lora of Authors (non Gray, 1847)
Turritomella Bartsch, 1941
pro Turritoma Bartsch, 1941
Belalora, Lorabela Powell, 1951
FUNITOMA Bartsch, 1941
Cestoma, Granotoma Bartsch, 1941
WIDALLI Bogdanov, 1986
PARACLATHURELLA Boettger, 1895
PARAMONTANA Laseron, 1954
PLATYCYTHARA Woodring, 1928
PSEUDOETREMA Oyama, 1953
PSEUDORAPHITOMA Boettger, 1895
PYRGOCITHARA Woodring, 1928
TENATURRIS Woodring, 1928
THELECYTHARA Woodring, 1928 (4)
TURELLA Laseron, 1954

DAPHNELLINAE (Raphitominae) (2a)
DAPHNELLA Hinds, 1844
Eudaphnella Bartsch, 1933
pro Eudaphne Bartsch, 1931 (non Reuss, 1922)
Paradaphne Laseron, 1954
DIAUGASMA Melvill, 1917
HEMIDAPHNE Hedley, 1918
ABYSSOBELA Kantor & Sysoev, 1986
ANTIMITRA Iredale, 1917
ASPERDAPHNE Hedley, 1922
pro Scabrella Hedley, 1918
(non Sacco, 1890)
ASPERTILLA Powell, 1944
AUSTRODAPHNELLA Laseron, 1954
BUCCINARIA Kittl, 1887 (fossil)
OOTOMELLA Bartsch, 1933 (recent)
pro Ootoma Koperberg, 1931 (non Dejean,1833)
Pionotoma Kuroda, 1952
COMARMONDIA Monterosato, 1884
pro Bellardia Bucquoy, Dautzenberg & Dollfus, 1883
(non Robineau-Desvoidy, 1863)
Bellatula Strand, 1928
pro Belardialla Fischer, 1883
CRYPTODAPHNE Powell, 1942
ACAMPTODAPHNE Shuto, 1971 (fossil ?)
EUBELA Dall, 1889
EUCYCLOTOMA Boettger, 1895
Turrhyssa Dall, 1924
EXOMILUS Hedley, 1918
FUSIDAPHNE Laseron, 1954
GYMNOBELA Verrill, 1884
Watsonaria F. Nordsieck, 1968
Majox F. Nordsieck, 1978
ISODAPHNE Laseron, 1954
KERMIA Oliver, 1915 (2)
Clathurina Melvill, 1917
LUSITANOPS F. Nordsieck, 1968
Pseudazorita F. Nordsieck, 1977
MAORIDAPHNE Powell, 1942 (fossil)
KUROSHIODAPHNE Shuto, 1965 (recent)
MICRODAPHNE McLean, 1971
MICROGENIA Laseron, 1954
NEOPLEUROTOMOIDES Shuto, 1971
NEPOTILLA Hedley, 1918
OTITOMA Jousseaume, 1898
PHILBERTIA Monterosato, 1884
Lineotoma F. Nordsieck, 1977
pro Cirilla Monterosato, 1884
(non Rondani, 1856)
Cordieria Monterosato, 1884
(non Rouault, 1848)
Peratoma Harris & Burrows, 1891

PHILBERTIA (Cont'd.)
pro Homatoma Bellardi, 1875
(non Guerin-Meneville, 1884)
Leufroyia Monterosato, 1884
GLYPHOSTOMOIDES Shuto, 1983 (fossil ?)
PSEUDODAPHNELLA Boettger, 1895 (2)
PLEUROTOMELLA Verrill, 1872
Azorilla, Azorita F. Nordsieck, 1968
ANOMALOTOMELLA Powell, 1966
PONTIOTHAUMA E. A. Smith, 1895
PSEUDEXOMILUS Powell, 1944
RAPHITOMA Bellardi, 1848
CYRTOIDES F. Nordsieck, 1968 ?
RIMSODAPHNELLA Cossmann, 1915
SPERGO Dall, 1895
STILLA Finlay, 1926
TASMADAPHNE Laseron, 1954
TELEOCHILUS Harris, 1897
Litachilus Powell, 1944
TERETIA Monterosato, 1890
THESBIA Jeffreys, 1867
THETA Clarke, 1959
Speoides Kuroda & Habe, 1962
TRITONOTURRIS Dall, 1924
TRUNCADAPHNE McLean, 1971
TYPHLODAPHNE Powell, 1951
VEPRECULA Melvill, 1917
Mordica Dall, 1924
XANTHODAPHNE Powell, 1942
ZENEPOS Finlay, 1928

THATCHERIINAE
THATCHERIA Angas, 1877
Cochlioconus Yokoyama, 1928

TEREBRIDAE
TEREBRA Bruguière, 1789
Terebrum Montfort, 1810
Microtrypetes Pilsbry & Lowe, 1933 ?
ABRETIELLA Bartsch, 1923
Abretia H. & A. Adams, 1853
(non Rafinesque, 1814)
ACUMINIA Dall, 1908
CLATHROTEREBRA Oyama, 1961
DECORIHASTULA Oyama, 1961
EUTEREBRA Cotton & Godfrey, 1932
FUSOTEREBRA Sacco, 1891
GRADATEREBRA Cotton & Godfrey, 1932
MYURELLA Hinds, 1845
Brevimyurella, Myurellina Bartsch, 1923
NODITEREBRA Cossmann, 1896
NOTOTEREBRA Cotton, 1947
OXYMERIS Dall, 1908 (8)
PANATEREBRA Olsson, 1967
PENTATEREBRA Olsson, 1967
PERIRHOE Dall, 1908
Dimidiacus Iredale, 1929
PUNCTOTEREBRA Bartsch, 1923
STRIOTEREBRUM Sacco, 1891
TEREBRINA Bartsch, 1923
TRIPLOSTEPHANIS Dall, 1908
DIPLOMERIZA Dall, 1919
Duplicaria Dall, 1908
(non Rafinesque, 1833)
MYURELLISCA Bartsch, 1923
EGENTELARIA Rehder, 1980
HASTULA H. & A. Adams, 1853
HASTULINA Oyama, 1961
IMPAGES E. A. Smith, 1873
PERVICACEA Iredale, 1924 (?)
TERENOLLA Iredale, 1929
SUBULA Schumacher, 1817
Acus Humphrey, 1797 (non Edwards, 1771)

References: (Conoidea)
Bouchet, P. & A. Warén "Revision of the Northeast Atlantic bathyal and abyssal Turridae" Jour.Mollusc.St. Supp.8:1-111 - 1980 (1)

Cernohorsky, W. O. "Taxonomy of some Indo-Pacific Mollusca" Rec.Auck.Inst.Mus. 12:213-234 - 1975 (2)
Ibid. 15:67-86 - 1978 (2a)
Ibid. 22:47-67 - 1985 (2b)

Garrard, T.A. "Genus Pervicacea in SE. Australia" J.Malac.Soc.Aust.3(3-4):179-185 - 1976 (3)

Gibson-Smith, J. W. "New Recent gastropod species from Venezuela" Veliger 25(3):177 - 1981 (4)

Kantor, Yu.I. & A.V. Sysoev "New genus: Turridae, Northern Pacific Ocean" Zool.Zh.65(4):485-498 - 1986 (5)

Marsh, J.A. & O.H. Rippingale "Cone Shells of the World" Jacaranda Press, Australia - 1968 (7)

McLean, J.A."A Revised Classification of: family Turridae: proposal of new subfamilies, genera and subgenera: Eastern Pacific" Veliger 14(1):114-130 - 1971 (8)

Powell, A. W. B. "The Molluscan Families Speightiidae and Turridae" Bull.Auck.Inst.Mus.5 - 1966 (10)

Sysoev, A.V. & Ivanov "New taxa: family Turridae" Zool.Zh.65(2):194-205 - 1985 (11)

---- & Y. I. Kantor "Deep sea gastropods of the genus Aforia in the Pacific --" Veliger 30(2):105-126 - 1987 (11a)

---- "New abyssal species of Turridae" Zool.Zh. 65(10):1457-1469 - 1986 (11b)

Walls, J. G. "Cone Shells" T.F.H. Publications, N.J. - 1978 (12)

RISSOELLOIDEA (1, 3)
OMALOGYRIDAE
OMALOGYRA Jeffreys, 1867 (Homalogyra)
AMMONICERA Vayssière, 1893

RISSOELLIDAE (1, 2, 3)
RISSOELLA Gray, 1847
Jeffreysia Alder in Forbes & Hanley, 1850

RISOELLA (Cont'd.)
Jeffreysiopsis Thiele, 1912
Jeffreysina Thiele, 1925
JEFFREYSILLA Thiele, 1925
Phycodrosus Rehder, 1943
JEFFREYSIELLA Thiele, 1912
Heterorissoa Iredale, 1912
ZELAXITAS Finlay, 1927

References:

Hazprunar, G. "The Heterobranchia; new concept of the higher Gastropoda" 1985 (1) Z.Zool.Syst.Evolut.forsch.23(1):15-37

Robertson, R. "A second West Atlantic Rissoella" Nautilus 74(4):131-136 - 1961 (2)

---- "Supplementary notes on the Rissoellidae" Notulae Nat.352 - 1962 (3)

Subclass Heterobranchia
Superorder Allogastropoda (3)

ARCHITECTONICOIDEA
ARCHITECTONICIDAE (1a)
ARCHITECTONICA Röding, 1798
Solarium Lamarck, 1799
Verticellus Jousseaume, 1888 (1)
(non Moquin-Tandon, 1848)
ADELPHOTECTONICA Bieler, 1987
ACUTITECTONICA Habe, 1961 (2)
Russetia Garrard, 1961
BASISULCATA Melone & Taviani, 1984
DISCOTECTONICA Marwick, 1931
GRANOSOLARIUM Sacco, 1892
Solariaxis Dall, 1892
Claraxis Iredale, 1936
HELIACUS Orbigny, 1842 (Heliacinae - 2)
Torinia Gray, 1847
GRANDELIACUS Iredale, 1957
GRANOHELIACUS Melone & Taviani, 1984
GYRISCUS Tiberi, 1867
PYRGOHELIACUS Bieler, 1987
REDIVIUS Melone & Taviani, 1984
TERETROPOMA Rochebrune, 1881
TORINISTA Iredale, 1936
Astronacus Woodring, 1959

ARCHITECTONICIDAE (Cont'd.)
PHILIPPIA Gray, 1847 (Philippiinae - 2)
PSILAXIS Woodring, 1928
PSEUDOMALAXIS Fischer, 1885 (4)
Discosolis Dall, 1892
Mangonuia Mestayer, 1930
PSEUDOTORINIA Sacco, 1892
Awarua Mestayer, 1930
Calodisculus Rehder, 1935
Punjabia Eames, 1952 ?
SOLATISONAX Iredale, 1931
SPIROLAXIS Monterosato, 1913 (2)
Paurodiscus Rehder, 1935
Aguayodiscus Jaume & Borro, 1946

MATHILDIDAE
MATHILDA Semper, 1865
BROOKESENA Finlay, 1927
FIMBRIATELLA Sacco, 1895
GEGANIA Jeffreys, 1884
Tuba Lea, 1838
Steirotuba Sacco, 1895
GRANULICHARILDA Kuroda, Habe & Oyama, 1971
MATHILDONA Iredale, 1929
OPIMILDA Iredale, 1929
VETERATOR Laws, 1944 (fossil)

References: (Architectonicoidea)
Bieler, R. "Die gattungen der Architectonicidae Gastropoda, Allogastropoda"
I. Arch.Moll. 115(1-3):53-103 - 1984 (1)
II. Ibid. 115(4-6):231-265 - 1984
III. Ibid. 116(1-3):89-117 - 1985
IV. Ibid. 117(4-6):203-215 - (1987)
---- "Revision of genera, and Indo-Pacific species in the family Architectonicidae" Amer.Malac.Bull.(4):108-109 - 1986 (1a)

Boss, K.J. & A.S. Merrill "Notes: Acutitectonica: radular configuration --" Occ.Pap.Mollusks, Harv.U. 4(65-66):333-411 - 1984 (2)

Haszprunar, G. "The Heterobranchia, a new concept for the phylogeny of the higher gastropoda" Z.zool.Syst.Evolut.forsch.23:15-37-1985 (3)

Ponder, W. F. "Review of the genera of Rissoidae" Rec. Aust.Mus.Supp. 4:1-221 - 1985 (4)

PYRAMIDELLOIDEA
PYRAMIDELLIDAE
PYRAMIDELLINAE
PYRAMIDELLA Lamarck, 1799
Pyramidellus Montfort, 1810
Obeliscus Gray, 1847
(non Beck, 1837)
CALLOLONCHAEUS Dall & Bartsch, 1904
COSSMANNICA Dall & Bartsch, 1904
Diptychus Cossmann, 1888
(non Milne-Edwards, 1880)
LONGCHAEUS Mörch, 1875
MILDA Dall & Bartsch, 1904
PHARCIDELLA Dall, 1889
SULCORINELLA Dall & Bartsch, 1904
VOLUSPA Dall & Bartsch, 1904
BACTERIDIELLA Saurin, 1983 (?)
CHARILDA Iredale, 1929
CONTRAXIALA Laseron, 1956 (3)
COSTABIETA Laseron, 1956 (3)
DERJUGINELLA Habe, 1958
EUCHARILDA Iredale, 1929
EULIMELLA Jeffreys, 1847
Loxoptyxis Cossmann, 1888
Visma Dall & Bartsch, 1904
Raoulostraca Oliver, 1915
BACTERIDIUM Thiele, 1929
BELONIDIUM Cossmann, 1892 ?
Aciculina Deshayes, 1864
(non A. Adams, 1853)
Baudonia Bayan, 1873
(non Mabille, 1868)
Raphium Bayan, 1875
(non Meigen, 1864)
EBALA Leach, 1847
Anisocycla Monterosato, 1880
Ebalina Thiele, 1929
FINLAYOLA Laws, 1937
HEREWARDIA Iredale, 1955 (3)
OTOPLEURA Fischer, 1885
PUPOSYRNOLA Cossmann, 1921
Hoonsyrnola Nomura, 1938 (nom.nud.)
STYLOPYRAMIS Thiele, 1931
TERELIMELLA Laws, 1938
TIBERIA Monterosato, 1875
Tiberiola Cossmann, 1900
OOPYRAMIS Thiele, 1930 ?
ORINA Agassiz, 1846

ODOSTOMIINAE
ODOSTOMIA Fleming, 1817
Odontostomia Jeffreys, 1839
Monoptaxis Clark in Jeffreys, 1867
Ptychostomon Locard, 1886
(non Ptychostomum Stein, 1860)
Turritodostomia Sacco, 1892
Heida Dall & Bartsch, 1904
AURISTOMIA Monterosato, 1884
BESLA Dall & Bartsch, 1904
BOONEA Robertson, 1978

ODOSTOMIA (Cont'd.)
BRACHYSTOMIA Monterosato, 1884
Brachystoma Meigen, 1822 ?
Zastoma Iredale, 1915
CINCTINGA Laseron, 1951 (?)
CYCLODOSTOMIA Sacco, 1892
DOLIELLA Monterosato, 1880
EGILA Dall & Bartsch, 1904
EULIMASTOMA Bartsch, 1904
Telloda Hertlein & Strong, 1951
Parodostomia Laseron, 1959
EVALEA A. Adams, 1860
Auriculina Gray, 1847
(non Grateloup, 1838)
Ondina de Folin, 1870
EVALINA Dall & Bartsch, 1904
FARGOA Robertson, 1978 ?
GUMINA Finlay, 1928
IOLAEA A. Adams, 1867
Iolea Pascoe, 1858
Iole A. Adams, 1860
(non Blythe, 1844)
Iolina Bailey, 1945
IVARA Dall & Bartsch, 1903
IVIDELLA Dall & Bartsch, 1909
Funicularia Monterosato, 1884
(non Lamarck)
Ividia Dall & Bartsch, 1904 ?
JORDANULA Chaster, 1901
Jordaniella Chaster, 1898
LINOPYRGA Laws, 1941
MEGASTOMIA Monterosato, 1884
Megastoma Swainson, 1837 (error)
Stomega Dall & Bartsch, 1904
NESIODOSTOMIA Pilsbry, 1918
NISOSTOMIA Cossmann, 1921
OBEXOMIA Laws, 1941
OOPRIAMUS Thiele, 1930
SALASSIA de Folin, 1885
SALASSIELLA Dall & Bartsch, 1909
SIOGAMAIA Nomura, 1936
STRIODOSTOMIA Laws, 1940
TURRIODOSTOMIA Habe, 1961
CHRYSALLIDA Carpenter, 1857
Elodia de Folin, 1873
(non Desvoidy, 1863)
Odostomella Bucquoy, Dautzenberg & Dollfus, 1883
Elodiamea de Folin, 1884
BABELLA Dall & Bartsch, 1906
FOLINELLA Dall & Bartsch, 1904
Amaura de Folin, 1873
(non Gray, 1847)
HALDRA Dall & Bartsch, 1904
MUMIOLA A. Adams, 1864
ODONTOSTOMIELLA Cossmann, 1921
PARTHENINA Bucquoy, Dautzenberg & Dollfus, 1883
PARTULIDA Schaufuss, 1869
Spirulina Chaster, 1898

ODOSTOMIINAE
CHRYSALLIDA (Cont'd.)
PRESTONIELLA Saurin, 1983 (= Pyrgulina ?)
PYRGULINA A. Adams, 1864
STANDENIELLA Saurin, 1983 ?
TRABECULA Monterosato, 1884
VILIA Dall & Bartsch, 1904
CINGULINA A. Adams, 1860
Polyspirella A. Adams, 1861
HERVIERA Melvill & Standen, 1897
Rugadentia Laseron, 1951
LIOSTOMIA Sars, 1878
Leiostoma Swainson, 1840 (err.)
Cremula Iredale, 1915
MYXA Hedley, 1903
PSEUDORISSOINA Tate & May, 1900
KOLOONELLA Laseron, 1959
MENESTHO Möller, 1842
NOEMIAMEA de Folin, 1886
Noemia de Folin, 1872
(non Pasco, 1857)
Oda Monterosato, 1901
ODETTA de Folin, 1870
MIRALDA A. Adams, 1864
Lia de Folin, 1870
(non Eschscholtz, 1829)
Euparthenia Thiele, 1929
HINEMOA Oliver, 1915
MIDDIELLA Cossmann, 1921
OSCILLA A. Adams, 1861
Jaminea de Folin, 1869
(non Gray, 1847)
MORRISONIETTA Brandt, 1968
PERISTICHIA Dall, 1889
SAYELLA Dall, 1885
SYRNOLA A. Adams, 1860
AGATHA A. Adams, 1860
Amathus A. Adams, 1861
ANGUSTISPIRA Pelseneer, 1912 ?
COLSYRNOLA Iredale, 1929
IPHIANA Dall & Bartsch, 1904
Evaletta Pilsbry, 1918
LYSACME Dall & Bartsch, 1904
PACHYSYRNOLA Cossmann, 1907
STYLOPSIS A. Adams, 1860 ?
STYLOPTYGMA A. Adams, 1860 ?
VAGNA Dall & Bartsch, 1904
ORINELLA Dall & Bartsch, 1904
Orina A. Adams, 1870
(non Oreina Chevrolat, 1834
= Orina Agassiz, 1846)
TRYPTYCHUS Mörch, 1875

TURBONILLINAE
TURBONILLA Risso, 1826
ASMUNDA Dall & Bartsch, 1904
BALDRA Dall & Bartsch, 1904
BARTSCHELLA Iredale, 1916
Dunkeria Dall & Bartsch,
(non Carpenter, 1865)
CARELIOPSIS Mörch, 1895
CHEMNITZIA Orbigny, 1839 ?

TURBONILLA (Cont'd.)
CYLINDRITURBONILLA F. Nordsieck, 1977
DUNKERIA Carpenter, 1857
Pyrgisculus Dall & Bartsch, 1909 ?
LANCELLA Dall & Bartsch, 1904
Lancea Pease, 1868
(non Lancia Walker, 1859)
GRACILITURBONILLA F. Nordsieck, 1977
MAGNITURBONILLA F. Nordsieck, 1977
MORMULA A. Adams, 1864
Pyrgostylus Monterosato, 1884
NISITURRIS Dall & Bartsch, 1906
PLANPYRGISCUS Laws, 1937
PSELLIOGYRA Dall & Bartsch, 1909
PTYCHEULIMELLA Sacco, 1892
PYRGISCILLA Laws, 1937
PYRGISCUS Philippi, 1841
PYRGOLAMPROS Sacco, 1892
PYRGOLIDIUM Monterosato, 1884
STRIARCANA Laws, 1937
STRIOTURBONILLA Sacco, 1892
TRAGULA Monterosato, 1884
Burkilla Iredale, 1915
VARITURBONILLA F. Nordsieck, 1977
KLEINELLA A. Adams, 1860
ACTAEOPYRAMUS Fischer, 1885 (Invalid ?)
pro Monoptygma A. Adams, 1851
(non Lea, 1833)
Monotygma Gray, 1840 (nom.nud.)
Myonia A. Adams, 1860
(non Dana, 1847)
Adelactaeon Cossmann, 1895
EUPARTHENIA Thiele, 1931 (pars)
Parthenia Lowe, 1840
(non Robineau-Desvoidy, 1830)
TROPAEAS Dall & Bartsch, 1904
Elusa A. Adams, 1861
(non Walker, 1858)
SYRNOLINA Dall & Bartsch, 1904
UGARTEA Bartsch, 1917

CYCLOSTREMELLINAE
CYCLOSTREMELLA Bush, 1897

Pyramidellidae - Placement ?
APHALISTA, EURATHEA, EXESILLA, INSTARELLA,
KUNOPIA, MORMULASTA, PARADOXELLA,
QUIRELLA, TATHRELLA, URAMBELLA, WINGENELLA,
ZAPHELLA, ZONELLA all Laseron, 1959
PLICIFER A. Adams, 1868 (4)
PSEUDOSKENELLA Ponder, 1973

AMATHINIDAE (3a)
AMATHINA Gray, 1842
AMATHINOIDES Sacco, 1896
CLATHRELLA Récluz, 1864
CYCLOTHYCA Stearns, 1891 (4)
Cyclotheca Petuch, 1987 (err.)
ISELICA Dall, 1918
Isapis H. & A. Adams, 1854
LEUCOTINA A. Adams, 1860
PHASIANEMA S. Wood, 1872 (3)

References:
Fretter, V. "Evolution: higher taxa gastropods" J.Malac.Soc.Aust.4(4):226-227 - 1980 (1)

Haszprunar, G. "The Heterobranchia: new concept -" Z.zool.Syst.Evolut.-forsch.23:15-37 -1985 (2)
—— "Fine morphology: Osphradial organs, Mollusca" Phil.Trans.R.Soc.Lond.B.307:497-505.1985 (2a)

Ponder, W. F. "Review of the genera of Rissoidae" Rec.Aust.Mus.Supp.4:1-221 - 1985 (3)
—— "The anatomy and relationships: Amathina" Asian Marine Biology 4:1-34 - 1987 (3a)

Warén, A. "Generic revision of the Eulimidae" Journ.Moll.Stud., Supp.13 - 1983 (4)

Subclass Opisthobranchia
Order Cephalaspidea

PHILINOIDEA ?
ACTEONIDAE (22a)
ACTEON Montfort, 1810 (Actaeon)
Tornatella Lamarck, 1816
Speo Risso, 1826
Kanilla Sowerby, 1838
Myosota Gray, 1847
Actaeon A. Adams, 1854
(non Oken, 1815)
Metacteon Thiele, 1931 (invalid) ?
Bullinopersilia F. Nordsieck, 1972
CRENILABIUM Cossmann, 1889
Lissacteon Monterosato, 1890
INOPINODON Bouchet, 1975
JAPONACTEON I. Taki, 1956 (22)
MAXACTEON Rudman, 1971
MICROGLYPHIS Dall, 1902
Hyporingicula Habe, 1952
MYSOUFFA Marcus, 1974
NEACTAEONINA Thiele, 1912
OBRUSSENA Iredale, 1930
pro Obrussa Iredale, 1925
(non Braun, 1915)
OMPHALOACTEONELLA Akopyan, 1972 ?
OVULACTEON Dall, 1889
PSEUDACTEON Thiele, 1925
PUNCTACTEON Kuroda & Habe, 1961
PUGNUS Hedley, 1896
PUPA Röding, 1798
Solidula Fischer, 1807 ? (3)
Dactylus Schumacher, 1817
Solidulus Férussac, 1829
Buccinulus H. & A. Adams, 1854
RICTAXIELLA Habe, 1958
STRIGOPUPA Habe, 1958
RICTAXIS Dall, 1871

RINGICULIDAE
RINGICULA Deshayes, 1838
Aptycha Meek, 1863
Plicatra, Ringactaeon F. Nordsieck, 1972
RINGICULINA Monterosato, 1884
Ringiculadda Iredale, 1936
RINGICULINOPSIS Lin, 1985

HYDATINIDAE
HYDATINA Schumacher, 1817
Hydatoria, Noalda Iredale, 1936
APLUSTRUM Schumacher, 1817
Aplustra Swainson, 1840
MICROMELO Pilsbry, 1895
PARVAPLUSTRUM Powell, 1981 ?

BULLINIDAE (22)
BULLINA Férussac, 1822
Bullinula Beck in Swainson, 1840 ?
Perbullina Iredale, 1929

SCAPHANDRIDAE (Cylichnidae ?)
SCAPHANDER Montfort, 1810
Bullaria Rafinesque, 1815
Assula Schumacher, 1817
Adamnestia Iredale, 1936
Mamillocylichna F. Nordsieck, 1972
BUCCANIA Dall, 1890
Bullocardia F. Nordsieck, 1972
EOSCAPHANDER Habe, 1952
ACTEOCINA Gray, 1847
Alicula Eichwald, 1830
Didontoglossa Annandale, 1924

ACTEOCINA (Cont'd.)
Neoacteocina Kuroda & Habe, 1952
Truncacteocina Kuroda & Habe, 1954
CYLICHNA Lovén, 1846
Bullina Risso, 1826
(non Férussac, 1822)
Cyclina Gray, 1857
(non Deshayes, 1850)
Bullinella R. B. Newton, 1891
ACROSTEMMA Cossmann, 1889
BROCKTONIA Iredale, 1915
CLISTAXIS Cossmann, 1895
Cryptaxis Jeffreys, 1883
(non Lowe, 1854)
CYLICHNOIDES Minichev, 1977
CYLICHNOPSIS Cossmann, 1904 (fossil ?)
EOCYLICHNA Kuroda & Habe, 1952 (Syn.?)
MNESTIA H. & A. Adams, 1854 (24)
Morestia Kobelt, 1879 (err.)
SPHAEROCYLICHNA Thiele, 1925
CYLICHNATYS Kuroda & Habe, 1952
CYLICHNELLA Gabb, 1873
MELOSCAPHANDER Schepman, 1913
MICRAENIGMA Berry, 1953
NIPPONOSCAPHANDER Kuroda & Habe, 1971
PARACTEOCINA Minichev, 1966
ROXANIA Leach in Gray, 1847 (Turton, 1834 ?)
Damoniella Iredale, 1918 ?
Abderospira Dall, 1896
SABATIA Bellardi, 1877
Sabatina Dall, 1908
TORNATINA A. Adams, 1850
UTRICULASTRA Thiele, 1925 ? (15)
Coleophysis P. Fischer, 1883
Coelophysis Habe, 1955 (err.)
TORNASTRA Marcus, 1977

PHILINIDAE (22b)
PHILINE Ascanius, 1772
Lobaria Müller, 1776
Bullaea Lamarck, 1801
Bullea Rafinesque, 1815
Colobocephalus, Utriculopsis Sars, 1870
Hermania Monterosato, 1884
Phillingwinia F. Nordsieck, 1972
Retusophiline F. Nordsieck, 1972
Philinorbis, Yokoyamaia Habe, 1950
Chosiphiline, Globophiline Habe, 1958
LAONA A. Adams, 1865 ?
Johania, Ossiania Monterosato, 1884
PSEUDOPHILINE Habe, 1976 ?
SPINIPHILINE Gosliner, 1988

AGLAJIDAE
AGLAJA Renier, 1807 (Aglaia auctt. - em.)
Doridium Meckel, 1809 (invalid)
Acera Cuvier, 1810
Bullidium Leue, 1813
Lobaria Blainville, 1825
(non Müller, 1776)
Eidothea Risso, 1826
Posterobranchaea Orbigny, 1837
Posteobranchus, Posteriobranchus,
Posterobranchia, Posterobranchea auctt.
NAVANAX Pilsbry, 1895
Strategus Cooper, 1863
(non Kirby & Spence, 1828)
Navarchus Cooper, 1863
(non Filippi & Verany, 1859)
CHELIDONURA A. Adams, 1850
Hirundella Gray, 1850
Chelinodura Fischer, 1883 (err.)
MELANOCHLAMYS Cheesman, 1881 (3)

AGLAJIDAE (Cont'd.)
NAKAMIGAWAIA Kuroda & Habe, 1961
ODONTOGLAJA Rudman, 1978
PHILINOPSIS Pease, 1860 (3)

GASTROPTERIDAE
GASTROPTERON Meckel in Kosse, 1813
Sarcopterus Rafinesque, 1814
Parthenopia Oken, 1815
Gasteroptera, Gastroptera,
Gasteropterum auctt. (err.)
SAGAMINOPTERON Tokioka & Baba, 1964

DIAPHANIDAE
DIAPHANA T. Brown, 1827 (26)
Amphisphyra Lovén, 1846
Physema A. Adams, 1854
AUSTRODIAPHANA Pilsbry, 1895
COLPODASPIS M. Sars, 1870
NEWNESIA E. A. Smith, 1902
Anderssonia Strebel, 1908
RHINODIAPHANA Lemche, 1967
TOLEDONIA Dall, 1902
Odostomiopsis Thiele, 1903
Ohlinia Strebel, 1905
Ptisanula Odhner, 1913
WOODBRIDGEA Berry, 1953
Clistaxis Dall, 1908
(non Cossmann,1895)
Brocktonia Dall, 1921
(non Iredale, 1915)

NOTODIAPHANIDAE
NOTODIAPHANA Thiele, 1917
Diaphanella Thiele, 1912
(non Clessin, 1880)
Osorattis Iredale, 1929

BULLIDAE (Bullariidae)
BULLA Linnaeus, 1758
Bullus Montfort, 1810
Bullites Schlotheim, 1813
(non Gmelin, 1793)
Bullaria auctt.
(non Rafinesque, 1815)
Bullea Blainville, 1825
(non Rafinesque, 1815)
Quibulla Iredale, 1929
Vesica Swainson, 1840
LEUCOPHYSEMA Dall, 1908

HAMINEIDAE (Atyidae) (5b, 6)
HAMINEINAE
HAMINOEA Turton & Kingston in Carrington, 1830
Haminea Leach in Gray, 1847
Haminella Thiele, 1925
Haminaea Zilch in Wenz, 1959-60
ALICULASTRUM Pilsbry, 1896
Alicula Ehrenberg, 1831
(non Eichwald, 1830)
Spissitydeus Iredale, 1936 ?
ATYS Montfort, 1810
Naucum Schumacher, 1817
AUSTROCYLICHNA Burn, 1974
CYLICHNIUM Dall, 1908
Volvulopsis Schepman, 1913
Volutopsis Habe, 1955 (err.)
DINIATYS Iredale, 1936
Dinia H. & A. Adams, 1854
(non Walker, 1854 - prior)
HALOA Pilsbry, 1921 (24)
Penthominea Iredale, 1929
Parahaminoea Kuroda & Habe, 1952 ?

HALOA (Cont'd.)
Vitreohaminoea Kuroda & Habe, 1952
SERICOHAMINOEA Habe, 1952
HAMINEOBULLA Habe, 1950
LAMPROHAMINOEA Kuroda & Habe, 1952
LILOA Pilsbry, 1921
LIMULATYS Iredale, 1936
Tepidatys Iredale, 1936
MICRATYS Habe, 1952
MIMATYS Habe, 1952
NIPPONATYS Kuroda & Habe, 1952
ROXANIELLA Monterosato, 1884
SPHAERATYS F. Nordsieck, 1972 ?
WEINKAUFFIA A. Adams, 1858

BULLACTINAE
BULLACTA Bergh, 1901
Atyscaphander Annandale, 1924

SMARAGDINELLIDAE (21)
SMARAGDINELLA A. Adams, 1848
Glauconella Gray, 1850
Linteria A. Adams, 1850
Nona H. & A. Adams, 1854 ?
Thecaphorus H. & A. Adams, 1854
PHANEROPHTHALMUS A. Adams, 1850
Xanthonella Gray, 1850
Lathophthalmus Pruvot-Fol, 1939
pro Cryptophthalmus Ehrenberg, 1831
(non Rafinesque, 1814)

RETUSIDAE
RETUSA Brown, 1827
Utriculus T. Brown, 1844
(non Schumacher, 1817)
Mamilloretusa F. Nordsieck, 1972
CYLICHNINA Monterosato, 1884
Acrotrema Cossmann, 1889
Cylichnania Marwick, 1931
Ventomnestia Iredale, 1936
DECORIFER Iredale, 1937 (29)
PYRUNCULUS Pilsbry in Tryon & Pilsbry, 1895
Sao H. & A. Adams, 1854
(non Billberg, 1820)
SEMIRETUSA Thiele, 1925
RELICHNA Rudman, 1971
SULCORETUSA Burch, 1945
pro Sulcularia Dall, 1921
(non Rafinesque, 1831)
VOLVULELLA Newton, 1891
Volvula A. Adams, 1862
(non Gistl, 1848)
Rhizorus Montfort, 1810 ?
PARAVOLVULELLA Harry, 1967

RUNCINOIDEA
RUNCINIDAE (Peltidae)
RUNCININAE
RUNCINA Forbes & Hanley, 1851
Pelta Quatrefages, 1844 (suppressed)
(non Beck, 1837)
RUNCINIDA Burn, 1963
RUNCINELLA Odhner, 1924
RUNNICA Miller & Rudman, 1968
METARUNCINA Baba, 1967

ILBINAE
ILBIA Burn, 1963
PSEUDOILBIA Miller & Rudman, 1968

ILDICIDAE
ILDICA Bergh, 1889
LAPINURA Marcus & Marcus, 1970

Order Acochlidioidea

EUACOCHLIDIOIDEA (?)
ACOCHLIDIIDAE (Freshwater)
ACOCHLIDIUM Strubell, 1892

HEDYLOPSIDAE
HEDYLOPSIS Thiele, 1931
Hedyle Bergh, 1895
(non Guenée, 1857)
PSEUDUNELA Salvini-Plawen, 1973
STRUBELIA Odhner, 1937

MICROHEDYLIDAE
MICROHEDYLE Hertling, 1930
MANCOHEDYLE Salvini-Plawan, 1973
PARAHEDYLE Thiele, 1931 ?
GANITUS Er. Marcus, 1953
PARAGANITUS Challis, 1968
UNELA Er. Marcus, 1953

PLATYHEDYLIDAE
PLATYHEDYLE Salvini-Plawen, 1973

TANTULIDAE ? (freshwater) (20)
TANTULUM Rankin, 1979

Placement ? (Various families) (20)
ANADEMARIA, ANTEMNELLA, ASPERSPINA,
GASTROHEDYLE, LIVORNIELLA,
MARAUNIBINA, PALLIOHEDYLE,
SABULINCOLA, STELLASPINA all Rankin, 1979

PHILINOGLOSSOIDEA (3)
PHILINOGLOSSIDAE
PHILINOGLOSSA Hertling, 1932
ABAVOPSIS Salvini-Plawan, 1973
SAPHA Er. Marcus, 1959

PLUSCULIDAE
PLUSCULA Er. Marcus, 1953

Order Sacoglossa (Ascoglossa)

CYLINDROBULLIDAE
CYLINDROBULLA Fischer, 1857
ASCOBULLA Ev. Marcus, 1972

VOLVATELLIDAE (Artessidae)
VOLVATELLA Pease, 1860
Artessa Evans, 1950 (part)

JULIIDAE
JULIINAE
JULIA Gould, 1862
Prasina Deshayes, 1886 (? 1863)

BERTHELINIINAE
BERTHELINIA Crosse, 1875 (= ? Edenttelina - 8a)
TAMANOVALVA Kawaguti & Baba, 1959
EDENTTELLINA Gatliff & Gabriel, 1911
MIDORIGAI Burn, 1960

OXYNOIDAE (Lobigeridae)
OXYNOE Rafinesque, 1814
Oxinoe Blainville, 1824
Icarus Forbes, 1844
Lophocercus Krohn, 1847
ARTESSA Evans, 1950 (Arthessa ?) (29)
LOBIGER Krohn, 1847
Pterygophysis Fischer, 1883
Dipterophysis Pilsbry, 1896 ?

OXYNOIDAE
LOBIGER (Cont'd.)
LOPHOPLEURELLA Zilch, 1956
pro Lophopleura Thiele, 1912
(non Ragonet, 1891)
ROBURNELLA Marcus, 1982

GASCOIGNELLIDAE ? (13)
GASCOIGNELLA Jensen, 1985

CALIPHYLLIDAE (Polybranchidae)
CALIPHYLLA Costa, 1867
Beccaria Trinchese, 1870
Lobifera Pease, 1860 (invalid)
CYERCE Bergh, 1871
LOBIANCOIA Trinchese, 1881
MOURGONA Marcus, 1970
PHYLLOBRANCHILLUS Pruvot-Fol, 1933
PHYLLOBRANCHOPSIS Eliot in
Cockerell & Eliot, 1905
Olea Agersborg, 1923 ?
POLYBRANCHIA Pease, 1860
Phyllobranchus Alder & Hancock, 1864
(non Girard, 1850)

STILIGERIDAE (Hermaeidae) (8)
STILIGERINAE
STILIGER Ehrenberg, 1831 (26)
Calliopaea Orbigny, 1837
Custiphorus Deshayes, 1864
Ercolania Trinchese, 1872
ALDERELLA Odhner in Franc, 1968
ALDERIA Allman, 1846
Canthopsis Agassiz, 1850
ALDERIOPSIS Baba, 1968
APLYSIOPSIS Deshayes, 1864
COSTASIELLA Pruvot-Fol, 1951
HERMAEA Lovén, 1844 (Aplysiopsis ?)
Physopneumon Da Costa, 1864
PLACIDA Trinchese, 1872
Laura Trinchese, 1873 (nom.nud.)
HERMAEINA Trinchese, 1874
HERMAEOPSIS Da Costa, 1869

LIMAPONTIINAE
LIMAPONTIA Johnston, 1836
Chalidis Quatrefages, 1844
Pontolimax Creplin, 1848
ACTEONIA Quatrefages, 1844
Ictis Alder & Hancock, 1847
(non Kaup, 1829)
Cenia Alder & Hancock, 1848
Lafontia Locard, 1886

ELYSIIDAE
ELYSIA Risso, 1818
Actaeon Oken, 1815
(non Montfort, 1810)
Aplysiopterus Delle Chiaje, 1829
Rhizobranchus Philippi, 1853 ?
Pterogasteron Pease, 1860
Hydropsyche Kelaart in Pease, 1871
Thuridilla Bergh, 1872
Elysiella Verrill, 1872
TRIDACHIA Deshayes, 1857
TRIDACHIELLA McFarland, 1924
ELYSIOBRANCHUS Pruvot-Fol, 1930
PATTYCLAYA Marcus, 1982
PLAKOBRANCHUS Hasselt, 1824

BOSELLIDAE (Marcus, 1982)
BOSELLIA Trinchese, 1891

Order Anaspidea

AKERIDAE (Aceridae)
AKERA O. F. Müller, 1776
Vitrella Swainson, 1840
Eucampe Leach in Gray, 1847
Aceras Locard, 1886

Order Aplysiomorpha (10)

APLYSIOIDEA (7)
APLYSIIDAE
APLYSIINAE
APLYSIA Linnaeus, 1767
Laplysia Linnaeus,1767 (Laplisia - err.)
Tethys Linnaeus, 1758 (ICZN 200) (29)
Tetis Renier, 1807 (err.)
Esmia Leach in Gray, 1847
Aclesia Adams, 1854 (7)
Aplysiopsis Bergh, 1898
(non Deshayes, 1864 ?)
Tullia Provot-Fol, 1933
NEAPLYSIA Cooper, 1863
PRUVOTAPLYSIA Engel, 1936
PHYCOPHILA A. Adams, 1861
Placobranchus A. Adams, 1854
VARRIA Eales, 1960
SYPHONATA H. & A. Adams, 1854
Siphonotus A. Adams & Reeve, 1848
(non Brandt, 1836)
Syphonpyge Bronn, 1863
Paraplysia Pilsbry, 1895

DOLABELLINAE
DOLABELLA Lamarck, 1801

DOLABRIFERINAE
DOLABRIFERA Gray, 1847
Thallepus Swainson, 1840 ?
Dolabrifer Fischer, 1883
PETALIFERA Gray, 1847
Aplysiella P. Fischer, 1872
PSEUDAPLYSIA Pilsbry, 1896
PHYLLAPLYSIA P. Fischer, 1872

NOTARCHIDAE ?
NOTARCHINAE
NOTARCHUS Cuvier, 1817

STYLOCHEILINAE
STYLOCHEILUS Gould, 1852

BURSATELLINAE
BURSATELLA Blainville, 1817
Aclesia Rang, 1828 ?
Barnardaclesia Eales & Engel, 1935 (7)
Ramosaclesia Cheesman & Iredale, 1929 ?

Order Notaspidea

UMBRACULOIDEA
UMBRACULIDAE
UMBRACULUM Schumacher, 1817
Patella Lightfoot, 1786
(non Linnaeus, 1758)
Acado Lamarck, 1801
(non Commercon, 1792)
Gastroplax Blainville, 1819
Umbrella Lamarck, 1819
Ombrella Blainville, 1824
? Spiricella Rang, 1827
Umbella Orbigny, 1841
Operculatum H. & A. Adams, 1854

UMBRACULUM (Cont'd.)
HYALOPATINA Dall, 1889 ? (24)
Bertinia Jousseaume, 1883

TYLODINIDAE (Willan, 1987)
TYLODINA Rafinesque, 1819
Parmophorus Cantraine, 1835
(non Blainville, 1817)
Joannisia Monterosato, 1884
Tylodinella Mazzarelli, 1897
ANIDOLYTA Willan, 1987
Tylodina Lovén, 1846
(non Rafinesque, 1819)
Roya Bertsch, 1980
(non Iredale, 1912)

PLEUROBRANCHOIDEA (27)
PLEUROBRANCHIDAE
Tribe Pleurobranchini
PLEUROBRANCHUS Cuvier, 1804
Discoides Renier, 1804
Gervisia, Westernia Quoy & Gaimard in Blainville, 1827
Oscaniella Bergh, 1897
Oscanius Leach in Gray, 1847
Osceanius Pruvot-Fol, 1954 ?
Susania Gray, 1857
Pseudolibania Stefani & Paytanelli, 1879
Isselia Bourguignat, 1877 ?
(non Schmeltz, 1874)
Arthuria Servain, 1891 ?
(non Dall, 1879)
Tribe Berthellini
BERTHELLA Blainville, 1825
Cleanthus Leach in Gray, 1847
Bouvieria Vayssière, 1896
Gymnotoplax Pilsbry, 1896
Berthellinops Burn, 1962
BATHYBERTHELLA Willan, 1983
BERTHELLINA Gardiner, 1936
Berthella Vayssière, 1896
(non Blainville, 1825)
PLEUREHDERA Marcus & Marcus, 1970

PLEUROBRANCHAEINAE
PLEUROBRANCHAEA Meckel in Leue, 1813
Pleurobranchidium Blainville, 1825
Cyanogaster Blainville, 1825
Koonsia Verrill, 1882
Pleurobranchillus Bergh, 1892
Macfarlandaea Ev. Marcus & Gosliner, 1984
EUSELENOPS Pilsbry, 1896
Neda H. & A. Adams, 1854
(non Mulsant, 1851)
Oscaniopsis Bergh, 1897
PLEUROBRANCHELLA Thiele, 1925
Gigantonotum Guang-Yu & Si, 1965
Pleurobranchoides O'Donoghue, 1930 (24)

Order Thecosomata
Suborder Euthecosomata

LIMACINIDAE (Spiratellidae)
LIMACINA Bosc, 1817
Spiratella Blainville, 1817
Limacina Lamarck, 1819
Heliconoides (pars) Orbigny, 1836
Heterofusus Fleming, 1823
Spirialis (pars) Eydoux & Souleyet, 1840
Helicophora Gray, 1842
Scaea Philippi, 1844
Valvatina Bornemann, 1855; Mörch, 1874

LIMACINIDAE
LIMACINA (Cont'd.)
Protomedea (pars) Costa, 1861
(non Blainville, 1830)
Embolus Jeffreys, 1869
(non Selenka, 1867)
MUNTHEA Van der Spoel, 1967
THILEA Strebel, 1908 (Thilia, Thielea -err.)

CAVOLINIIDAE (Cuvieriidae)
CAVOLINIINAE
CAVOLINIA Abildgaard, 1791
Cavolina Bruguière, 1791
Anomia Forskål, 1775
(non Linnaeus, 1758)
Rheda Humphrey, 1797
Hyalaea Lamarck, 1799
Archonta Montfort, 1810
Tricla Oken, 1815
Pleuropus Eschscholtz, 1825
(non Rafinesque, 1815)
Cavolinia Menke, 1844
Potamius Gistl, 1848
Fabius Gistl, 1848
(non Duncan, 1837)
Orbignyia A. Adams, 1859
DIACAVOLINIA Van der Spoel, 1987 (26x)
DIACRIA J. E. Gray, 1842
Pleuropus Pfeffer, 1879
(non Eshscholtz, 1825)

CLIONAE
CLIO Linnaeus, 1767
Cleodora Péron & Leseuer, 1810
Balantium Children in Gray, 1823
Flabellum, Poculina Bellardi, 1872
Euclio Bonnevie, 1914
Proclio Hubendick, 1951 ?
CRESEIS Rang, 1828
Crisia Menke, 1844
Boasia C. W. Johnson, 1934
HYALOCYLIS Fol, 1875 (Hylocyclis - err.)
Hyalocylix P. Fischer, 1883
STYLIOLA Lesueur, 1825 (? Blainville, 1827)

CUVIERININAE
CUVIERINA Boas, 1886
Cuvieria Rang, 1825
(non Lesueur, 1807)
Herse Gistl, 1848
(non Oken, 1815)
Hyperia Gistl, 1848
(non Latreille,1829)
Rangistela Pruvot-Fol, 1946

Suborder Pseudothecosomata

PERACLIDAE (Procymbuliidae)
PERACLE Forbes, 1844
Campylonaus Gray, 1847
(non Benson, 1835)
Euromus H. & A. Adams, 1858
Peraclis Pelseneer, 1888
PROCYMBULIA Meisenheimer, 1905

CYMBULIIDAE
CYMBULIINAE
CYMBULIA Péron & Lesueur, 1810

GLEBINAE
GLEBA Forskål, 1776 (29)
Tiedemannia Chiaje in Benedin, 1839
(non Leuckart, 1830)

GLEBINAE (Cont'd.)
COROLLA Dall, 1871
Cymbuliopsis Pelseneer, 1888

DESMOPTERIDAE
DESMOPTERUS Chun, 1889

Order Gymnosomata

PNEUMODERMATIDAE
PNEUMODERMA Roissy, 1805
Cirrifer Pfeffer, 1879
ABRANCHAEA Zhan Fu-Sui, 1975
CRUCIBRANCHAEA Pruvot-Fol, 1942
PLATYBRACHIUM Minichev, 1976
PNEUMODERMOPSIS Bronn in
Bronn & Keferstein, 1862
Dexiobranchaea Boas, 1885
PRUVOTELLA Pruvot-Fol, 1932 ?
SCHIZOBRACHIUM Meisenheimer, 1903 (24, 3)
SPONGIOBRANCHAEA Orbigny, 1836
Cliodita (pars) Quoy & Gaimard, 1824
Trichocyclus Eschscholtz, 1825 ?

NOTOBRANCHAEIDAE
NOTOBRANCHAEA Pelseneer, 1886
MICRODONTA Bonnevie, 1914

CLIOPSIDAE
CLIOPSIS Troschel, 1854
Clionopsis Bronn, 1862

CLIONIDAE
CLIONE Pallas, 1774
Clionina Provot-Fol, 1924 ?
FOWLERINA Pelseneer, 1906
PAEDOCLIONE Danforth, 1907
PARACLIONE Tesch, 1903 (24)
PRIONOGLOSSA Tesch, 1950 (1951 ?)
THALASSOPTERUS Kwietniewski, 1910

THLIPTODONTIDAE
THLIPTODON Boas, 1886
Pteroceanus Meisenheimer, 1902
CEPHALOBRANCHIA Bonnevie, 1914

HYDROMYLIDAE (Anopsiidae)
HYDROMYLES Gistl, 1848
Anopsia Gistl, 1848 ?
Psyche Rang, 1825
(non Linnaeus, 1758)
Euribia Rang, 1827
(non Meigen, 1800)
Philopsuedes Gistl, 1848
Halopsyche Bronn, 1862
Theceurybia Bronn, 1862
Verrillopsyche Cossmann, 1900

LAGINIOPSIDAE
LAGINIOPSIS Pruvot-Fol, 1922

Order Nudibranchia
Suborder Doridoidea

GNATHODORIDOIDEA (16)
DORIDOXIDAE
DORIDOXA Bergh, 1899
DORIDOIDES Eliot & Evans, 1908 ? (24)
Doridomorpha Eliot,1906
(non Audouin & Milne-Edwards,1832)

BATHYDORIDIDAE
BATHYDORIS Bergh, 1884

ANADORIDOIDEA

CORAMBIDAE

CORAMBE Bergh, 1869
Hypobrancheiaea A. Adams, 1847 ?
DORIDELLA Verrill, 1870
Corambella Balch, 1899

GONIODORIDIDAE

GONIODORIS Forbes & Goodsir, 1839
Pelagella Gray, 1850
LOPHODORIS G. O. Sars, 1878
TESHIA Edmunds, 1966

OKENIIDAE

OKENIINAE

OKENIA Leuckart & Bronn in Menke, 1830
Idalia Leuckart, 1828
(non Hübner, 1820)
Idalina Norman, 1890
Cargoa Vogel & Schultz, 1970
IDALIELLA Bergh, 1881
Idalla Oersted, 1844 (suppr.)

ANCULINAE

ANCULA Lovén, 1846
Miranda Alder & Hancock, 1847
BERMUDELLA Odhner, 1941
SPHARIA Risbec, 1928
TRAPANIA Pruvot-Fol, (June) 1931 (22d)
Drepanida MacFarland, (July) 1931
pro Drepania Lafont, 1874
(non Hübner, 1816)
Drepaniella Burn, 1961 ?

HOPKINSIINAE

HOPKINSIA MacFarland, 1905
HOPKINSIELLA Baba, 1938

ONCHIDORIDIDAE (Lamellidoridae) (18)

ONCHIDORIS Blainville, 1816
Onchidiorus Férussac, 1821
Onchidora Cuvier, 1830
Villiersia Orbigny, 1837
(non Locard, 1886)
Oncodoris Agassiz, 1847
Oncidiodoris Agassiz, 1847
Oncidoris Herrmannsen, 1847
Oicodespina Gistl, 1848
Proctaporia Mörch, 1857
ATALADORIS Iredale & O'Donoghue, 1923
DIAPHORODORIS Iredale & O'Donoghue, 1923
ACANTHODORIS Gray, 1850
ACIODORIS Bergh, 1879
ADALARIA Bergh, 1878
ARCTADALARIA Roginskaya, 1971 (18)
ARMODORIS Minichev, 1972
CALYCIDORIS Abraham, 1876
DORIDUNCULUS G. O. Sars, 1878
LAMELLIDORIS Alder & Hancock, 1876
Ancylodoris W. Dybowski, 1900 (24)
PRODORIDUNCULUS Thiele, 1912

TRIOPHIDAE

TRIOPHINAE

TRIOPHA Bergh, 1880
CRIMORA Alder & Hancock, 1862
HOLOPLOCAMUS Odhner, 1926
JOUBINIOPSIS Risbec, 1928
KALOPLOCAMUS Bergh, 1893 (Caloplocamus, err.)
Euplocamus Philippi, 1836
(non Latreille, 1809)
Heteroplocamus Oliver, 1915
LAILA MacFarland, 1905 (Polyceridae, 6x)

KALINGINAE

KALINGA Alder & Hancock, 1864
BRACHYCHLANIS Ehrenberg, 1831
Brachychlamys Gray, 1847
Brachylanis O'Donoghue, 1929

NOTODORIDIDAE (Aegiretidae) (25)

NOTODORIS Bergh, 1875
AEGIRES Lovén, 1844
SERIGEA F. Nordsieck, 1984 ?
TRIOPELLA G. O. Sars, 1878

POLYCERIDAE (Polyceratidae)

POLYCERA Cuvier, 1817
Themisto Oken, 1815 (non 1807)
Cufaea Leach, 1852
EUPHURUS Rafinesque, 1815 ?
Triopa Johnston, 1838
FOLIETTA F. Nordsieck, 1972 ?
GALACERA Risso-Dominguez, 1960 (24)
GREILADA Bergh, 1894
Ohola Bergh, 1883
ISSENA Iredale & O'Donoghue, 1923 ?
? Colga Bergh, 1880 (Valid genus - 6x)
(non Kolga Danielssen
& Koren, 1879)
Issa Bergh, 1880
(non Walker, 1867)
LIMACIA O. F. Möller, 1781 ?
PALIO GRAY, 1857
PLOCAMOPHERUS Leuckart,1928 (10)(Triophinae, 6x)
Plocamocerus Cuvier, 1830
Peplidia Lowe, 1842
Histiophorus Pease, 1860
Plocamophorus Rüppell & Leuckart, 1931
POLYCERELLA Verrill, 1881
THECACERA Fleming, 1828

GYMNODORIDIDAE

GYMNODORIS Stimpson, 1855
Trevelyana Kelaart, 1858
Trevelyania P. Fischer, 1883
RHODIGINA Bergh, 1880 ?
ANALOGIUM Risbec, 1928
ANGASIELLA Crosse, 1804
LECITHOPHORUS Macnae, 1958
NEMBROTHA Bergh, 1877
Angasiella Crosse, 1864 (Invalid)
PALIOLLA Burn, 1958
ROBOASTRA Bergh, 1877
TAMBJA Burn, 1962

VAYSSIEREIDAE (Okadaiidae)

VAYSSIEREA Risbec, 1928
OKADAIA Baba, 1930

EUDORIDOIDEA (Cryptobranchia ?)

HEXABRANCHIDAE

HEXABRANCHUS Ehrenberg, 1831
Heptabranchus A. Adams, 1849 ?

CADLINIDAE

CADLINA Bergh, 1878
Acanthochila Mörch, 1868 (suppressed)
Ectinochila Mörch, 1869 (suppr.)
Juanella Odhner, 1921
CADLINELLA Thiele, 1931

CHROMODORIDIDAE

CHROMODORIS Alder & Hancock, 1855 (nom.cons.)
Felimida Marcus, 1971
ARDEADORIS Rudman, 1984
BABAINA Odhner in Franc, 1968 (= Thorunna ?)

CHROMODORIDIDAE (Cont'd.)
CERATODORIS Gray, 1850
Echinodoris Bergh, 1878 ?
CERATOSOMA Adams & Reeve, 1850 ?
CHROMODORIDELLA Eliot, 1905 ?
DIGIDENTIS Rudman, 1984
DURVILLEDORIS Rudman, 1984
FELIMARE Marcus & Marcus, 1967
GLOSSODORIS Ehrenberg, 1831 (nom.cons.) (22f)
Actinodoris Ehrenberg, 1831 (Suppr.)
Doriprismatica Orbigny, 1839
Casella H. & A. Adams, 1854
Goniobranchus Pease, 1866 ?
Crepidoris Pagenstecher, 1877
Rosodoris Pruvot-Fol, 1954
Chromolaichma Bertsch, 1977
GRUVELIA Risbec, 1928
HYPSELODORIS Stimpson, 1855 (nom.cons.)
Pterodoris Ehrenberg, 1831 (Suppr.)
MEXICHROMIS Bertsch, 1977
NOUMEA Risbec, 1928 ?
OTINODORIS White, 1948
PECTENODORIS Rudman, 1984
THORUNNA Bergh, 1878
TYRINNA Bergh, 1898 ?

ACTINOCYCLIDAE
ACTINOCYCLUS Ehrenberg, 1831
Sphaerodoris Bergh, 1877
HALLAXA Eliot, 1909
Halla Bergh, 1878
(non A. Costa, 1844)

MIAMIRIDAE
MIAMIRA Bergh, 1874
ORODORIS Bergh, 1875
GRAVIERIA Vayssière, 1912

CONUALEVIIDAE (16)
CONUALEVIINAE
CONUALEVIA Collier & Farmer, 1964

INUDINAE
INUDA Marcus & Marcus, 1967

ALDISIDAE
ALDISA Bergh, 1878
APORODORIS Ihering, 1886 (3)

ROSTANGIDAE
ROSTANGA Bergh, 1879
Rhabdochila P. Fischer, 1883 (part)
BOREODORIS Odhner, 1939

DORIDIDAE
DORIDINAE
DORIS Linnaeus, 1758
Doridigitata Orbigny, 1839
Staurodoris Bergh, 1878
AUSTRODORIS Odhner, 1926
DORIORBIS Kay & Young, 1969
PERCUNAS Marcus, 1970
PUPSIKUS Marcus, 1970
SIRAIUS Marcus, 1955
THORYBOPUS Bouchet, 1977

ARTACHAEINAE
ALLIODORIS Bergh, 1904
ARTACHAEA Bergh, 1882
GUYONIA Risbec, 1928 ?

ARCHIDORIDIDAE
ARCHIDORIS Bergh, 1878

ARCHIDORIDIDAE (Cont'd.)
ATAGEMA Gray, 1850
CTENODORIS Eliot, 1907
ETIDORIS Ihering, 1886
PERONODORIS Bergh, 1904
PETELODORIS Bergh, 1882 ?
TRIPPA Bergh, 1878
Phlegmodoris Bergh, 1878

GEITODORIDIDAE
GEITODORIS Bergh, 1892
VERILLIA Ortea, 1981

HOMOIODORIDIDAE
HOMOIODORIS Bergh, 1882

DISCODORIDIDAE
DISCODORIDINAE
DISCODORIS Bergh, 1877
ANISODORIS Bergh, 1898
Montereina MacFarland, 1905
CARRYODORIS Vayssière, 1919
CARMINODORIS Bergh, 1889 (22d)
DIAULULA Bergh, 1879
DYSTYLODORIS Odhner, 1959 (22d)
FRACASSA Bergh, 1878
NIRVA Bergh, 1878
NUVUCA Marcus & Marcus, 1967
PARADORIS Bergh, 1884 (22d)
PELTODORIS Bergh, 1880
PHIALODORIS Bergh, 1876
SEBADORIS Marcus & Marcus, 1960
THORDISA Bergh, 1877

TARINGINAE
TARINGA Er. Marcus, 1955
TAYUVA Marcus & Marcus, 1967

NEODORIDINAE
NEODORIS Baba, 1938

KENTRODORIDIDAE
KENTRODORIS Bergh, 1876 (Centrodoris)
Awuka Er. Marcus, 1955
AUDURA Bergh, 1878
JORUNNA Bergh, 1876

HALGERDIDAE (Asteronotidae) (16)
HALGERDA Bergh, 1880
Dictydoris Bergh, 1880
APHELODORIS Bergh, 1879
ASTERONOTUS Ehrenberg, 1831
SCLERODORIS Eliot, 1904 (16)

PLATYDORIDIDAE
PLATYDORIDINAE
PLATYDORIS Bergh, 1877
Argus Bohadsch, 1761 ?
BAPTODORIS Bergh, 1884 (22d)
GARGAMELLA Bergh, 1894

HOPLODORIDINAE
HOPLODORIS Bergh, 1880
XENODORIS Odhner (?)

PHYLLIDIOIDEA (Porodoridoidea)
PHYLLIDIIDAE (9c)
PHYLLIDIA Cuvier, 1797
PHYLIDIELLA Bergh, 1869 (?)
CERATOPHYLLIDIA Eliot, 1903
PHYLLIDIOPSIS Bergh, 1875
REYFRIA Yonow, 1986
Fryeria J. E. Gray, 1853

DENDRODORIDIDAE
DENDRODORIS Ehrenberg, 1831
Doriopsis Pease, 1860 ?
Rhacodoris Mörch, 1863
Doridopsis Alder & Hancock, 1864
Haustellodoris Pease, 1871
DORIOPSILLA Bergh, 1880

IDULIIDAE ? (34)
IDULIANA O'Donoghue, 1932

Suborder Dendronotoidea

DENDRONOTOIDEA
TRITONIIDAE (Duvaucellidae) (25)
TRITONIA Cuvier, 1798
Tritonia Cuvier, 1803 (non 1798)
Duvaucellia Risso, 1826
Liriope, Necromantes Gistl, 1848
Candiella Gray, 1850
TRITONIDOXA Bergh, 1907
CANDELLISTA Iredale & O'Donoghue, 1923 (24)
MICROLOPHUS Rochebrune & Mabille, 1889 (24)
SPHAEROSTOMA MacGillivray, 1843 (24)
Cloelia Lovén, 1841
(non Fitzinger,1833)
MARIONIA Vayssière, 1877
TOCHUINA Odhner, 1963
TRITONIELLA Eliot, 1907
TRITONOPSIS Eliot, 1905

MARIANINIDAE
MARIANINA Pruvot-Fol, 1930
ARANUCUS Odhner, 1936
DIPLOPELYCIA Mörch, 1872

LOMANOTIDAE
LOMANOTUS Verany, 1844
Eumenis Alder & Hancock, 1845
(non Hübner, 1818)

SCYLLAEIDAE
SCYLLAEA Linnaeus, 1758
NOTOBRYON Odhner, 1936
CROSSLANDIA Eliot, 1902
Nerea Lesson, 1830 ?

HANCOCKIIDAE
HANCOCKIA Gosse, 1877
Govia Trinchese, 1885

DENDRONOTIDAE
DENDRONOTUS Alder & Hancock, 1845
Amphitrite Ascanius,1774
(non Möller, 1771)
Amphitritidea Kroyer, 1847
Campaspe Bergh, 1863

BORNELLIDAE
BORNELLA Adams & Reeve, 1848
BORNELLOPSIS O'Donoghue, 1929
PSEUDOBORNELLA Baba, 1932 (24)

TETHYIDAE
TETHYS Linnaeus, 1758
Fimbria Bohadsch, 1761 (invalid)
Phoenicurus Rudolphi, 1819 ? (24)
(non Förster, 1817)
Vertumnus Otto, 1823 ?
MELIBE Rang, 1829
Chioraera Gould, 1852
Jacunia Filippi, 1867
PROPEMELIBE Allan, 1932

DOTIDAE
DOTO Oken, 1815
Idulia Leach, 1852
Dotona Iredale, 1918
Iduliella Thiele, 1931
pro Dotilla Bergh, 1878
Heromorpha Bergh, 1873
Caecinella Bergh, 1870
GELLINA Gray, 1850
MIESEA Er. Marcus, 1961

PHYLLIROIDAE
PHYLLIROE Péron & Lesueur, 1810 (Phyllirrhoe, em.)
Eurydice Eschscholtz, 1825
Phylirine Menke, 1844
ACURA H. & A. Adams, 1855
CEPHALOPYGE Hanel, 1905
Boopsis Pierantoni, 1923
CTILOPSIS Andre, 1906 ? (24)
NECTOPHYLLIRRHOE H. Hoffmann, 1922 (24)
Bonneviia Pruvot, 1929
pro Dactylopus Bonnevie, 1921
(non Gill, 1859)

Suborder Arminoidea

ARMINOIDEA (Euarminoidea) (25)
HETERODORIDAE
HETERODORIS Verrill & Emerton, 1882
ARCTOLEURA Odhner ? 1968 (3)
ATTHILA Bergh, 1899 (10) ?

ARMINIDAE ?
ARMINA Rafinesque, 1814
Pleurophyllidia Mekkel, 1816
Diphyllidia Cuvier, 1817
LINGUELLA Férussac, 1822
SANCARA Bergh, 1861
HISTIOMENA Mörch, 1859
Camarga Bergh, 1866
PLEUROPHYLLIDIELLA Eliot, 1903
DERMATOBRANCHUS Hasselt, 1924
PLEUROLEURA Bergh, 1874
Dermatobranchopsis Baba, 1949

LEMINDIDAE (10)
LEMINDA Griffiths, 1985

METARMINOIDEA
MADRELLIDAE
MADRELLA Alder & Hancock, 1864
ELIOTIA Vayssière, 1909

DIRONIDAE
DIRONA MacFarland in Cockerell & Eliot, 1905

ZEPHYRINIDAE (18x) (Janolidae)
ZEPHYRINA Quatrefages, 1844
Venilia Alder & Hancock, 1844
(non Godart, 1829)
PROCTONOTUS Alder, 1844
JANOLUS Bergh, 1884
Janus Verany, 1844
(non Stephens, 1835)
Antiopella Hoyle, 1902
Antiopa Alder & Hancock, 1848
(non Meigen,1800)
Bonisa Gosliner, 1981
CALDUKIA Burn & Miller, 1969 (5c)
GALEOJANOLUS Miller, 1971

GONIAEOLIDIDAE
GONIAEOLIS M. Sars, 1861

CHARCOTIIDAE
CHARCOTIA Vayssière, 1906
PSEUDOTRITONIA Thiele, 1912
TELARMA Odhner, 1934

HEROIDAE
HERO Alder & Hancock, 1855

Suborder Aeolidioidea
EUAEOLIDIOIDEA
NOTAEOLIDIIDAE
NOTAEOLIDIA Eliot, 1905

CORYPHELLIDAE
CORYPHELLA Gray, 1850
RIOSELLEOLIS Ortea, 1979
CHLAMYLLA Bergh, 1886
CORYPHELLINA O'Donoghue, 1929
HIMATINA Thiele, 1931 ?
Himatella Bergh, 1891 (non Zittel, 1879)

PARACORYPHELLIDAE ?
PARACORYPHELLA Miller, 1971

NOSSIDAE
NOSSIS Bergh, 1902
Eocoryphella, Hespereolis Odhner (n.nd.)

FLABELLINIDAE (? Aeolidiidae)
FLABELLINA Voigt, 1834
CALMELLA Eliot, 1906
SAMLA Bergh, 1900
TULARIA Burn, 1966
FLABELLINOPSIS MacFarland, 1966

PROTOAEOLIDIELLIDAE
PROTAEOLIDIELLA Baba, 1955

PLEUROLIDIIDAE
PLEUROLIDIA Burn, 1966

EUBRANCHIDAE
EUBRANCHINAE
EUBRANCHUS Forbes, 1838
Amphorina Quatrefages, 1844
Galvina Alder & Hancock, 1855
Capellinia Trinchese, 1874 ? (2)
Dunga Eliot, 1902
Egalvina Odhner, 1929
Eubranchopsis Baba, 1949
GALVINELLA Eliot, 1907

CUMANOTINAE
CUMANOTUS Odhner, 1907

PSEUDOVERMIDAE
PSEUDOVERMIS Periaslavzeff, 1891

TERGIPEDIDAE (1) (Cuthonidae) (16)
TERGIPES Cuvier, 1805
Psiloceros Menke, 1844
CATRIONA R. Winkworth, 1941
EURYCATRIONA Burn, 1963
Montagua Fleming, 1822 (non Leach, 1814)
CUTHONA Alder & Hancock, 1855
Selva Edmunds, 1964
Guyvalvoria Vayssière, 1906
Precuthona Odhner, 1929
Indocratena Odhner, 1940
Diaphorealis Iredale & O'Donoghue, 1923
Cratenopis Lemche, 1936
Hervia Bergh, 1871 (2)
Cuthonella Bergh, 1884 (18a)

TERGIPEDIDAE (Cont'd.)
EMBLETONIA Alder & Hancock, 1851
Pterochilus Alder & Hancock, 1844
(non Klug, Weber & Mohr, 1805)
MYJA Bergh, 1896
PHESTILLA Bergh, 1874
PISEINOTECUS Er. Marcus, 1955 (23)
TENELLIA A. Costa, 1866
TRINCHESIA Ihering, 1879 (1)
Xenocratena Odhner, 1940
Subcuthona Baba, 1947
Njurja Ev. & Er. Marcus, 1960
Narraeolidia Burn, 1961
Toorna Burn, 1964
ZATTERIA Eliot, 1902

FIONIDAE
FIONA Alder & Hancock 1855
Oithona Forbes & Hanley, 1851
(non Baird, 1843)
Hymenaeolis A. Costa, 1866
DOLICHEOLIS Finlay, 1927 ?

CALMIDAE
CALMA Alder & Hancock, 1855
Forestia Trinchese, 1881

GLAUCIDAE (3)
GLAUCINAE
GLAUCUS Förster, 1777
Eucharis Péron, 1807
Laniogerus Blainville, 1816
Dadone, Nausimacha Gistl, 1848
GLAUCILLA Bergh, 1860

BABAKININAE
BABAKINA Roller, 1973
pro Babaina Roller, 1972
(non Franc, 1968)

FACELINIDAE (16)
FACELININAE
FACELINA Alder & Hancock, 1855 (25)
Acanthopsole Trinchese, 1888
Learchis Bergh, 1896
Liopsole Odhner, ?
ANTIONETTA Schmekel, 1966
PHIDIANA Gray, 1850 (3)
Hermissenda Bergh, 1879 ?
Caloria Trinchese, 1888 (25)
Moridilla Bergh, 1888
Facelinella Baba, 1949
Rolandia Pruvot-Fol, 1951
Facelinopsis Pruvot-Fol, 1954
Palisa Edmunds, 1964
Pruvotfolia Tardy, 1969
Emarcusia Roller, 1972

FAVORININAE (Facalaninae)
FAVORINUS M. E. Gray, 1850
Matherena Bergh, 1850
ARMANDA Macnae, 1954
AUSTRAEOLIS Burn, 1962
BABIELLA Risso-Dominguez, 1965
BAJAEOLUS Gosliner & Behrens, 1986 (9b)
DICATA Schmekel, 1967
DONDICE Er. Marcus, 1958
ECHINOPSOLE Macnae, 1954
FACALANA Bergh, 1888
GLOBIFERINA Risbec, 1937
GODIVA Macnae, 1954
HERMOSITA Gosliner & Behrens, 1986 (9b)
JASON Miller, 1974

FAVORININAE (Cont'd.)
NANUCA Er. Marcus, 1957
NOUMEAELLA Risbec, 1937
PHYLLODESMIUM Ehrenberg, 1831
PHYLLODESMOPSIS Risso-Dominguez, 1964
SAKURAEOLIS Baba in Baba & Hamatani, 1965
SETOEOLIS Baba in Baba & Hamatani, 1965
SHINAOELIS Baba, 1937

PTERAEOLIDIINAE
PTERAEOLIDIA Bergh, 1875

CRATENINAE ?
CRATENA Bergh, 1864
Cavolina Bruguière, 1792
(non Abildgaard, 1791)
Rizzolia Trinchese, 1877 (3)
ENNOIA Bergh, 1896

HERVIELLINAE (5a)
HERVIELLA Baba, 1949
Muessa Er. Marcus, 1965
MARCIELLA Burn, 1967

AEOLIDIIDAE (Aeolidioidea)
AEOLIDIA Cuvier, 1797
Eolia Cuvier, 1800
Eolis Cuvier, 1805
Eolida Cuvier, 1816
AEOLIDIELLA Bergh, 1867
EOLIDINA Quatrefages, 1843 ?
Ethalion Risso, 1826
(non Aethalion Peletier, 1825)
BAEOLIDIA Bergh, 1888
CERBERILLA Bergh, 1873
LIMENANDRA Haefelfinger & Stamm, 1958 ?
Placement ?
PINUFIUS Marcus, 1959 (Pinufiidae)
FENRISIA Bergh, 1888 (24)
ANTONIETTA Schmekel, 1966

MYRRHINIDAE
MYRRHINE Bergh, 1905

SPURILLIDAE
SPURILLA Bergh, 1864
BERGHIA Trinchese, 1877

References:

Behrens, D. W. "—Tergipedid nudibranch: N.E. Pacific" Veliger 27(1):65-71 - 1984 (1)

Bertsch, H. "Review of Euthyneura in American (2) Seashells" (2nd Ed.) Veliger 18:104-106.1973
—— "The Chromodoridine nudibranchs from the Pacific coast of America" I, II: Chromodoris Veliger 20(2):107-118 - 1977 (2a)
—— "Authorship and Date Citations: California Taxa" Veliger 24(1):49-50 - 1981 (2b)
—— & Kerstich, A. "Distribution and Radular Morphology of various Nudibranchs: Gulf of Mexico" Veliger 26(4):264-273 - 1984 (2c)

Boss, K. J. in Sybil Parker "Synopsis and Classification of Living Organisms" 1982 (3)

Brown, G. H. "British species of Tergipedidae" Zool.J.Lin.Soc.69:225-255 - 1980 (4)

Burn, R. "Description of Australian Eolidacea" (5) J.Malac.Soc.Aust.7:12-20 - 1963
—— Ibid. 9:25-35 - 1966
—— "Revision of the genus Herviella" Malacologia 6(1-2):223-230 - 1967 (5a)
—— "Review: Australian Austrocylichna,Nipponatys etc." J.Mal.Soc.Aust.4(1-2):93-112 -1978 (5b)
—— & M. C. Miller "New genus; Caldukia" J.Mal.Soc.Aust.12:23-31 - 1969 (5c)

Cattaneo-Vietti, R. & M. Sordi "New — Triophidae: Mediterranean" Basteria 52:49-59 - 1988 (6x)

Cernohorsky, W. O."Taxonomy: Indo-Pac. Mollusca" Part 12 Rec.Auck.Mus.22:47-67 - 1985 (6)

Eales, W. "Notes on Cephalaspideans" Opisthobranch Vol.16(3):26 - 1984 (7)

Gascoigne, T. "— Classification: Stiligeridae" J.Mal.Soc.Aust. 3(3-4):157-172 - 1976 (8)
—— "Provisional classification: Order Ascoglossa" Jour.Moll.Stud.51(1):8-22 - 1985 (8a)

Gosliner, T.M."Review:systematics of Cylichnella" Veliger 93(2-3):85-92 - 1979 (9)
—— Philinacea:Aldabra Atoll; new genus" Biol.Soc.Wash.Bull.8:79-100 - 1988 (9a)
—— & D.W. Behrens "New aeolid Nudibranch; trop. E. Pacific" Veliger 29(1):101-113 -1986 (9b)
—— & D. W. Behrens "Generic divisions within Phyllidiidae" Ibid.30(3):305-314 -1988 (9c)

Hamann, J.C. & W.M. Farmer "New species: Plocamopherus: W.Atl".Veliger 31(1-2):68-74.1988(10)

Jaeckle, A. "The Opisthobranchs; Humboldt County, California" Veliger 26(3):207-213 -1984 (12)

Jensen, K. R. "Hong Kong Ascoglossa" in Morton & Dudgern (Eds.)"Malacofauna: H.K. & S.China" Hong Kong U. Press, Hong Kong, 1985 (13)

Kay, E. A. & D. K. Young "Hawaiian Doridacea" Pac.Sci.23:172-231 - 1969 (14)

Marcus, Ev. de B. "Tornatina and related forms" Jour.Mollusc.Stud.Supp.2:1-35 - 1977 (15)
—— & Marcus "Some gastropoda; Madagascar & West Mexico" Malacologia 10(1):181-223-1970 (15a)

McDonald, G. R. "Rev. of Nudibranchs; California coast" Malacologia 24(1-2):114-276 1983 (16)

Mikkelsen, P. & P. "Comparison; Acteocina (17) canaliculata" Veliger 27(2):164-192 - 1984

Millen, S.V. "Onchidoris, Diapodoris: Northeast. Pacific" Veliger 28(1):80-93 - 1985 (18)
—— "Northern — tergiped nudibranchs Can. Pac." Can.Journ.Zool.64(6):1356-1362 - 1986 (18a)

Miller, M.C. & R.C. Willan "Rev.:N.Z.nudibranchs" N.Z.Jour.Zool.13(3):377-408 - 1986 (18x)

Minichev, Yu.S. "Cylichna: Franz Joseph Land"(19) IFML 14(22):428-434 (Veliger Review, 1977)
—— "Pelagic Pneumodermatidae: Antarctic" IFML 18(26:102-106 - 1979 (19a)

Rankin, J.J. "Freshwater shell-less mollusk from the Caribbean: new understanding,Acochlidia" Life Sci.Cont.Roy.Ont.Mus.116:1-123.1979(20)

References: (Opisthobranchia, Cont'd.)
Rudman, W. B. "The genus Bullina in New Zealand" Jour.Mal.Soc.Aust.2(1):195-203 - 1971 (22)
---- "The family Acteonidae in New Zealand" Ibid.2(1):205-214 - 1971 (22a)
---- "The genus Philine" Proc.Malac.Soc.London 40:171-187 -1972 (22b)
---- "Comparison of the genera of Aglajidae" Zool.J.Linn.Soc.London 54:185-212 -1973 (22c)
---- "The Chromodorididae of - Indo-West Pacific" Ibid. 83(3):241-299 - 1985 (22d)
---- "Chromodoridae: genus Glossodoris = Casella" Ibid. 86(2):101-184 - 1986 (22e)
---- "Genus Trapania in the Indo-West Pacific" J.Moll.Stud.53:189-212 - 1987 (22f)

Schmekel, L. "Synopsis of the genus Piseinotecus" Veliger 23(1):21-24 - 1980 (23)
Thiele, J. "Weichtierkunde" - 1936 (24)

Thompson, F.C. & G.H. Brown "British- Nudibranchia" Ray Soc."Biol:Opisthobranch Mollusks"1984 (25)

Thompson, T.E. "Sacoglossan gastropods of E. Aust." Proc.Malac.Soc.London 40:239-251 - 1973 (26)

Van der Spoel, S. "Diacavolinia nov. gen. separated from Cavolinia" 1987 (26x) Bull.Zool.Mus.U.of Amsterdam 11(8)77-79

Willan, R. C. "New Zealand Pleurobranchidae" Malacologia 23(2):221-270 - 1983 (27)

---- "Phylogenetic systematics of Notaspidea" Am.Malac.Bull.5(2):215-241 - 1987 (27a)

Zilch, A. in W. Wenz "Gastropoda" Tiel 2 "Euthyneura" pp.1-63 - 1960 (29)

Subclass Gymnomorpha (24a)
Order Systellommatophora (45h)

ONCHIDIOIDEA (6)
ONCHIDIIDAE
ONCHIDIUM Buchanan, 1800
Onchidion, Onchydium, Orchidium,
Oncida, Oncidium, Oncidia auctt. (err.)
Peronia Blainville, 1824 (nom.nud.)
HOFFMANNOLA Strand, 1932
pro Watsonella Hoffmann, 1928
(non Berg, 1898)
LABELLA Starobogatov, 1971
pro Elophilus Labbe, 1933
(non Meigen, 1803)
LESSONINA Starobogatov, 1976
pro Lessonia Labbe, 1934
(non Swainson, 1832)
ONCHIDELLA Gray, 1850
(Oncidiella auctt.- err.)
Arctonchis Dall, 1905
Occidentella Hoffmann, 1929
Peroniella Starobogatov, 1976 ?
pro Peronella Mörch, 1863
(non Gray, 1855)
ONCHIDINA Semper, 1885 (Oncidina)
PARAONCIDIUM Labbe, 1934
PERONIA Fleming, 1822
Onchis Férussac,1822
Oncus, Oncis auctt. (err.)
Eudrastus Gistl, 1848
Paraperonia, Scaphis Labbe, 1934
PERONINA Plate, 1893
PLATEVINDEX H. B. Baker, 1938
pro Oncis Plate, 1893
(non Herrmannsen, 1847)
QUOYELLA Starobogatov, 1976
pro Quoya Labbe, 1934
(non Deshayes, 1843)
SEMPERONCIS Starobogatov, 1976
pro Semperella Labbe, 1934
(non Gray, 1868)

Order Soleolifera (45h)

VERONICELLIDAE (Vaginulidae)
VERONICELLA Blainville, 1817
ANGUSTIPES Colosi, 1922
Monocaulis Colosi, 1921
Belocaulis H. Hoffmann, 1925
COLOSIUS Thome, 1976

VERONICELLIDAE (Cont'd.)
DIPLOSOLENODES Thome, 1976
FILICAULIS Simroth, 1913
Vanigula Grimpe & Hoffmann, 1925
ELEUTHEROCAULIS Simroth, 1913
Leonardia Tapparone-Canefri, 1889 ?
Annulicaulis Simroth, 1913
Laevicaulis Simroth, 1913
Meisenheimeria Grimpe & Hoffmann, 1924
TENACIPES H. B. Baker, 1931
FORCARTULUS Thome, 1976
HETEROVAGINULA Krauss, 1953
IMERINIA Cockerell, 1891
Rhopalocaulis Simroth, 1913
DESMOCAULIS Simroth, 1913 ?
Curticaulis Simroth, 1913
DREPANOCAULIS Simroth, 1913
Prismatocaulis Simroth, 1913
FLAGELLICAULIS Simroth, 1913
Spirocaulis Simroth, 1913
LATIPES Colosi, 1922 ?
LEIDYULA H. B. Baker, 1925
Cylindrocaulus H. Hoffmann, 1925
(non Fairmaire, 1880)
Cylidrocaulidea Strand, 1928
MICROVERONICELLA Thome, 1976
NOVOVAGINULA Thiele, 1931
PHYLLOCAULIS Colosi, 1922
Phyllocaulus H. Hoffmann, 1925 (err.)
POTAMOJANUARIUS Thome, 1976
PSEUDOVERONICELLA Germain, 1908
Cycloprocta, Drepanoprocta,
Pleuroprocta Simroth, 1913
SIMROTHULA Thome, 1976
VAGININA Simroth, 1897
Vaginulopsis Simroth, 1897
VAGINULA Férussac, 1821
SARASINULA Grimpe & Hoffmann, 1924
ZILCHULUS Thome, 1976

RATHOUISIIDAE
RATHOUISIA Heude, 1883
ATOPOS Simroth, 1891
PADANGIA Babor, 1900
Podangia Ghosh, 1912 (err.)
PARAPODANGIA Ghosh, 1915
PRISMA Simroth, 1891

RHODOPIDAE ?
RHODOPE Kolliker, 1847
Sidonia M. Scrultze, 1854

Subclass Pulmonata
Order Archaeopulmonata

ELLOBIOIDEA (Auriculoidea) (Brackish)
ELLOBIIDAE
ELLOBIINAE
ELLOBIUM Röding, 1798
Auricula Lamarck, 1799
Auriculus Montfort, 1810
Marsyas Oken, 1815
Geovula Swainson, 1840
AURICULODES Strand, 1928
pro Auriculina Kobelt, 1898
(non Grateloup, 1838)
Auricella Möllendorff, 1898
(non Jurine, 1817)
ALLOCHROA Ancey, 1887
AURICULASTRA von Martens, 1880
Auriculstra Yen, 1952 (err.)
CASSIDULA Férussac, 1821
Cassidulus Barthold, 1827
(non Lamarck, 1801)
Rhodostoma Swainson, 1840
Sidula Gray, 1840
CASSIDULTA Strand, 1942 (53)
pro Cassidulella Thiele, 1936
(non Orbigny, 1826)
pro Cassidulina Thiele, 1931
(non Strand, 1928)
CYLINDROTIS Möllendorff, 1895
LEUCOPHYTIA Winckworth, 1949 (53)
Jaminia Brown, 1827
(non Risso, 1826)
Leuconia Gray, 1840
(non Grant, 1833)
Jaminea Gray, 1847
Leucopepla Peile, 1926
(non Kirkaldy, 1907)
OPHICARDELUS Beck, 1837
Melodisula Iredale, 1936
PYTHIA Röding, 1798 (Phytia Gray, 1821 - err.)
Polydonte Fischer, 1807
Scarabus Montfort, 1810
Strigula Perry, 1811
Scarabaea Oken, 1817
Scarabaeus Blainville, 1821
(non Linnaeus, 1758)
Polydonta Férussac, 1829
(non Megerle, 1811)
TRIGONOPYTHIA Kobelt, 1898
Holcomphalia Möllendorff, 1898
HENDRIKIA Preston, 1913
SARNIA H. & A. Adams, 1855
Siona H. & A. Adams, 1855
(non Duponchel, 1829)

MELAMPODINAE (Melampinae)
MELAMPUS Montfort, 1810
Conovulus Lamarck, 1816
Melampa Schweigger, 1820
Conovulum Sowerby, 1842
Pira H. & A. Adams, 1855
DETRACIA Gray in Turton, 1840
Tifata H. & A. Adams, 1855
Eusiphorus Conrad, 1863
Microtralia Dall, 1894
AUTONOELLA Wenz, 1947 (53)
pro Autonoe Guppy, 1868
(non Leach, 1852)
MICROMELAMPUS Möllendorff, 1898 ?
SIGNIA H. & A. Adams, 1855
TRALIA Gray in Turton, 1840
Persa H. & A. Adams, 1855

PEDIPEDINAE
PEDIPES Bruguière, 1792 (Férussac, 1821 ?)
Carassa Gistl, 1848 (1850)
APODOSIS Pilsbry & McGinty, 1949
BLAUNERIA Shuttleworth, 1854
LAEMODONTA Philippi, 1846
Lirator Beck, 1837 (nom.nud.) ?
Plecotrema H. & A. Adams, 1853 (1854)
Enterodonta Sykes, 1894
pro Laimodonta H. & A. Adams, 1855
(non Bronn, 1847)
BULLAPEX Haas, 1950
LEUCONOPSIS Hutton, 1884
MARINULA King & Broderip, 1832
Cremnobates Swainson, 1855
Maripythia Iredale, 1936
OVATELLA Bivona, 1832
Ovatilla Gray, 1847 (err.)
MYOSOTELLA Monterosato, 1906
Alexia Leach in Gray, 1847
(non Stephens, 1835)
Monica H. & A. Adams, 1855
Kochia Pallary, 1900
(non Frech, 1888)
Nealexia Wenz, 1920
PSEUDOMELAMPUS Pallary, 1900 ?
RANGITOTOA Powell, 1933

CARYCHIIDAE
CARYCHIUM O. F. Müller, 1774
Auricella Brard in Jurine, 1817
Auriculina Moquin-Tandon, 1856
(non Grateloup, 1838)
SARAPHIA Risso, 1826
ZOSPEUM Bourguignat, 1856
Speozoum Hamann, 1896

OTINIDAE
OTINA Gray, 1847

Order Basommatophora

SIPHONARIOIDEA (Marine)
TRIMUSCULIDAE
TRIMUSCULUS Schmidt, 1818
Gadinia Gray, 1824
Clypeus Scacchi, 1833
Mouretia Sowerby, 1842
Muretia Orbigny, 1843
Gardinia Pictet, 1855
Rowellia Carpenter, 1864
Gadinalea Iredale, 1940

SIPHONARIIDAE
SIPHONARIA Sowerby, 1823
Liria Gray, 1847
Hebesiphon, Legosiphon, Mallorsiphon,
Mestosiphon, Parellsiphon, Torquisiphon,
Triellsiphon all Iredale,1940
Ductosiphonaria Hubendick, 1945
BENHAMINA Finlay, 1927
ELLSIPHON Iredale, 1940
HETEROSIPHONARIA Hubendick, 1945
HUBENDICKULA McAlpine, 1952 (53)
LIRIOLA Dall, 1870
PACHYSIPHONARIA Hubendick, 1945
Talisiphon Iredale, 1940 ?
PATELLOPSIS Nobre, 1886
PLANESIPHON Iredale, 1940
PUGILLARIA Iredale, 1924
Kerguelenia Rochebrune & Mabille, 1889
(non Stebbing, 1888)
Kerguelenella Powell, 1951 (Kergulenella)

SIPHONARIA (Cont'd.)
SACCULOSIPHONARIA Hubendick, 1945
SIMPLISIPHONARIA Hubendick, 1945
ANTHOSIPHONARIA Kuroda & Habe, 1971
WILLIAMIA Monterosato, 1884 [ICZN 1410]
Brondelia Bourguignat, 1862 (suppr.)
Scutulum Monterosato, 1877
(non Tournouer, 1869)
Allerya Mörch, 1877
(non Bourguignat, 1876)
Parascutum Cossmann, 1891
Aporemodon Robson, 1913
SIPHONACMEA Habe, 1958

AMPHIBOLOIDEA (Brackish)
AMPHIBOLIDAE (26)
AMPHIBOLA Schumacher, 1817
Ampullacera Quoy & Gaimard, 1832
Thallicera Swainson, 1840
Ampullarina Sowerby, 1842
SALINATOR Hedley, 1900

CHILINOIDEA (Freshwater)
CHILINIDAE
CHILINA Gray, 1828
Diplicaria Rafinesque, 1833
Ida Jay, 1836
Potomophila Swainson, 1840 =
Potamophila Agassiz, 1846
(non Sowerby, 1822)
Dombeia Orbigny, 1836
Dombeya Beck, 1837
Pseudochilina Dall, 1870
Acyrogonia Rochebrune & Mabille, 1889

LATIIDAE
LATIA Gray, 1850
Pelex Gould, 1852

ACROLOXOIDEA (Freshwater limpets)
ACROLOXIDAE
ACROLOXUS Beck, 1837
Ancylus Geoffroy, 1767 (invalid)
Ancyclus, Ancyllus auctt. (err.)
Velletia Gray, 1840
Acroxus Bourguignat, 1853
Calytra Westerlund, 1885
Helicon Westerlund, 1885

LYMNAEOIDEA (Freshwater)
LYMNAEIDAE
LYMNAEA Lamarck, 1799
Limneus, Lymnea, Lymnus, Lymneus,
Lymnaeus, Lymnula, Limnea,
Limnaea auctt. (err.)
Leachia Jeffreys, 1833
(non Lesueur, 1821)
Bulimnea Haldeman, 1841 (14)
Eulimneus Sandberger, 1875
Stagnaliana Servain, 1881
Limnus, Omphalolimnus Dybowski, 1903
Peregriana Gundrizer & Starobogatov,1985
(non Servain, 1881)
Fossarina Gundrizer & Starobogatov, 1985
(non Adams & Angas, 1864)
ACELLA Haldeman, 1841
AFROGALBA, LENAGALBA, LIMNOBULLA, PACIFIMYXAS,
ORIENTOGALBA, SIBIRIGALBA, SPHAEROGALBA,
WALTERIGALBA all Kruglov
& Starobogatov, 1985 (29)
BAKERILYMNAEA Weyrauch, 1964
pro Nasonia F. C. Baker, 1928
(non Ashmead, 1864)

LYMNAEA (Cont'd.)
MICROLIMNAEA Dybowski, 1908 ?
PECTINIDENS Pilsbry, 1911
PSEUDISIDORA Thiele, 1931
PSEUDOGALBA F. C. Baker, 1913
pro Simpsonia F. C. Baker, 1911
(non Rochebrune, 1904)
PSEUDOSUCCINEA F. C. Baker, 1908
WALTERLYMNAEA Starobogatov, 1976
AENIGMOMPHISICOLA Kruglov, 1981
ERINNA H. & A. Adams, 1855 (1858)
OMIA Annandale & Prashad, 1919
LANTZIA Jousseaume, 1872
GALBA Schrank, 1803 (Lymnaea Subgenus - 29)
Truncatuliana Servain, 1881
Tanousia Bourguignat in Servain, 1881
Fossaria Westerlund, 1885
Sandria Brusina, 1886
MYXAS Leach in Sowerby, 1822 ?
Myxus auctt. (error)
Amphipeplea Nilsson, 1822
Lutea Gray, 1840
Bullastra Bergh, 1901 ?
Austropeplea Cotton, 1942 ?
CANEFRIA Issel, 1874
RADIX Montfort, 1810
Gulnaria Leach in Turton, 1831
Neritostoma Klein in H.&.A. Adams, 1855
Auriculariana, Limnosiana,
Peregriana all Servain, 1881
Cyclolimnea Dall, 1905
CERASINA Kobelt, 1880
STAGNICOLA Leach in Jeffreys, 1830
Limnophysa Fitzinger, 1833
Entochilius Sandberger, 1880
Corvusiana Servain, 1881
Palustrisiana Servain, 1881
Palustria Dybowski, 1908
Turrilimnea Dybowski, 1908
BERLANIANA Kruglov & Starobogatov, 1986
COSTOLIMNAEA Dybowski, 1912
FENZIANA Kruglov & Starobogatov, 1986 (29a)
HINKLEYIA F. C. Baker, 1928 ?
LADISLAVELLA Kruglov & Starobogatov, 1986
OMPHISCOLA Beck, 1837
Leptolimnaea Swainson, 1840
Leptolimnaeus Sandberger, 1875
Glabriana Servain, 1881
Ladislavella Dybowski, 1912
POLYRHITIS Meek, 1876

LANCIDAE
LANX Clessin, 1882
Walkerola Hannibal, 1912
FISHEROLA Hannibal, 1912

GLACIDORBOIDEA (39a)
GLACIDORBIDAE
GLACIDORBIS Iredale, 1943
GONDWANORBIS Ponder, 1986

PHYSOIDEA
PHYSIDAE
PHYSA Draparnaud, 1801
Exydra Hübner in Férussac, 1821 ?
Phyza Risso, 1826
Rivicola Fitzinger, 1833
Physella, Physodon Haldeman, 1842
Echemythes Gistl, 1848
Alampetista Zilch, 1956
pro Alampetis von Martens, 1898
(non Thompson, 1878)
BERINOPHYSA Starobogatov, 1976

PHYSIDAE
PHYSA (Cont'd.)
COSTATELLA Dall, 1870
Costella Meek, 1876 (err.)
PETROPHYSA Pilsbry, 1926
APLEXA Fleming, 1820
Nauta Leach in Turton, 1831
Aplexus, Amplexa, Aplecta auctt. (err.)
STENOPHYSA Von Martens, 1898

PLANORBIOIDEA
PLANORBIDAE
PLANORBINAE (53)
Tribe Planorbeae
PLANORBIS Müller, 1774
Tropidiscus Stein, 1850 ?
Trophidiscus H. & A. Adams, 1855
Tropodiscus Clessin, 1877
AFROGYRUS Brown & Mandahl-Barth, 1973 ?
HAVORBIS Brown & Mandahl-Barth, 1979
ANISUS Stüder, 1820 (= Gyraulus subgenus ?)
Spirorbis Swainson, 1840
(non Daudin, 1800)
Planorbina Haldeman, 1842
Omalodiscus Benson, 1855
Gyrorbis Moquin-Tandon, 1856
(non Fitzinger, 1833)
Paraspira Dall, 1905
Wustia Honigmann, 1909
Gyrostoma Haas, 1923
(non Hesse, 1908)
COSTORBIS Lindholm, 1926
DISCULIFER Boettger, 1944
Diplodiscus Westerlund, 1897
(non Diesing, 1835)
Spiralina von Martens, 1899
(non Chaster, 1898)
BATHYOMPHALUS Agassiz in Charpentier, 1837
Polygyrus Gray, 1847
Discoidina Stein, 1850
CHOANOMPHALUS Gerstfeldt, 1859
ACHOANOMPHALUS Lindholm, 1909
Valvatomphalus Dybowski
& Grochmalicki, 1925
ANOMPHALODES Tomlin, 1929
pro Anomphalus Dybowski
& Grochmalicki, 1925
(non Meek & Worthen, 1866)
ANTICHOANOMPHALUS Beckman, 1975
BAICALOPLANORBIS Beckman, 1975
BIANGULATUS Dybowski & Grochmalicki, 1925
KOZHOVISULCIFER Beckman, 1975
OMPHALOCRYPTA Tomlin, 1929
pro Cryptomphalus Dybowski
& Grochmalicki, 1925
(non Charpentier, 1837)
SULCIFER Lindholm, 1909
Platybasalis Dybowski & Grochmalicki,1925
GYRAULUS Agassiz in Charpentier, 1837 (32)
Planaria T. Brown, 1827
(non Müller, 1776)
Trochlea Haldeman, 1841
Poecilospira Mörch, 1852
Giraulus Moquin-Tandon, 1856
Glyptanisus, Plananisus Iredale, 1943
ARMIGER Hartmann, 1844
Nautilus Gray, 1847
(non Linnaeus,1758)
Nautilina Stein, 1850
Spiniformis Germain, 1903
Atropoarmiger, Cristoarmiger,
Nautiloarmiger all Dybowski
& Grochmalicki, 1922

GYRAULUS (Cont'd)
CAILLAUDIA Bourguignat, 1883
CARINOGYRAULUS Polinski, 1929 (1932 ?)
CERATOPHALLUS Brown & Mandahl-Barth, 1973
CHOANOMPHALODES Lindholm, 1927
LAMORBIS Starobogatov, 1967
NAUTILINUS Mousson, 1872 ?
TORQUIS Dall, 1905
Tribe Amerianneae
AMERIANNA Strand, 1928
pro Ameria H. Adams, 1861
(non Walker, 1854)
Ameriella Cotton, 1943
GLYPTOPHYSA Crosse, 1872
Glyptamoda Iredale, 1943
Tribe Physastreae
PHYSASTRA Tapparone-Canefri, 1883
Pulmobranchia Pelseneer, 1894
Lenameria, Tasmadora Iredale, 1943
MIRATESTA P. & F. Sarasin, 1897
Tribe Camptocerateae
CAMPTOCERAS Benson, 1843
Culminella Clench, 1927
Tribe Segmentineae
SEGMENTINA Fleming, 1822
Hemithalamus Leach in Turton, 1831
Segmentaria Swainson, 1840
Appendiculata Finicus, 1867 (part)
Hemithalasmus Paetel, 1889
POLYPYLIS Pilsbry, 1906
TROCHORBIS Benson, 1855
ACRORBIS Odhner, 1937 ?
Segnitilia Cotton & Godfrey, 1938
HIPPEUTIS Agassiz in Charpentier, 1837
Hippeutes Gray, 1850
HELICORBIS Benson, 1855
INTHA Annandale, 1922
LENTORBIS Mandahl-Barth, 1954
PINGIELLA F. C. Baker, 1945
SEGMENTORBIS Mandahl-Barth, 1954
ACUTORBIS Mandahl-Barth, 1954
Carinorbis Mandahl-Barth, 1954
(non Conrad, 1862)
Tribe Drepanotremeae
DREPANOTREMA Crosse & Fischer, 1880
ANTILLORBIS Harry & Hubendick, 1964
FOSSULORBIS Pilsbry, 1934
Tribe Biomphalarieae
BIOMPHALARIA Preston, 1910
AFROPLANORBIS Thiele, 1931
Planorbulina Germain, 1923
(non Orbigny, 1826)
ARMIGERUS Clessin, 1884
Obstructio Haas, 1939
LATEORBIS F. C. Baker, 1945
TROPICORBIS Brown & Pilsbry, 1914
AUSTRALORBIS Pilsbry, 1934
pro Planorbina Dall, 1905
(non Haldeman, 1842)
PLATYTAPHIUS Pilsbry, 1924
SYRIOPLANORBIS F. C. Baker, 1945
Heterodiscus Westerlund, 1902
(non Sharp, 1886)
TAPHIUS H. & A. Adams, 1855

HELISOMATINAE (14)
HELISOMA Swainson, 1840
Helicosoma Agassiz, 1846
CARINIFEX Binney, 1863 (1865)
Megasystropha Lea, 1864 (invalid)
Megastropha Walker, 1917
MENETUS H. & A. Adams, 1855
MICROMENETUS F. C. Baker, 1945

HELISOMATINAE (Cont'd.)
PLANORBARIUS Froriep, 1806 ?
Coretus Gray, 1847
Spirodiscus Stein, 1850
(non Ehrenberg, 1830)
PLANORBELLA Haldeman, 1842 (10)
Hypsogyra Lindholm, 1927
pro Adula H. Adams, 1861
(non H. & A. Adams, 1857)
pro Ancaeus H. Adams, 1869
(non Agassiz, 1846)
PIEROSOMA Dall, 1905
SEMINOLINA Pilsbry, 1934
Thomsonia Ancey, 1886
(non Signoret, 1879)
PLANORBULA Haldeman, 1842
pro Discus Haldeman, 1840
(non Fitzinger, 1833)
Dentalis Gray, 1847
Planorbulina von Martens, 1899
(non Jickel, 1874)
HALDEMANINA Dall, 1905 (10)
PROMENETUS F. C. Baker, 1935
Phreatomenetus Taylor, 1960
VORTICIFEX Meek in Dall, 1870
PARAPHOLYX Hanna, 1922 (10)
Pompholyx Lea, 1856
(non Gosse, 1851)
Pompholycodea Lindholm, 1927

BULLINIDAE
BULLININAE
BULINUS O. F. Müller, 1781
Bullinus Oken, 1815
Scaevola Scudder, 1882
Isidora Ehrenberg, 1831 (53)
Isodora Mousson, 1874 (err.)
DIASTROPHA Gray in Turton, 1840
PYRGOPHYSA Crosse, 1879
Pyrgobullinus Pallary, 1923
INDOPLANORBIS Annandale & Prashad, 1920
ISIDORELLA Tate, 1896 (51)
Oppletora Iredale & Whitley, 1938
MANDAHLBARTHIA Biocca, 1980 ?
PHYSOPSIS Krauss, 1848

PLESIOPHYSINAE (53)
PLESIOPHYSA P. Fischer, 1883

PROTANCYLIDAE
PROTANCYLUS P. & F. Sarasin, 1897

PATELLOPLANORBIDAE
PATELLOPLANORBIS Hubendick, 1957

NEOPLANORBIDAE (Subfamily ?) (10)
NEOPLANORBIS Pilsbry, 1906
AMPHIGYRA Pilsbry, 1906
Payettia Dall, 1924

ANCYLIDAE (10) (Rhodacmeidae)
ANCYLINAE
ANCYLUS O. F. Müller, 1774
Pseudancylus Walker, 1921
RHODACMAEA Walker, 1917
RHODOCEPHALA Walker, 1917

FERRISSIINAE (10)
FERRISSIA Walker, 1903
Haldemania Clessin, 1880
(non Tryon, 1862)
WATSONULA Mirolli, 1960 ?

FERRISSINAE (Cont'd.)
ANCYLASTRUM Bourguignat, 1853
Legrandia Legrand, 1879
Cumingia Clessin, 1880
(non Sowerby, 1833)
Tasmancylus Iredale, 1926
ANISANCYLUS Pilsbry, 1924
BURNUPIA Walker, 1912
HEBETANCYLUS Pilsbry, 1914
UNCANCYLUS Pilsbry, 1914
GUNDLACHIA Pfeiffer, 1849
Poeyia Bourguignat, 1862
KINCAIDILLA Hannibal, 1912
Kincaidella auctt. (err.)
LAEVAPEX Walker, 1903
PETTANCYLUS Iredale, 1943
PSEUDANCYLASTRUM Lindholm, 1909 ?

Order Stylommatophora (36b)
Suborder Orthurethra

ACHATINELLOIDEA (Terrestrial)
ACHATINELLIDAE (17)
PITYSINAE
PITYS Mörch, 1852
Pitys Beck, 1837 (nom.nud.)
Diaglyptus Pilsbry, 1892
Diglyptus Pilsbry, 1893
ANTONELLA Cooke & Kondo, 1960
APOPITYS Cooke & Kondo, 1960
CELTICOLA Cooke & Kondo, 1960
MERYTICOLA Cook & Kondo, 1960
NESONOICA Cooke & Kondo, 1960
LAMELLOVUM Pilsbry, 1910
MAITUA Cooke & Kondo, 1960
MANGAOA Cooke & Kondo, 1960
MITIPERUA Cooke & Kondo, 1960
PERAHUA Cooke & Kondo, 1960
PUKINIA Cooke & Kondo, 1960
STROBILUS Anton, 1839
Strombilus Anton, 1839 in Gray, 1847
Elasmatina Petit, 1843
TANGA Cooke & Kondo, 1960
TAUTAUTUA Cooke & Kondo, 1960
TAITAA Cooke & Kondo, 1960
TAIREVA Cooke & Kondo, 1960
TARAIA Cooke & Kondo, 1960
TUBUAIA Cooke & Kondo, 1960

LAMELLIDEINAE
LAMELLIDEA Pilsbry, 1910
Lamellina Pease, 1860
(non Bory, 1824)
ATEA Pilsbry & Cooke, 1933
ELAMELLIDEA Cooke & Kondo, 1960
ELASMIAS Pilsbry, 1910
FERNANDEZIA Pilsbry, 1911
TORNATELLINA L. Pfeiffer, 1842
ELASMATINA Petit, 1843 ?
TORNATELLINOPS Pilsbry & Cooke, 1915
Pacificella Odhner, 1922
Tornatellinops Iredale, 1937 ?
Tornelasmias Iredale, 1944

TORNATELLIDINAE
TORNATELLIDES Pilsbry, 1910
AIDITUANS Cooke & Kondo, 1960
WAIMEA Pilsbry & Cooke, 1915
AURICULELLA L. Pfeiffer, 1855
Frickella L. Pfeiffer, 1855
GULICKIA C. M. Cooke, 1915
PHILOPOA Cooke & Kondo, 1960
TORNATELLARIA Pilsbry, 1910

ACHATINELLINAE
ACHATINELLA Swainson, 1828
Helicteres Férussac in Beck, 1837
Helicter Pease, 1862
Apex von Martens, 1860
Helicterella Gulick, 1873
ACHATINELLASTRUM L. Pfeiffer, 1854
BULIMELLA L. Pfeiffer, 1854
NEWCOMBIA L. Pfeiffer, 1854
Newcombiana Thwing, 1907
PARTULINA L. Pfeiffer, 1854
BALDWINIA Ancey, 1899
EBURNELLA Pease, 1869
PARTULINELLA Hyatt, 1914
PERDICELLA Pease, 1869

TEKOULININAE
TEKOULINA Solem, 1972

COCHLICOPOIDEA (Cionelloidea) (36b)
COCHLICOPIDAE (Cionellidae)
COCHLICOPA Férussac, 1821 in Risso, 1826
Cionella Jeffreys, 1829
Zua Leach in Turton, 1831
Styloides Fitzinger, 1833 (part)
Folliculus Agassiz in Charpentier, 1837
Chionella Jeffreys, Swainson, 1840 (err.)
Hydastes Parreyss, 1849
Ferussacia Risso in Binney, 1878
(non Risso, 1826)
AZECA Leach in Fleming, 1828
Odontalus Parreyss, 1850
Azecastrum Bourguignat, 1858
CRYPTAZECA de Folin & Berillon, 1897 (22x)
GOMPHROA Westerlund, 1902
HYPNOPHILA Bourguignat, 1858

AMASTRIDAE
AMASTRA H. & A. Adams, 1855
AMASTRELLA Sykes, 1900
HETERAMASTRA Pilsbry, 1911
METAMASTRA Hyatt & Pilsbry, 1911
PARAMASTRA Hyatt & Pilsbry, 1911
ARMSIA Pilsbry, 1911
CARELIA H. & A. Adams, 1855
CYCLAMASTRA Pilsbry & Vanatta, 1905
KAUAIA Sykes, 1900
Carinella L. Pfeiffer, 1875
(non Sowerby, 1839)
ARMIELLA Hyatt in Pilsbry, 1911
LAMINELLA L. Pfeiffer, 1854
LEPTACHATINA Gould, 1847
ANGULIDENS Pilsbry & Cooke, 1914
ILIKALA Cooke, 1911
PAUAHIA Cooke, 1911
LABIELLA L. Pfeiffer, 1854
THAANUMIA Ancey, 1899
PLANAMASTRA Pilsbry, 1911
TROPIDOPTERA Ancey, 1889
Pterodiscus Pilsbry, 1893
Helicamastra Pilsbry & Vanatta, 1905

PUPILLOIDEA
PUPILLIDAE
PUPILLINAE
PUPILLA Leach in Fleming, 1828
Pupa Draparnaud, 1801 (part)(non Röding)
Torquatella Held, 1837
Omegapilla Iredale, 1937 (10a)
AFRIPUPILLA Pilsbry, 1921
FRAGIPUPILLA Shileyko, 1984
GIBBULINOPSIS Germain, 1919
Primipupilla Pilsbry, 1921

PUPILLA (Cont'd.)
STRIOPUPILLA Pilsbry, 1921
GLYPTOPUPOIDES Pilsbry, 1926
Famarinia Iredale, 1933
MICROSTELE Boettger, 1886
PUPOIDES L. Pfeiffer, 1854
Leucochila von Martens, 1860
Leucochiloides L. Pfeiffer, 1879
Themapupa Iredale, 1930
ISCHNOPUPOIDES Pilsbry, 1926
PUPOIDOPSIS Pilsbry & Cooke, 1921

LAURIINAE (44)
LAURIA Gray in Turton, 1840
Eruca Swainson, 1840
Gastrodon Lowe, 1852
(non Rafinesque, 1815)
Reinhardtia O. Boettger, 1879
SENILAURIA Pilsbry, 1928
Petrarca Pilsbry, 1922
(non Fowler, 1899)
AGARDHIA Gude, 1911
Sphyradium Hartman, 1844
(non Charpentier, 1837)
Coryna Westerlund, 1887
(non Billberg, 1833)
Rhytidochasma A. J. Wagner, 1914
ARGNA Cossmann, 1889 (Argninae ?)
AGARDHIELLA P. Hesse, 1923
EUXINOLAURIA Lindholm, 1924
CAUCASIPUPA Pilsbry, 1926
Caucasica Caziot & Margier, 1909
(non Boettger, 1877)
MATSCHACHELIA Shileyko, 1976
NEOLAURIA Shileyko, 1975
HEMILAURIA Walden, 1983 ?
LEIOSTYLA Lowe, 1852
Alvearella Lowe, 1852
Charadrobia Albers, 1854
Liostyla von Martens in Albers, 1860
AZORIPUPA Pilsbry, 1923
CRATICULA Lowe, 1852
Eryma Albers, 1854
MASTULA Lowe, 1852
SCARABELLA Lowe, 1852
WOLLASTONULA Pilsbry, 1922
PHOLEOTERAS Sturany, 1904 ? (53)

VALLONIIDAE (Spelaeodiscinae)
VALLONIA Risso, 1826
Amplexis T. Brown, 1827
Amplexus Brown, 1844
(non Sowerby, 1814)
Zurama Leach in Turton, 1831
Circinaria Beck, 1837
Lucena Moquin-Tandon,1855
(non Oken,1815)
Lucena "Hartmann" Gray in Turton, 1840
(non Hartmann, 1821)
Glaphyra Albers, 1850
Lurama Leach in Honigmann, 1910 (err.)
Zuramia Leach in Kennard & Woodward,1926
PLANIVALLONIA Shileyko, 1984
ACANTHINULA Beck, 1847
Achantinella Betta, 1870
Euacanthinula Westerlund, 1889
Aulaca Westerlund, 1902
Imputegla Iredale, 1937
SPERMODEA Westerlund, 1902
KLEMMIA Gittenberger, 1969
PLANOGYRA Morse, 1864
SPELAEODISCUS Brusina, 1886
ASPASITA Westerlund, 1869

VALLONIIDAE (Cont'd.)
VIRPAZARIA Gittenberger, 1969
ZOOGENETES Morse, 1864
Zoogenites Morse, 1864 (err.)
Acanthinula Binney, 1878
(non Beck, 1847)

VERTIGINIDAE
VERTIGININAE
VERTIGO O. F. Müller, 1774
Isthmia Gray, 1821
Alaea Jeffreys, 1830
Staurodon Lowe, 1855 (non 1852)
Dexiogyra Stabile, 1864
Dexiogira Betta, 1870
Nearctula Sterki, 1892
Haplopupa Pilsbry, 1898
ALLOPTYX Pilsbry, 1953
ANGUSTULA Sterki, 1888
VERTILLA Moquin-Tandon, 1855
VERTILLARIA Pilsbry, 1919

TRUNCATELLININAE
TRUNCATELLINA Lowe, 1852
Isthmia Reinhardt, 1897
(non Gray, 1821)
Laurinella P. Hesse, 1915
BOTHRIOPUPA Pilsbry, 1898
BOYSIA Pfeiffer, 1849
Hypostoma Albers, 1850
(non Rudolphi, 1809)
Hypotrema von Martens, 1860
Hypoma L. Pfeiffer, 1879
Hypostrema L. Pfeiffer, 1879
COLUMELLA Westerlund, 1878
Paludinella Lowe, 1852
(non L. Pfeiffer, 1841)
Edentulina Clessin, 1876
(non L. Pfeiffer, 1855)
Paludellina Tryon, 1884 (err.)
Sphyradium Westerlund, 1887
(non Charpentier, 1837)
NEGULUS Boettger, 1889
PRONESOPUPA Iredale, 1913
EDENTULOPUPA Pilsbry & Cooke, 1920
SERICIPUPA Pilsbry & Cooke, 1920

GASTROCOPTINAE (Hypselostomatinae)
GASTROCOPTA Wollaston, 1878
Bifidaria Sterki in Pilsbry, 1891
Eubifidaria Sterki, 1893
Falsopupa Germain, 1919
ALBINULA Sterki, 1892
Leucochilus Boettger, 1880 (ICZN 115)
(non Leucochila von Martens, 1860)
AUSTRALBINULA Pilsbry, 1916
Gyrodaria Iredale
GEMINIDENS Pilsbry, 1930
GIBBULINA Beck, 1837
Infundibularia Pfeiffer, 1876
IMMERSIDENS Pilsbry & Vanatta, 1900
PRIVATULA Sterki, 1893
SINALBINULA Pilsbry, 1916
STAUROTREMA Pilsbry, 1948
VERTIGOPSIS Cockerell in Sterki, 1893
AFRIBOYSIA Zilch, 1939 ?
ANAUCHEN Pilsbry, 1917
BENSONELLA Pilsbry & Vanatta, 1900
BOYSIDIA Ancey, 1881
Gredleriella Möllendorff, 1884
DASYPUPA Thompson & Dance, 1983
PARABOYSIDIA Pilsbry, 1917 ?
CAVIPUPA Pilsbry, 1934

GASTROCOPTINAE (Cont'd.)
CHAENAXIS Pilsbry & Ferriss, 1906
GYLIOTRACHELA Tomlin, 1930 ? (45f)
Gyliauchen Pilsbry, 1917
(non Nicoll, 1915)
Gyliotrachea Pilsbry, 1931
HYPSELOSTOMA Benson, 1856 (45f)
Tanystoma Benson, 1856
(non Latreille, 1829)
Hyperostoma Simroth, 1909
ULPIA Scott, 1955

AULACOSPIRINAE
AULACOSPIRA Möllendorff, 1890
Micropetasus Möllendorff, 1890
PSEUDOSTREPTAXIS Möllendorff, 1890
SYSTENOSTOMA Bavay & Dautzenberg, 1909

NESOPUPINAE
NESOPUPA Pilsbry, 1900
Ptychochilus O. Boettger, 1880
(non Jordan, 1877)
Westralcopta Iredale, 1939
AFRIPUPA Pilsbry & Cooke, 1920
COCOPUPA Pilsbry & Cooke, 1920
HELENOPUPA Pilsbry & Cooke, 1920
INDOPUPA Pilsbry & Cooke, 1920
INFRANESOPUPA Cooke & Pilsbry, 1929
INSULIPUPA Pilsbry & Cooke, 1920
Pagodella H. Adams, 1867 ?
(non Swainson, 1840)
LIMBATIPUPA Cooke & Pilsbry, 1920
NESODAGYS Cooke & Pilsbry, 1920
NESOPUPARIA Pilsbry, 1926
NESOPUPILLA Pilsbry & Cooke, 1920
CAMPOLAEMUS Pilsbry, 1892
COSTIGO O. Boettger, 1891
CYLINDROVERTILLA O. Boettger, 1880 (10a)
WALLIVERTILLA Iredale, 1937
LYROPUPA Pilsbry, 1900
LYROPUPILLA Pilsbry & Cooke, 1920
MIRAPUPA Cooke & Pilsbry, 1920
PUPISOMA Stoliczka, 1873
Parazoogenetes Habe, 1956
PTYCHOPATULA Pilsbry, 1889
SALPINGOMA Haas, 1937
PTYCHALAEA O. Boettger, 1889
SOMNIOPUPA Iredale, 1937
STAURODON Lowe, 1852
STERKIA Pilsbry, 1898
METASTERKIA Pilsbry, 1920

ORCULIDAE
ORCULA Held, 1837
Pupula Leach in Mörch, 1852
(non Charpentier, 1837)
Scyphus Monterosato, 1908
PILORCULA Germain, 1912
SPHYRADIUM Charpentier, 1837
ORCULELLA Steenberg, 1925
MESORCULELLA Shileyko, 1976
PAGODULINA Clessin, 1876
Pagodina Stabile, 1864
(non Beneden, 1853)
Pagodula P. Hesse, 1916
(non Monterosato, 1884)
Pagodinella Thiele, 1917
CRYSTALLIFERA Shileyko, 1976
SCHILEYKULA Gittenberger, 1983
SPELEODENTORCULA Gittenberger, 1985 ?

PLEURODISCIDAE
PLEURODISCUS Wenz, 1919

STROBILOPSIDAE
STROBILOPS Pilsbry, 1893
Pro Strobila Morse, 1864
(non M. Sars, 1835)
Strobilus Sandberger, 1872-1875
(non Anton, 1839)
COELOSTROBILOPS Pilsbry, 1931
DISCOSTROBILOPS Pilsbry, 1927
EOSTROBILOPS Pilsbry, 1927
NESOSTROBILOPS Pilsbry, 1931
ENTEROPLAX Gude, 1899 ?

PYRAMIDULIDAE
PYRAMIDULA Fitzinger, 1833
Pyramidulops Habe, 1956

CHONDRINIDAE
CHONDRINA Reichenbach, 1828
Torquilla Stüder, 1820 (part)
(non Brisson, 1760)
Alloglossa Lindstrom, 1868
MODICELLA H. & A. Adams, 1855
SOLATOPUPA Pilsbry, 1917
ABIDA Leach in Turton, 1831
Pupa Draparnaud, 1801 (part)
(non Röding, 1798)
Torquilla Stüder, 1820 (part)
(non Brisson, 1760)
Jaminia Leach in Risso, 1926 (part)
Granaria Held, 1837
Pupella Swainson, 1840
Stomodonta Mermet, 1843 (53) ?
Deloplecta Agassiz, 1846
FAUXULUS Schaufuss, 1869
Faula H. & A. Adams, 1855
(non Blanchard, 1850)
ANISOLOMA Ancey, 1901
FAUXULELLA Pilsbry, 1917
TOMIGERELLA Pfeiffer, 1879
GRANOPUPA Boettger, 1889
RUPESTRELLA Monterosato, 1894
ODONTOCYCLAS Schlüter, 1838
Scopelophila Albers, 1850
SANDAHLIA Westerlund, 1887

PUPILLOIDEA; Family uncertain:
GITTENBERGIA Giusti & Mangonelli, 1986

BULIMINOIDEA (36b)
BULIMINIDAE (Enidae - 45b)
BULIMININAE
BULIMINUS Beck, 1837
Bulimina Ehrenberg, 1831
(non Orbigny, 1826)
Petraeus Albers, 1850
PENE Pallary, 1929
AKRAMOVSKIELLA Shileyko, 1984
AMIMOPINA Iredale, 1933
BONINENA Habe, 1955
CHONDRUS Cuvier, 1817
ANTICHONDRUS Lindholm, 1925
Clausilia Blainville, 1825 (part)
(non Draparnaud, 1805)
CLAUSILIOIDES Lindholm, 1925
COCCODERMA Möllendorff, 1901
HEUDIELLA Annandale, 1924
DIFFERENA Shileyko, 1984
DRAPARNAUDIA Montrouzier, 1859 (41d)
ENA Leach in Turton, 1831
Bulinus Stüder, 1820
(non Müller, 1781)
Napaeus von Martens in Albers, 1860
(non Albers, 1850)

ENA (Cont'd.)
PERISTOMA Krynicki, 1833
CAUCASICOLA Hesse, 1917
Medea Boettger, 1883
(non Eschscholtz, 1829)
Helle Hesse, 1916
(non Osten-Sacken, 1896)
PARAMASTUS Hesse, 1933
LJUDMILENA Shileyko, 1984
LOPHAUCHEN Möllendorff, 1901
LUCHUENA Habe, 1955
MERDIGERA Held, 1837 (Merdigerinae ?)
Merdigerus Albers, 1850
MIRUS Albers, 1850
NAPAEUS Albers, 1850
Macaronapaeus Kobelt, 1899
NAPAEINUS Hesse, 1933
OMPHALOCONUS Westerlund, 1887
CIRNA Pallary, 1928
KABYLIA Pallary, 1928
MAURONAPAEUS Kobelt, 1899
PSEUDONAPAEUS Westerlund, 1887
Pseudopetraeus Westerlund, 1902
(non 1896)
ARIDENUS Shileyko, 1984
CAULICULUS Lindholm, 1925
LABROENA Lindholm, 1927
LAEONAPAEUS Lindholm, 1925
SIRAPHORUS Lindholm, 1925
PSEUDOCHONDRULA Hesse, 1933 ?
PUPINIDIUS Möllendorff, 1901
PETRAEOMASTUS Möllendorff, 1901
PUPOPSIS Gredler, 1898
CLAUSILIOPSIS Möllendorff, 1901
HOLCAUCHEN Möllendorff, 1901
RACHISPECULUM Iredale, 1933 ?
RETOWSKIA O. Boettger, 1881
SERINA Gredler, 1898
SESTERIA Bourguignat, 1884
SIRAPHOROIDES Shileyko, 1977
TRIANGUSTOMA Shileyko, 1984
TURANENA Lindholm, 1922
YAKUENA Habe, 1955

JAMINIINAE
JAMINIA Risso, 1826
Chondritortus Monterosato, 1908
BOLLINGERIA Forcart, 1940
CHONDRULOPSINA Lindholm,1925
(Chondrulopsinae ?)
Chondrulopsis Westerlund,1902 (non 1887)
EUCHONDRUS O. Boettger, 1883
Multidentinia Lindholm, 1925
MASTUS Beck, 1837
Pseudomastus Boettger, 1889
TOKATIA Hüdec, 1972
IMPARIETINIA Lindholm, 1925
IMPARIETULA Lindholm, 1925
MULTIDENTULA Lindholm, 1925
SENARIDENTA Shileyko, 1984 ?

CHONDRULINAE
CHONDRULA Beck, 1837
Gonadon Held, 1837
Eucore Charpentier, 1837
Chondrulus Westerlund, 1887
Dentistomus Kimakowicz, 1890
AMPHISCOPUS Westerlund, 1887
pro Euchondrus Retowski, 1886
(non O. Boettger, 1883)
EUBREPHULUS Wagner, 1927
GEMINULA Lindholm, 1925
SPANIODONTA Kobelt, 1902

CHONDRULINAE (Cont'd.)
ADZHARIA Hesse, 1933
ANDRONAKIA Lindholm, 1913
CHONDRULOPSIS Westerlund, 1887 ?
CYRENAEUS Heller, 1971
AMPHITRORSUS Kimakowicz, 1890 (Mastus ?)
SEWERTZOWIA Kobelt, 1888
Severtzowia, Severtzovia Kobelt (err.)
PARACHONDRULA Lindholm, 1925
SUBZEBINUS Westerlund, 1887 (Ena Subgenus ?)
Pseudopetraeus, Mastoides Westerlund,1896
LAEVOZEBRINUS Lindholm, 1925
OEDICHILUS Lindholm, 1925
STYLOPTYCHUS Lindholm, 1925
ZEBRINA Held, 1837
Brephulus Beck, 1837
Zebrinus Westerlund, 1887
Euzebrina Kobelt, 1899
Rhabdoena Kobelt & Möllendorff, 1902
Leucomastus Wagner, 1927
Aschera Pallary, 1939
BREPHULOPSIS Lindholm, 1925
Brephulus Westerlund, 1902
(non Beck, 1837)
NAPAEOPSIS Sturany & Wagner, 1914
RAMUSCULUS Lindholm, 1925
THOANTEUS Lindholm, 1925

SPELAEOCONCHINAE
SPELAEOCONCHA Sturany, 1912

CERASTUIDAE
CERASTUA Strand, 1928
pro Cerastus Albers, 1860
(non DeJean, 1821)
PARACERASTUS Thiele, 1933
ALTENAIA Zilch, 1972
APOECUS Kobelt, 1902 ?
CONULINUS von Martens, 1895
EDOUARDIA Gude, 1914
EURYPTYXIS Fischer, 1883
Petraeocerastus Kobelt, 1902
LIMICENA Connolly, 1925
Krapfiella Connolly, 1925 ?
Xerocerastus Connolly, 1925
MABILLIELLA Ancey, 1886 (53)
PACHNODES Albers in von Martens, 1860
Pachnodus auctt. (err.)
NESIOCERASTUS Van Mol et al, 1980
PASSAMAELLA Pfeiffer, 1878
ACHATINELLOIDES Nevill, 1878
Ovella Pfeiffer, 1879
POLYCHORDIA Connolly, 1941
RHACHIDINA Thiele, 1911
RHACHIS Albers, 1850 (Rachis auctt. - err.)
Rachisellus Bourguignat, 1889
PLEURORHACHIS Connolly, 1939
RHACHISTIA Connolly, 1925
Eorrachis Tomlin & Peile, 1930
ZEBRINOPS Thiele, 1931

Suborder Mesurethra

CLAUSILIOIDEA (36)
CLAUSILIIDAE
PHAEDUSINAE
PHAEDUSA H. & A. Adams, 1855
Macrenoica Wagner, 1920 (part)
Calcariclavis Lindholm, 1924
PSEUDOPHAEDUSA Tomiyama, 1984
CASPIOPHAEDUSA Lindholm, 1924
Phaedusa Westerlund, 1902
(non H. & A. Adams, 1855)

PHAEDUSINAE (Cont'd.)
EUPHAEDUSA O. Boettger, 1877 ?
RENSCHIPHAEDUSA Loosjes, 1973
HEMIPHAEDUSA O. Boettger, 1877 ?
Synprosphyma Wagner, 1920
Symprosphyma Thiele, 1931
ADVERSARIA Lindholm, 1925
CYLINDROPHAEDUSA O. Boettger, 1877
EXCUSSISPIRA Lindholm, 1925
FORMINIGERA Lindholm, 1925
NOTOPTYCHIA Ehrmann, 1927
SINIGENA Lindholm, 1925
TOSAPHAEDUSA Ehrmann, 1929
TYRANNOPHAEDUSA Pilsbry, 1901
OOSPIRA Blanford, 1872
Pseudonenia O. Boettger, 1877
Acrophaedusa O. Boettger, 1877
JUTTINGIA Loosjes, 1973
PARAPHAEDUSA O. Boettger, 1899
REINIA Kobelt, 1876
PROREINIA Thiele, 1931
STEREOPHAEDUSA O. Boettger, 1877
Stereostoma Kobelt, 1906
(non Murray, 1857)
Sterophaedusa Annandale, 1924

GARNIERIINAE ?
GARNIERIA Bourguignat, 1877
TROPIDAUCHENIA Lindholm, 1924
SYMPTYCHIA Ehrmann, 1927

SERRULININAE
SERRULINA Mousson, 1873
SERRULINELLA H. Nordsieck, 1984
BOFILLIELLA Ehrmann, 1927 (recent)
DOBATIA H. Nordsieck, 1973
GRAECOPHAEDUSA Rahle, 1982
LAEVIPHAEDUSA Likharev & Steklov, 1965
LAMINIFERA O. Boettger, 1863 (fossil)
NENIATLANTA Bourguignat, 1876 (recent)
Pyrenaica O. Boettger, 1877
Tortula Westerlund, 1878
LAMNIPLICA H. Nordsieck, 1975
MICROPHAEDUSA H. Nordsieck, 1978
PONTOPHAEDUSA Lindholm, 1924
PRAVISPIRA Lindholm, 1924
SCIOCOCHLEA C. Boettger, 1935
Phygas Wagner, 1914
(non Treitschke, 1833)

MEGALOPHAEDUSINAE
MEGALOPHAEDUSA O. Boettger, 1877
Aprosphyma Wagner, 1920 (part)
VASTINA Ehrmann, 1929
BATHYPTYCHIA Lindholm, 1925
DAUTZENBERGIELLA Lindholm, 1924
FORMOSANA O. Boettger, 1877
Macrophaedusa Möllendorff, 1883
Polyptychophora Wagner, 1920 (part)
DEXTROFORMOSANA Boettger & Schmacker, 1894
FUCHSIANA Gredler, 1887
LEPTACME Ehrmann, 1927
LINDHOLMIELLA Ehrmann, 1927
LIPAROPHAEDUSA Lindholm, 1924
ATRACTOPHAEDUSA Ehrmann, 1927
SIPHONOPHAEDUSA Lindholm, 1924
Syphonophaedusa Ehrmann, 1927 (err.)
LUCHUPHAEDUSA Pilsbry, 1905
OOPHAEDUSA Pilsbry, 1905
MESOPHAEDUSA Ehrmann, 1929
MUNDOPHAEDUSA Kuroda & Habe, 1958
VITRIPHAEDUSA Kuroda & Habe, 1958
NESIOPHAEDUSA Pilsbry, 1905

MEGALOPHAEDUSINAE (Cont'd.)
PAUCIPHAEDUSA Minato & Habe, 1983
PROTOPHAEDUSA Szekeres, 1970
RUFOSPIRA Möllendorff, 1883
Rufispira Paetel, 1889 (err.)
SELENOPHAEDUSA Lindholm, 1924
STREPTODERA Lindholm, 1925

ZAPTYCHINAE
ZAPTYX Pilsbry, 1901
PROZAPTYX Loosjes, 1950
METAZAPTYX Pilsbry, 1905
PARAZAPTYX Pilsbry, 1905
STEREOZAPTYX Pilsbry, 1905
OLIGOZAPTYX Pilsbry, 1905
DICERATOPTYX Pilsbry, 1905
IDIOZAPTYX Pilsbry, 1909
SELENOPTYX Pilsbry, 1908
Stenoptyx Kobelt, 1908
Selonoptyx Lindholm, 1924
THAUMATOPTYX Pilsbry, 1908
TYRANNOZAPTYX Kaufel, 1930
ZAPTYCHOPSIS Ehrmann, 1927
HEMIZAPTYX Pilsbry, 1905
HETEROZAPTYX Pilsbry, 1906

ALOPIINAE (Cochlodininae)
ALOPIA H. & A. Adams, 1855
Baleo-Clausilia Bielz, 1863
Eualopia Westerlund, 1884
Transsilvanica Westerlund, 1890
Nixa Szekeres, 1976
KIMAKOWICZIA Szekeres, 1969
AGATHYLLA H. & A. Adams, 1855
AGATHYLLINA H. Nordsieck, 1969
ALBINARIA Vest, 1867
Archipelagica, Bigibbosa, Corrugata, Cerigana, Cretica, Egana, Filumna, Graja, Interstriata, Laconica, Mirabellina, Naevosa, Scopulosa, Teres
all O. Boettger, 1878
Munda O. Boettger, 1878 ?
(non Stal, 1877)
Pistillum O. Boettger, 1892
Diadoma Westerlund, 1901
Mundaria Lindholm, 1924
(non Kerremans, 1896)
Catharella Lindholm, 1925
GRAECA O. Boettger, 1877 ?
Profuga O. Boettger, 1878
Protalbinaria Wagner, 1923 (part)
STRIATA O. Boettger, 1878 ?
Protalbinaria Wagner, 1923 (part)
BARCANIA Brandt, 1956
TORBANA Brandt, 1956
CARINIGERA Möllendorff, 1873
ANGITICOSTA H. Nordsieck, 1977
CHARPENTIERIA Stabile, 1864
Pedemontiana Pini, 1884
ITALA O. Boettger, 1877
Pollinia Betta, 1870 ?
(non Targioni-Tozetti, 1868)
Tirolica O. Boettger, 1877
Alpidelima Wagner, 1924 ?
COCHLODINA Férussac, 1821
Marpessa Gray, 1821 ?
Clausiliastra L. Pfeiffer, 1855 ?
Stabilea Betta, 1870 ?
Marpressa Clessin, 1876
Clausilia Lindholm, 1910 ?
(non Draparnaud, 1805)
COCHLODINASTRA H. Nordsieck, 1977

COCHLODINA (Cont'd.)
PARACOCHLODINA H. Nordsieck, 1969
PROCOCHLODINA H. Nordsieck, 1969
CRISTATARIA Vest, 1867
Strangulata Westerlind, 1884
Cristaria Westerlund, 1884
(non Schumacher, 1817)
Rhipsauchenia O. Boettger, 1890
Callopleura, Delessertia, Florienia
all Pallary, 1939 ?
Westerlundella Lindholm, 1924
DELIMA Hartmann, 1842
Albanica, Binodata O. Boettger, 1877
Dalmatica O. Boettger, 1877
Laevissima O. Boettger, 1877 ?
Piceata O. Boettger, 1877
Substricta O. Boetteger, 1877
Binotata Römer, 1891 (err.)
SEMIRUGATA O. Boettger, 1878
Robusta O. Boettger, 1877
Adriaca Westerlund, 1884
Adriatica Westerlund, 1901 (err.)
DILATARIA Vest, 1867
HERILLA H. & A. Adams, 1855
Bosnica Westerlund, 1884
ISABELLARIA Vest, 1867
Venusta O. Boettger, 1877
Euclista O. Boettger, 1878
Nymphogena Sajo, 1968
LAMPEDUSA O. Boettger, 1877
Lopadusaria Monterosato, 1892
Priodelima Wagner, 1924
IMITATRIX Westerlund, 1884
LEUCOSTIGMA Wagner, 1919
MACEDONICA O. Boettger, 1877
Attica O. Boettger, 1877
Serbica O. Boettger, 1877
Turcica O. Boettger, 1877
(non H. & A. Adams, 1854)
Neoserbica Wagner, 1919
Balcanica Kennard & Woodward, 1923
MAURITANICA O. Boettger, 1879
MEDORA H. & A. Adams, 1855
Medorella Woodward in Lindholm, 1924
MONTENEGRINA O. Boettger, 1877
Heteroptycha Westerlund, 1884
Heteroptychia Kennard & Woodward, 1923
Albanodelima Wagner, 1924 ?
Beieriella Klemm, 1962
MUTICARIA Lindholm, 1925
Lamellifera Monterosato, 1892 ?
(non Westerlund, 1890)
PAPILLIFERA Hartmann, 1842
Papillina Moquin-Tandon, 1855
PROTOHERILLA Wagner, 1921
TARACLAUSILIA H. Nordsieck, 1972
SERICATA O. Boettger, 1878
Bathyclista O. Boettger, 1885
Charites Westerlund, 1901
OLYMPICOSTA H. Nordsieck, 1972
SICILIARIA Vest, 1867
Sicania Tomlin, 1929
Trinacria O. Boettger, 1877
(non Mayer, 1868)
GIBBULARIA Monterosato, 1908
Gibbula O. Boettger, 1877
(non Risso, 1826)
STIGMATICA O. Boettger, 1877
Paestana Westerlund, 1884
STRIGILODELIMA Wagner, 1925
TRILOBA Vest, 1867
Eutriloba Westerlund, 1884
Trilobella Woodward in Lindholm, 1924

MENTISSOIDINAE
MENTISSOIDEA O. Boettger, 1877
Polinskia Wagner, 1920 ?
ACROTOMA O. Boettger, 1881
Acrostoma Westerlund, 1901
[non Le Sauvage, 1826]
Acrotomella Woodward in Lindholm, 1924 ?
ACROTOMINA H. Nordsieck, 1977
BZYBIA H. Nordsieck, 1977
ARMENICA O. Boettger, 1877
Sprattia O. Boettger, 1883
Armeniaca Westerlund, 1884 ?
Creniclavis Lindholm, 1924
Astrogena Szekeres, 1970
AKRAMOWSKIA H. Nordsieck, 1975
BITORQUATA O. Boettger, 1877
BOETTGERIA O. Boettger, 1863
ELIA H. & A. Adams, 1855
Maesta Kennard & Woodward, 1923 ?
ACROEUXINA O. Boettger, 1877
CAUCASICA O. Boettger, 1877
MEGALEUXINA O. Boettger, 1877
EUXINA O. Boettger, 1877
Hetaera O. Boettger, 1877
Peneptychia Westerlund, 1844
Wagneria Hesse, 1912
[non Robineau-Desvoidy, 1830]
Thraciella Hesse, 1916
Euxenia Wagner, 1919
Illunellaria Lindholm, 1924
EUXINASTRA O. Boettger, 1888
EUXINELLA H. Nordsieck, 1973
FILOSA O. Boettger, 1877
GALEATA O. Boettger, 1877
Pleioptychia Wagner, 1913 ?
Plistoptychia Lindholm, 1924
IDYLA H. & A. Adams, 1855
Hellenica O. Boettger, 1877
Oligoptychia O. Boettger, 1877
Olygoptycha, Oligothychia auctt. [err.]
Crucita Westerlund, 1878
Pleioptychia Kennard & Woodward, 1923
[non Wagner, 1913]
MACROPTYCHIA O. Boettger, 1877
ABBADIA Bourguignat, 1883
AEQUINOCTIA Lindholm, 1925
SABAEOLA Lindholm, 1925
AUSTROBALEA Pilsbry, 1924
ODONTEUXINA H. Nordsieck, 1975
OLYMPICOLA Hesse, 1916
Olympia Vest, 1867
[non Risso, 1826]
ROSENIELLA Thiele, 1931
Thalestris Lindholm, 1913
[non Claus, 1863]
Rosenia P. Hesse, 1916
[non Schepman, 1913]
SCROBIFERA O. Boettger, 1877
STRIGILEUXINA H. Nordsieck, 1975
STRUMOSA O. Boettger, 1877

NENIINAE
NENIA H. & A. Adams, 1855
Neniastrum Bourguignat, 1876
NENIOPS Pilsbry, 1926
NENISCA Rehder, 1939
PARANENIA Rehder, 1939
ANDINIA Polinski, 1922
EHRMANNIELLA Zilch, 1949
BEQUAERTINENIA Weyrauch, 1964 ?
COLUMBINIA Polinski, 1924
GIBBONENIA Zilch, 1954
GONIONENIA Pilsbry, 1929 [? 1916] [53]

NENIINAE [Cont'd.]
GRACILINENIA Polinski, 1921
GRANDINENIA Minato & Cheu, 1984
HEMICENA Pilsbry, 1949
INCAGLAIA Pilsbry, 1949
INCANIA Polinski, 1922
INDONENIA Ehrmann, 1927 ?
NENIATRACTA Pilsbry, 1926
PERUINIA Polinski, 1922
PFEIFFERIELLA Weyrauch, 1957
PSEUDOGRACILINENIA Loosjes, 1984
STEATONENIA Pilsbry, 1926
STEERIANA Jousseaume, 1900
Steereana Pilsbry, 1926
ANDINIELLA Weyrauch, 1958
CYLINDRONENIA Ehrmann, 1949
TEMESA H. & A. Adams, 1855
Parabalea Ancey, 1882
Exbalea Jousseaume, 1900
ZILCHIELLA Weyrauch, 1957

CLAUSILIINAE
CLAUSILIA Draparnaud, 1805
Stomodonta Mermet, 1843 [53] ?
Andraea Hartman in L. Pfeiffer, 1848
Pseudocerva Schaufuss, 1869
Kuzmicia Brusina, 1870
Cusmicia, Kuzmichia,
Cusmischia auctt. [err.]
STROBIELLA H. Nordsieck, 1977
ERJAVECIA Brusina, 1870
Trigonostoma Vest, 1867
[non Blainville, 1825]
FUSULUS Fitzinger, 1833
Dyodonta Hartmann, 1842
Diodonta auctt. [error]
ERJAVECIELLA H. Nordsieck, 1977
GRACILIARIA Bielz, 1867
Neostyria Wagner, 1920
Rupicola Hartman, 1844
[non Brisson, 1760]
JULICA H. Nordsieck, 1963
MACROGASTRA Hartmann, 1841
pro Iphigenia Gray, 1840
[non Schumacher, 1817]
PSEUDOVESTIA H. Nordsieck, 1977
Iphigena Gray, 1821 [nom.nud.]
MICRIDYLA H. Nordsieck, 1973
NEOSTYRIACA Wagner, 1920
PSEUDOFUSULUS H. Nordsieck, 1977
RUTHENICA Lindholm, 1924

BALEINAE
BALEA Gray, 1824
Baleastra L. Pfeiffer, 1855
Balaea, Balia, Bolaea, auctt. [err.]
ALINDA H. & A. Adams, 1855
Spoliata Frankenberger, 1912
PSEUDALINDA O. Boettger, 1877
Banatica O. Boettger, 1877 ?
BULGARICA O. Boettger, 1877
Idylopsina Lindholm, 1924
Idyla Westerlund, 1902
[non Vest, 1867]
DENTICULARIA Lindholm, 1924
PAVLOVICIA H. Nordsieck, 1973
LACINIARIA Hartmann, 1844
Rhodopiella H. Nordsieck, 1973
LIKHAREVIA H. Nordsieck, 1975
MENTISSA H. & A. Adams, 1855
Mantissa Chenu, 1859 [err.]
MENTISSELLA H. Nordsieck, 1973

BALEINAE (Cont'd.)
MICROPONTICA O. Boettger, 1881
BALEOPSINA Lindholm, 1924
MUCRONARIA O. Boettger, 1877
INDEX O. Boettger, 1877
MULTIPLICARIA Lindholm, 1924
Polyptychia O. Boettger, 1877
Euxina Wagner, 1920
(non O. Boettger, 1877)
QUADRIPLICATA O. Boettger, 1878
STRIGILECULA Kennard & Woodward, 1923
Strigillaria Vest, 1867
(non Rafinesque, 1815)
VESTIA Hesse, 1916
Uncinaria Vest, 1867
(non Froelich, 1789)
BRABENECIA H. Nordsieck, 1974
VESTIELLA H. Nordsick, 1977

Incertae Sedis (H. Nordsieck, 1977)
INOBSERATELLA Lindholm, 1924
KUESTERIELLA Lindholm, 1924
MIRALINDA Lindholm, 1924
Mira O. Boettger, 1877
(non Schellenberg, 1803)

(Achatinid Sigmurethra) (36b)

PARTULOIDEA (45h)
PARTULIDAE
PARTULA Férussac, 1821
Partulus Beck, 1837
Nenia Hartman, 1881
(non H. & A. Adams, 1855)
Astraea Hartman, 1881
(non Röding, 1798)
Clytia Hartman, 1881
(non Lamouroux, 1812)
Ilia Hartman, 1881
(non Leach, 1817)
Oenone Hartman, 1881
(non Savigny, 1817)
Helena Hartman, 1881
(non Risso, 1826)
Matata Hartman, 1881
Pasithea Hartman, 1881
(non Lamouroux, 1812)
CAROLINELLA Pilsbry, 1909
EUA Pilsbry & Cooke, 1934
LEPTOPARTULA Pilsbry, 1909
Echo Hartman, 1881
(non Selys, 1853)
MARIANELLA Pilsbry, 1909
Harmonia Hartman, 1881
(non Mulsant, 1846)
MARQUESANA Pilsbry, 1909
Aega Hartman, 1881
(non Leach, 1815)
Latia Hartman, 1881
(non Gray, 1849-1850)
MELANESICA Pilsbry, 1909
Sterope Hartman, 1881
(non Goodsir, 1845)
NESANASSA Pilsbry & Cooke, 1934
PALAOPARTULA Pilsbry, 1909
RENNELLIA Clench, 1941
SAMOANA Pilsbry, 1909
Evadne Hartman, 1881
(non Lovén, 1835)
THAKOMBAUA Pilsbry, 1909
Thacombaua Thiele, 1931 (err.)

ORTHALICOIDEA (36b) (Bulimulacea)
ORTHALICIDAE (7)
BULIMULINAE (7b)
BULIMULUS Leach, 1814
Leptomerus Albers, 1850
Loboa Ihering, 1917
COCHLEORNATA Scott, 1976
DENTAXIS Pilsbry, 1902
ITABORAHIA Maury, 1935
BOSTRYX Troschel, 1847
Kionoptyx Haas, 1966
Floreziellus, Naesiotellus Weyrauch,1967
ATAXELLUS Dall, 1912
ATAXUS Albers, 1850
BILAMELLIFERUS Weyrauch, 1958 ?
ELATIBOSTRYX Weyrauch, 1958
MULTIFASCIATUS Weyrauch, 1958
PAMPASINUS Weyrauch, 1958
PSEUDOPERONAEUS Weyrauch, 1958
DISCOBOSTRYX Pilsbry & Olsson, 1949
GEOCERAS Pilsbry, 1896
LISSOACME Pilsbry, 1896
PLATYBOSTRYX Pilsbry, 1896
GEOPYRGUS Pilsbry, 1896
Pyrgus Albers, 1850
(non Hübner, 1816)
PERONAEUS Albers, 1850
PHENACOTAXUS Dall, 1912
VERMETELLUS Haas, 1951
BOTHRIEMBRYON Pilsbry, 1894
Liparus von Martens in Albers, 1860
(non Albers, 1850)
Hartogembryon, Larapintembryon,
Satagembryon all Iredale, 1933
Celatembryon, Dialembryon, Ponembryon,
Telembryon all Iredale, 1939
PLECTOSTYLUS Beck, 1837 ?
Plecostylus Wurtz, 1947 (err.)
SPIROSCUTALUS Pilsbry, 1932 ?
TASMANEMBRYON Iredale, 1933
LEIOSTRACUS Albers, 1850
Liostracus von Martens in Albers, 1860
(non Angelin, 1854)
GRAPTOSTRACUS Pilsbry, 1939
PSEUDOXYCHONA Pilsbry, 1931
SCANSICOHLEA Pilsbry, 1930 ?
Scansicochlea Thiele, 1931 (em.)
NAESIOTUS Albers, 1850
Nesiotes, Nesiotus auctt. (err.)
Adenodia, Granella, Granitza, Granucis,
Nucisus, Ochsneria, Olinodia, Reclasta,
Saeronia, Stemmodiscus all Dall, 1920
Pelecostoma Reibisch, 1893
Pleuropyrgus von Martens, 1860
RHAPHIELLUS Pfeiffer, 1856
RABDOTUS Albers, 1850
Rhabdotus von Martens, 1860
Globulinus Fischer & Crosse, 1874
Orthotomium Fischer & Crosse, 1874 ?
HANNARABDOTUS Emerson & Jacobson, 1964
LEPTOBYRSUS Fischer & Crosse, 1875
Sonorina Pilsbry, 1896
Puritana Jacobson, 1958
MARANHONIELLUS Weyrauch, 1958 ?
PLICOCOLUMNA J. G. Cooper, 1895
Columna J. G. Cooper, 1892
(non Perry, 1811)
Pseudorhodea Dall, 1895
PURITANINA Jacobson, 1958
RHINUS Albers, 1860
SCUTALUS Albers, 1850
Bocourtia Rochebrune, 1882
Xenothauma Fulton, 1896

BULIMULINAE

SCUTALUS (Cont'd.)
SUNIELLUS Breure, 1978
VERMICULATUS Breure, 1978

BULIMULINAE s.l.

ASPASTUS Albers, 1850
CALLISTOCHARIS Pilsbry, 1900
Charis Albers, 1850
(non Hübner, 1816)
LEUCOCHARIS Pilsbry, 1900
POECILOCHARIS Kobelt, 1891
AURIS Spix, 1827
Pachyotus Beck, 1837 (part)
ANTITRAGUS Oberwimmer, 1931
BERENDTIA Crosse & Fischer, 1869 ?
COCHLORINA Jan, 1830
Navicula Spix, 1827
(non Blainville, 1825)
Zaplagius Pilsbry, 1896
DIPLOMORPHA Ancey, 1884
Hebridaria Pilsbry, 1900
DRYMAEUS Albers, 1850
Hamadryas Albers, 1850
(non Hübner, 1806)
Semiclausaria Pfeiffer, 1856
Mormus Albers, 1860
Otostomus von Martens, 1873 (37)
(non Beck, 1837)
Goniognathmus Fischer & Crosse, 1875
ANTIDRYMAEUS Germain, 1907
LEPTODRYMAEUS Pilsbry, 1946 (37)
LEPTOMORMUS Weyrauch, 1958
MESEMBRINUS Albers, 1850 (8a)
METADRYMAEUS Pilsbry, 1926
ORODRYMAEUS Pilsbry, 1926
ORNATIMORMUS Weyrauch, 1958
DRYPTUS Albers, 1860 (53)
KUSCHELENIA Hylton Scott, 1951
LOPESIANUS Weyrauch, 1958
NEOPETRAEUS von Martens, 1885
NEWBOLDIUS Pilsbry, 1932
OTOSTOMUS Beck, 1837
OXYCHONA Mörch, 1852
PAENISCUTALUS Wurtz, 1947 ?
PLACOSTYLUS Beck, 1837
BASILEOSTYLUS Haas, 1935
MAORISTYLUS Haas, 1935
EUMECOSTYLUS Albers, 1860
EUPLACOSTYLUS Crosse, 1875
MALAITELLA Clench, 1941
Acrostylus Clench, 1935
(non Cossmann, 1896)
PLACOCHARIS Pilsbry, 1900
PROASPASTUS Clench, 1941
SANTACHARIS Iredale, 1927
PLECOCHILUS Guilding, 1827
Pelecocheilus, Pelecychilus,
Pelekocheilus, Pleocheilus,
Plecocheilus, Plekocheilus auctt.(err.)
Caprella Guilding, 1824
(non Lamarck, 1801)
AEROPICTUS Weyrauch, 1967
Orcesiellus Weyrauch, 1967
EUDOLICHOTIS Pilsbry, 1896
EURYTUS Albers, 1850
SPARNOTION Pilsbry, 1944
PROTOGLYPTUS Pilsbry, 1897
OBSTRUSSUS Parodiz, 1946
RIMATULA Parodiz, 1946
SPHAEROCONCHA Breure, 1978
STENOSTYLUS Pilsbry, 1898
THAUMASTUS Albers, 1860

THAUMASTUS (Cont'd.)
Orphnus Albers, 1850
(non MacLeay, 1819)
Orphaicus Schaufuss, 1869
Tatutor Jousseaume, 1887
ATAHUALPA Strebel, 1910
Tholus, Pachytholus Strebel, 1909
KARA Strebel, 1910
QUECHUA Strebel, 1910
SCHOLVIENA Strebel, 1910
Thomsenia Strebel, 1910
THAUMASTIELLA Weyrauch, 1956

ORTHALICINAE

ORTHALICUS Beck, 1837
Oxystyla Schlüter, 1838
Zebra Shuttleworth, 1856
Ortalichus von Martens, 1893
CORONA Albers, 1850
Paraliguus Pilsbry, 1899
HEMIBULIMUS von Martens, 1885
MYIORTHALICUS Strebel, 1909
LIGUUS Montfort, 1810
Chersina Beck, 1837
(non Férussac, 1821)
Pseudotrochus Herrmannsen, 1847
Lignus Nevill, 1878
(non Gray, 1834)
Ilotia L. Pfeiffer, 1881
LIGUELLUS Clench, 1946
OXYSTROMBUS Mörch, 1852
Orthalicinus Fischer & Crosse, 1873
PORPHYROBAPHE Shuttleworth, 1856
OXYORTHALICUS Strebel, 1909
SULTANA Shuttleworth, 1856
Orthaliscus Gray, 1847
Orthalicus Pilsbry, 1899
(non Beck, 1837)
CLATHRORTHALICUS Strebel, 1909
LAEORTHALICUS Strebel, 1909
Laeiorthalicus Strebel, 1909
MELANIORTHALICUS Strebel, 1909
METORTHALICUS Pilsbry, 1899
TRACHYORTHALICUS Strebel, 1909

ODONTOSTOMIDAE

ODONTOSTOMUS Beck, 1837
Macrodontes Swainson, 1840
Odontostoma Forcart, 1946 (em.)
(non Turton, 1829)
MACRODONTOPSIS Thiele, 1931
ANCTUS von Martens, 1860
ANOSTOMA Fischer von Waldheim, 1807
Tomogeres Montfort, 1810
Angystoma Schumacher, 1817
Tomogerus Blainville, 1824
Anastoma Cristofori & Jan, 1832
Tomogerina Jousseaume, 1877
RINGICELLA Gray, 1847
CLESSINIA Döring, 1874
Scalarinella Döring in Dohrn, 1874
CYCLODONTINA Beck, 1837
BAHIENSIS Jousseaume, 1877
BARRINGTONIA Parodiz, 1944
MORICANDIA Pilsbry & Vanatta, 1898
PANTAGRUELINA Forcart, 1946
HYPERAULAX Pilsbry, 1897
BONNANIUS Jousseaume, 1900
PLAGIODONTES Döring, 1876
PILSBRYLIA Hylton Scott, 1952
SPIXIA Pilsbry & Vanatta, 1898
Eudontostomus Holmberg, 1912
Ventania Parodiz, 1940

ODONTOSTOMIDAE (Cont'd.)
TOMIGERUS Spix, 1827
Cearella Ihering, 1906
pro Pilsbryella Ihering, March, 1905
(non Nierstrasz, January, 1905)
DIGERUS Haas, 1937

AMPHIBULIMIDAE
AMPHIBULIMA Lamarck, 1805
Amphibulia Rafinesque, 1815 (em.)
Brachyspira L. Pfeiffer, 1855
Rowsoni Bland, 1875
Mastogyra Ancey, 1881
RHODONYX Fischer, 1873
GAEOTIS Shuttleworth, 1854
PELLICULA Fischer, 1856
PELTELLA Webb & Beneden, 1836
PRESTONELLA Connolly, 1929
SIMPULOPSIS Beck, 1837
Simulopsis Gray, 1847
Miersi L. Pfeiffer, 1856 ?
Dominicensis L. Pfeiffer, 1858 ?
Bulimulopsis Pilsby, 1899
EUDIOPTUS Albers, 1860
Pseudoglandina Weyrauch, 1967

UROCOPTIDAE
TETRENTODONINAE (28)
TETRENTODON Pilsbry, 1903
Trachelia L. Pfeiffer,1853
(non Scopoli, 1777)
ANAFECOPTIS Jaume & de la Torre, 1972
CYLINDROCOPTIS Jaume & de la Torre, 1972
GRAVICOPTIS Jaume & de la Torre, 1972
SCALARICOPTIS Jaume & de la Torre, 1972
HETEROCOPTIS Jaume & de la Torre, 1972
ASTHENOCOPTIS Jaume & de la Torre, 1972
BERMUDEZICOPTIS Jaume & de la Torre, 1972
CAVICOPTIS Jaume & de la Torre, 1972
CLENCHICOPTIS Jaume & de la Torre, 1972
TORRECOPTIS Bartsch, 1943

BRACHYPODELLINAE (28)
BRACHYPODELLA Beck, 1837
Cylindrella L. Pfeiffer, 1840 (part)
(non Swainson, 1840)
Siphonostoma Swainson, 1840
(non Otto, 1821)
Brachypus Guilding, 1828
(non Swainson, 1824)
Brachypodisca Agassiz, 1846
Spiraxis Pilsbry, 1903 ?
(non C. B. Adams, 1850)
ANGULICERVIX Pilsbry, 1933
BREVIPEDELLA Pilsbry, 1903
Katastoma Clench, 1935 ?
CYCLAUCHEN Pilsbry, 1930
GEOSCALA Pilsbry & Vanatta, 1898
GYRAXIS Pilsbry, 1903
Spiraxilla Thiele, 1931
ISCHNOSTROPHINA Pilsbry, 1933
SIMPILICERVIX Pilsbry, 1903
SIPHONOLAEMUS Pilsbry, 1903

UROCOPTINAE
UROCOPTIS Beck, 1837
Cylindrella L. Pfeiffer, 1840 (part)
(non Swainson, 1840)
BACTROCOPTIS Pilsbry, 1902
SPIROCOPTIS Pilsbry, 1902
ACROCOPTIS Jaume & de la Torre, 1972
ALLOCOPTIS Thompson, 1976
AMPHICOSMIA Pilsbry & Vanatta, 1898

UROCOPTINAE (Cont'd.)
AMPHISTEMMA Jaume & de la Torre, 1972
ANOMA Albers, 1850
Leia Albers, 1850 (non Meigen, 1818)
Lia Mörch, 1852 (non Eschscholtz, 1829)
Inliaculus Schaufuss, 1869
Vendryesia Simpson, 1894
Vendrysia Pilsbry, 1903 (err.)
APOMA Beck, 1837
Casta Albers, 1850
ARANGIA Pilsbry & Vanatta, 1898
AUTOCOPTIS Pilsbry, 1902
UROCOPTOLA Clench, 1935
BADIOFAUX Pilsbry, 1941
BIALASMUS Jaume & de la Torre, 1972
CALLOCOPTIS Jaume & de la Torre, 1972
CALLONIA Crosse & Fischer, 1870
CAPILLACEA Jaume & de la Torre, 1972
CARCINOSTEMMA Jaume & de la Torre, 1972
CENTRALIA Jaume & de la Torre, 1972
MIMIAXIS Jaume & de la Torre, 1972
COCHLODINELLA Pilsbry & Vanatta, 1898
Cylindrella Pfeiffer (part)
& authors prior to 1902
BLAINCOPTIS Jaume & de la Torre, 1972
ORIENTICOPTIS Jaume & de la Torre, 1972
VENTRICOCHLIS Jaume & de la Torre, 1972
GEMINICOPTIS Jaume & de la Torre, 1972
GONGYLOSTOMA Albers, 1850
GONGYLOSTOMELLA Pilsbry, 1941
ARANGOCOPTIS Jaume & de la Torre, 1972
JUANOCOPTIS Jaume & de la Torre, 1972
LEVICOPTIS Jaume & de la Torre, 1972
PALENQUECOPTIS Jaume & de la Torre, 1972
IDIOSTEMMA Pilsbry & Vanatta, 1898
FIBRICUTIS Pilsbry, 1903
MACEO Pilsbry & Vanatta, 1898
LEVISTEMMA Jaume & de la Torre, 1972
LIOCALLONIA Pilsbry, 1902
CENOCOPTIS Jaume & de la Torre, 1972
JAUMEA Jaume & de la Torre, 1972
LIPAROTES Pilsbry, 1903
MACROCERAMUS Guilding, 1828 (Macroceraminae ?)
Leptospira Swainson, 1840 (part)
MYCHOSTOMA Albers, 1850
NESOCOPTUS Pilsbry, 1941
NODULIA Jaume & de la Torre, 1972
ORGANOCOPTIS Jaume & de la Torre, 1972
PARACALLONIA Pilsbry, 1903
PFEIFFERICOPTIS Jaume & de la Torre, 1972
GARCIACOPTIS Jaume & de la Torre, 1972
PINERIA Poey, 1854
PLANOSTEMMA Jaume & de la Torre, 1972
PLEUROSTEMMA Pilsbry, 1941
POECILOCOPTIS Pilsbry, 1941
POEYCOPTIS Jaume & de la Torre, 1972
PYCNOPTYCHIA Pilsbry & Vanatta, 1898
ESOCHARA Pilsbry & Vanatta, 1898
SAGRACOPTIS Jaume & de la Torre, 1972
SECTILUMEN Pilsbry & Vanatta, 1898
SPIROSTEMMA Pilsbry & Vanatta, 1898
STEATOCOPTUS Pilsbry, 1941
STROPHINA Mörch, 1852
TENERIA Jaume & de la Torre, 1972
TENUISTEMMA Jaume & de la Torre, 1972
TOMELASMUS Pilsbry & Vanatta, 1898
Ternelasmus auctt. (err.)
Scopulospica Pilsbry, 1941
TRILAMELLAXIS Jaume & de la Torre, 1972
UNCINICOPTIS Jaume & de la Torre, 1972

JOHANICERAMINAE (28)
JOHANICERAMUS Jaume & de la Torre, 1972

EUCALODIINAE
EUCALODIUM Crosse & Fischer, 1868
Thaumasia Albers, 1850
(non Perty, 1830)
Spartina Harris & Burrows, 1891
OLIGOSTYLUS Pilsbry, 1895
RESUPINATA von Martens, 1897
ANISOSPIRA Strebel & Pfeffer, 1880
DISSOTROPIS Bartsch, 1906
ARCHEGOCOPTIS Pilsbry, 1903
COELOCENTRUM Crosse & Fischer, 1872
CROSSOSTEPHANUS Dall, 1908
ELASMOCENTRUM Pilsbry, 1902
GYMNOCENTRUM Pilsbry, 1942
Liocentrum Pilsbry, 1902
(non Karsch, 1890)
PTYCHOCENTRUM Bartsch, 1943
PTYCHODONTA Bartsch, 1906
SCHIZOPYLE Pilsbry, 1939
EPIROBIA Strebel & Pfeffer, 1880
GYROCION Pilsbry, 1904
HENDERSONIELLA Dall, 1905
Hendersonia Dall, 1905
(non Wagner, 1905)
SPARTOCENTRUM Dall, 1895
Teneritia Mabille, 1897

HOLOSPIRINAE
HOLOSPIRA von Martens, 1860
Acera Albers, 1850
(non Cuvier, 1810)
ALLOCORYPHE Pilsbry, 1946
BOSTRICHOCENTRUM Strebel & Pfeffer, 1880
Eudistemma, Distomospira Dall, 1895
Haplostemma Dall, 1895
HAPLOCION Pilsbry, 1902
Liostemma Bartsch, 1906
MALINCHEA Bartsch, 1945
pro Tristemma Bartsch, 1906
(non Brandt, 1835)
PRIONOPLAX Pilsbry, 1953
COELOSTEMMA Dall, 1895
Megaxis Pilsbry, 1946
APERTAXIS Pilsbry, 1953
CRYCORYNE Thompson, 1971
STYLOPTYX Thompson, 1971
METASTOMA Strebel & Pfeffer, 1880
PROPILSBRYA Bartsch, 1906
STALACTELLA Bartsch, 1906
PECTINISTEMMA Rehder, 1940

MICROCERAMIDAE (?)
MICROCERAMUS Pilsbry & Vanatta, 1898
Colobus Albers, 1850
(non Illiger, 1811)
SPIROCERAMUS Pilsbry & Vanatta, 1898

CERIONIDAE (Ceriidae)
CERION Roding, 1798
Cerium Link, 1807
Puparia Rafinesque, 1814
Cochlodonta Férussac, 1821
pro Pupa Lamarck, 1801
(non Röding, 1798)
Puppa Montfort, 1810
Canistrum Fabricius, 1823
Cochlodon Sowerby, 1825 (part)
Cereales Férussac (nom.nud.)
Pulpa Poey, 1858
CYCLOCERION Bartsch, 1952
DIACERION Dall, 1894
Paracerion Pilsbry & Vanatta, 1895
Tridentistrophia Maynard, 1896

CERIONIDAE
CERION (Cont'd.)
STROPHIOPS Dall, 1894
Strophia Albers, 1850
(non Meigan, 1832)
Maynardia Dall, 1894
Longidens, Multistrophia,
Pinguitia, Seniculus,
Umbonis all Maynard, 1896

MEGASPIRIDAE (?)
MEGASPIRA Jay, 1836
Pyrgelix Beck, 1837
CALLIONEPION Pilsbry & Vanatta, 1899

COELOCIIDAE (36b) (?)
COELOCION Pilsbry, 1904
PERRIERIA Tapparone-Canefri, 1878

ACHATINOIDEA
FERUSSACIIDAE
FERUSSACIA Risso, 1826
Ferrusacia Auctt. (err.)
Vediantius Risso, 1826
Euferussacia Bourguignat, 1864
Folliculiana Bourguignat, 1864
Pseudostreptostyla Nevill, 1881
Folliculina Westerlund, 1887
(non Lamarck, 1816)
PEGEA Risso, 1826
Pseudazeca L. Pfeiffer, 1877
Proceruliana Bourguignat, 1864
Phylacus Westerlund, 1887
AMPHORELLA Lowe, 1852
Agraulina Bourguignat, 1858
Lovea B. Watson, 1876
FUSILLUS Lowe, 1852
Alsobia Bourguignat, 1858
HYPSELIA Lowe, 1854
PYRGELLA Lowe, 1854
CALAXIS Bourguignat, 1887
Elasmophora Westerlund, 1887
PSEUDOCALAXIS Pallary, 1912
CECILIOIDES Férussac, 1814
Acicula Risso, 1826
(non Hartmann, 1821)
Styloides Fitzinger, 1833 (37)
Caecilioides Herrmannsen, 1846
Caecilianella Bourguignat, 1856
Aciculina Westerlund, 1887
(non A. Adams, 1853)
Belonis Hartmann, 1840 ?
GEOSTILBIA Crosse, 1867
KAROLUS de Folin in de Folin & Perrier, 1870
Caecilianopsis Pilsbry, 1907
RHAPHIDIELLA Maltzan, 1886
TEREBRELLA Maltzan, 1886
COILOSTELE Benson, 1864
Francesia Paladilhe, 1872
Coelostele Crosse, 1876
Coelestele Bourguignat, 1880
CONNOLLYA Odhner, 1932
CYLICHNIDIA Lowe, 1852
DIGONIAXIS Jousseaume, 1889
HOHENWARTIANA Bourguignat, 1864
Hohenwarthia Bourguignat, 1887
SCULPTIFERUSSACIA Germain, 1911

SUBULINIDAE
SUBULININAE (Subulini)
SUBULINA Beck, 1837
Macrospira Swainson, 1840
ITIOPIANA Preston, 1910

SUBULININAE (Cont'd.)

BACILLUM Theobald, 1870
BOCAGEIA Girard, 1893
LIOBOCAGEIA Pilsbry, 1919
PETRIOLA Dall, 1905
Trichodina Ancey, 1888
(non Ehrenberg, 1830)
CERAS Dupuis & Putzeys, 1901
CHILONOPSIS Fischer von Waldheim, 1848
CLEOSTYLA Dall, 1896
CURVELLA Chaper, 1885
Hapalus Albers, 1850
(non Billberg, 1820)
HYPOLYSIA Melvill & Ponsonby, 1901
GLESSULA Von Martens, 1860
Electra Albers, 1850
(non Lamouroux, 1816)
Dikrangia Godwin-Austen, 1920
Jadukamia Godwin-Austen, 1920
Rishetia Godwin-Austen, 1920
HOMORUS Albers, 1850
STRIOSUBULINA Thiele, 1933
KRAPFIELLA Preston, 1911
LAMELLAXIS Strebel & Pfeffer, 1882
ALLOPEAS H. B. Baker, 1935
LEPTOPEAS H. B. Baker, 1927
EUTOMOPEAS Pilsbry, 1946
Tomopeas Pilsbry, 1906
(non Miller, 1900)
LEPTINARIA Beck, 1837
Nothus Albers, 1850
(non Olivier, 1811)
LUNTIA E. A. Smith, 1898
NEOSUBULINA E. A. Smith, 1898
PELATRINIA Pilsbry, 1907
NEOGLESSULA Pilsbry, 1909
NOTHAPALUS von Martens, 1897
Kenia Preston, 1911
OCHRODERMA Ancey, 1885
OCHRODERMATINA Thiele, 1931
OCHRODERMELLA Pilsbry, 1907
OPEAS Albers, 1850
(non Foerster, 1856)
MICROPEAS Connolly, 1923
NOTHAPALINUS Connolly, 1923
PLICAXIS Sykes, 1903
Rhodina Morgan, 1885
(non Guenée, 1854)
PROSOPEAS Mörch, 1876
BALFOURIA Crosse, 1884
COMOROPEAS Pilsbry, 1906
PAROPEAS Pilsbry, 1906
PSEUDOGLESSULA Boettger, 1892
ISCHNOGLESSULA Pilsbry, 1919
KEMPIOCONCHA Preston, 1913
Kempia Preston, 1913
(non Mathews, 1912)
Pseudocerastus Germain, 1918
Pseudoceratus Fischer-Piette, 1947
PSEUDOPEAS Putzeys, 1899
Beccaria Bourguignat, 1883
(non Trinchese, 1870)
DYSOPEAS H. B. Baker, 1927
EREMOPEAS Pilsbry, 1906
RIEBECKIA von Martens, 1883
SUBULINISCUS Pilsbry, 1919
SUBULONA von Martens, 1889
DICTYOGLESSULA Pilsbry, 1919
OREOHOMORUS Pilsbry, 1919
TORTAXIS Pilsbry, 1906
TRISTANIA Boettger, 1878 ?

RUMININAE

RUMINA Risso, 1826
Stenogyra authors (non Shuttleworth)
Orbitina Risso, 1826
Cylindrina Schlüter, 1838
Sira A. Schmidt, 1855
NAMIBIELLA Zilch, 1954
OBELISCELLA Jousseaume, 1899
XEROCERASTUS Kobelt & Möllendorff, 1902
Eburnea Mousson, 1887
(non Fleming, 1828)
LUBRICETTA Haas, 1928
ZOOTECUS Westerlund, 1887
Chilogymnus Jousseaume, 1894

OBELISCINAE

OBELISCUS Beck, 1837
DOLICHOLESTES Pilsbry, 1906
ISCHNOCION Pilsbry, 1907
LYOBASIS Pilsbry, 1903
PROTOBELISCUS Pilsbry, 1906
PSEUDOBALEA Shuttleworth, 1854
RECTOBELUS H. B. Baker, 1937
STENOGYRA Shuttleworth, 1854
CUPULELLA Aguayo & Jaume, 1948
EUONYMA Melvill & Ponsonby, 1896
NEOBELISCUS Pilsbry, 1896
PSEUDOCOLUMNA Wenz, 1920
RHODEA H. & A. Adams, 1855
SYNAPTERPES Pilsbry, 1896
Oxycheilus Albers, 1850
Oxychilus L. Pfeiffer, 1856
(non Fitzinger, 1833)
CHRYSERPES Pilsbry, 1906
PROMOUSSONIUS Pilsbry, 1906
ZONIFERELLA Pilsbry, 1906

CRYPTELASMINAE (?)

CRYPTELASMUS Pilsbry, 1907
EUCRYPTELASMUS Jaume & Sanchez, 1943

ACHATINIDAE

ACHATINA Lamarck, 1799
Chersina Humphrey, 1797 (part)
Ampulla Röding, 1798 (part)
Cochlitoma Férussac, 1807
Agathina Férussac, 1807
Achatium Link, 1807 (part)
Achathina, Achantina, Achatinum, Aghatina, Achatinus Auctt. (err.)
Oncaea Gistl, 1848
(non Philippi, 1843)
Geodes Gistl, 1848
Urceus Mörch, 1857
Parachatina, Sepaea Bourguignat, 1889
EUAETHIOPINA Bequaert, 1950
Euaethiops Clench & Archer, 1930
(non Hampson, 1926)
LEPTOCALA Ancey, 1888
Petitia Jousseaume, 1884
(non Chitty, 1857)
Leptocola Kobelt, 1910
(non Gerstaecker, 1883)
LEPTOCALINA Bequaert & Clench, 1934
LEPTOCALLISTA Pilsbry, 1904
LISSACHATINA Bequaert, 1950
PINTOA Bourguignat, 1889 ?
TRIPACHATINA Bourguignat, 1889
ARCHACHATINA Albers, 1850
CALACHATINA Pilsbry, 1919
Megachatina Bequaert & Clench, 1936
MEGACHATINOPSIS Bequaert & Clench, 1936
THOLACHATINA Bequaert, 1950

ACHATINIDAE (Cont'd.)
ATOPOCOCHLIS Crosse & Fischer, 1888
Eutaxis Ancey, 1888
BURTOA Bourguignat, 1889
Burtopsis Bourguignat, 1889
Livinhacia Crosse, 1889
CALLISTOPEPLA Ancey, 1888
Ganomidos Ailly, 1896
COLUMNA Perry, 1811
LIGNUS Gray, 1834
Pseudotrochus Mörch, 1852
(non Herrmannsen, 1847)
Perideris Shuttleworth, 1856
(non Brandt, 1835)
LIMICOLARIA Schumacher, 1817
Limicularia Schumacher, 1817
Limicolarius Beck, 1837
Omphalostyla Schlüter, 1838 ?
LIMICOLARIOPSIS Ailly, 1910
Rebmanniella Preston, 1911
METACHATINA Pilsbry, 1904
PERIDERIOPSIS Putzeys, 1898
PSEUDACHATINA Albers, 1850

COELIAXIDAE
COELIAXIS Adams & Angas, 1865
Bathyaxis Ancey, 1887
Sphalerostoma Girard, 1892
PYRGINA Greef, 1882
THOMEA Girard, 1893

THYROPHORELLIDAE (36a -?)
THYROPHORELLA Graeff, 1882

AILLYOIDEA
AILLYIDAE
AILLYA Odhner, 1927

OLEACINOIDEA
SPIRAXIDAE
SPIRAXINAE
SPIRAXIS C. B. Adams, 1850
BIANGULAXIS Pilsbry, 1907
DIGNAXIS H. B. Baker, 1939
EUSPIRAXIS L. Pfeiffer, 1854
RAVENIA Crosse, 1873
REPRESSAXIS H. B. Baker, 1939
MICROMENA H. B. Baker, 1939
MIRARADULA H. B. Baker, 1939
PSEUDOSUBULINA Strebel & Pfeffer, 1882
TORNAXIS von Martens, 1898
RECTAXIS H. B. Baker, 1926
VOLUTAXIS Strebel & Pfeffer, 1882
MIRAPEX H. B. Baker, 1939
VERSUTAXIS H. B. Baker, 1939

STREPTOSTYLINAE
STREPTOSTYLA Shuttleworth, 1852
CHERSOMITRA von Martens, 1860
EUSTREPTOSTYLA H. B. Baker, 1927
PETENIELLA Pilsbry, 1907
Petenia Crosse & Fischer, 1869
(non Gunther, 1862)
SALASIELLA Strebel, 1878
PERPUSILLA H. B. Baker, 1941
STREPTOSTYLOPS Pilsbry, 1933

EUGLANDININAE
EUGLANDINA Fischer & Crosse, 1870 (2, 3)
Pfaffia Behn, 1845
Glandina Binney, 1878
(non Schumacher, 1817)
COSMOMENUS H. B. Baker, 1941

EUGLANDINA (Cont'd.)
LAEVIGLANDINA Pilsbry, 1908
SINGLEYA H. B. Baker, 1941
GUILLARMODIA H. B. Baker, 1941
PROAMERIA H. B. Baker, 1941
PITTIERIA von Martens, 1901 ?
SHUTTLEWORTHIA H. B. Baker, 1941
POIRETIA Fischer, 1883
PSEUDOLEACINA Wenz, 1914
VARICOGLANDINA Pilsbry, 1908
VARICOTURRIS Pilsbry, 1907
GHIESBREGHTIA H. B. Baker, 1941
STREPTOSTYLELLA Pilsbry, 1907

OLEACINIDAE
OLEACININAE
OLEACINA Röding, 1798
Polyphemus Montfort, 1810
Glandina Schumacher, 1817
Boltenia L. Pfeiffer, 1878
(non Savigny, 1821)
FLAVOLEACINA Pilsbry, 1908
LAEVOLEACINA Pilsbry, 1907
PLICOLEACINA Pilsbry & Vanatta, 1928
ORYZOSOMA Pilsbry, 1891
Orizosoma Pilsbry, 1891
RECTOLEACINA Pilsbry, 1907
STREBELIA Crosse & Fischer, 1868
Physella L. Pfeiffer, 1861
(non Haldeman, 1842)
Spirobulla Ancey, 1881

VARICELLINAE
VARICELLA L. Pfeiffer, 1856
Melia Albers, 1850
(non Billberg, 1820)
VAGAVARIX H. B. Baker, 1941
VARICELLARIA Pilsbry, 1907
VARICELLINA Pilsbry, 1907
VARICELLIDEA Pilsbry, 1907
VARICELLOPSIS Pilsbry, 1907
GLANDINELLA L. Pfeiffer, 1878
LAEVARICELLA Pilsbry, 1907
BORIQUENA H. B. Baker, 1941
MELANIELLA L. Pfeiffer, 1857 (37)
Pichardiella P. Fischer, 1887
VARICELLULA Pilsbry, 1907
SIGMATAXIS Pilsbry, 1907

TESTACELLIDAE (36b)(45h)
TESTACELLA Cuvier, 1800
Testacellus Férussac, 1819
TESTACELLOIDES Wagner, 1914

STREPTAXOIDEA
STREPTAXIDAE
STREPTAXINAE
STREPTAXIS Gray, 1837
Artemon Beck, 1837
Alcidia Bourguignat, 1889
(non Westwood, 1879)
AUGUSTULA Thiele, 1931
CORENA A. Adams, 1870
DISCARTEMON L. Pfeiffer, 1856
EDENTULINA L. Pfeiffer, 1856
EUSTREPTAXIS L. Pfeiffer, 1878
Afristreptaxis Thiele, 1932
GLYPTOCONUS Möllendorff, 1894
GONAXIS Taylor, 1877
HAPLOPTYCHIUS Möllendorff, 1905
Pseudartemon Mabille, 1887 ?
IMPERTURBATIA von Martens, 1898

STREPTAXINAE (cont'd.)
INDOARTEMON Forcart, 1946
Odontartemon Möllendorff, 1905
(non L. Pfeiffer, 1856)
MARCONIA Bourguignat, 1889
STENOMARCONIA Germain, 1934
MARTINELLA Jousseaume, 1887
MICRARTEMON Möllendorff, 1890
ODONTARTEMON L. Pfeiffer, 1856 ?
Lamelliger Ancey, 1884
OOPHANA Ancey, 1884
PERROTTETIA Kobelt, 1905
PLATYCOCHLIUM Laidlaw, 1950
RECTARTEMON H. B. Baker, 1925
HYPSELARTEMON Wenz, 1947
SAIROSTOMA Haas, 1938
STEMMATOPSIS Mabille, 1887
STREPTARTEMON Kobelt, 1905
TAYLORIA Bourguignat, 1889
ARTEMONOPSIS Germain, 1908
COLPANOSTOMA Bourguignat, 1889
GIGANTAXIS Tomlin, 1930 ?
Gibbonsia Bourguignat, 1889
(non Cooper, 1864)
MACROGONAXIS, PSEUDOGONAXIS Thiele, 1932
TONKINIA Mabille, 1887

ENNEINAE (Ptychotrematinae) (53)
ENNEA H. & A. Adams, 1855
Enneastrum L. Pfeiffer, 1856
Carychiopsis von Martens, 1895
(non Sandburger, 1872)
ACANTHENNA von Martens, 1898
BRUGGENNEA Dance, 1972
DIAPHERA Albers, 1850
Diaphora von Martens, 1860
ELMA A. Adams, 1866
GIBBULINELLA Wenz, 1920
Oppenheimiella Pfeffer, 1929
(non Meunier, 1893)
Webbia Odhner, 1931
GIBBUS Montfort, 1810
Gibbulina Beck, 1837 (part)
GONIDOMUS Swainson, 1840
Idolum L. Pfeiffer, 1856
PLICADOMUS Swainson, 1840
GONOSPIRA Swainson, 1840
MICROSTROPHIA Möllendorff, 1887
Nevillia von Martens, 1880
(non H. Adams, 1868)
ORTHOGIBBUS Germain, 1919
GULELLA L. Pfeiffer, 1856
AENIGMIGULELLA Pilsbry & Cockerell, 1933
AVAKUBIA, CONOGULELLA, COSTIGULELLA,
PLICIGULELLA, PRIMIGULELLA, SILVIGULELLA,
TORTIGULELLA all Pilsbry, 1919
DIGULELLA Haas, 1934
HUTTONELLA, UNIPLICARIA L. Pfeiffer, 1856
MIRAGULELLA Pilsbry & Cockerell, 1933
MOLARELLA Connolly, 1922
PAUCIDENTELLA Thiele, 1933
PAUCIDENTINA von Martens, 1897
Paucidentata Preston, 1916
RHABDOGULELLA Haas, 1934
SPHINCTEROCOCHLION Verdcourt, 1985
PARENNEA Pilsbry, 1919
WILMATTINA Pilsbry & Cockerell, 1933
PSEUDELMA Kobelt, 1904
Fultonelma Haas, 1951
PTYCHOTREMA L. Pfeiffer, 1853
ADJUA Chaper, 1885
EXCISA Ailly, 1896
HAPLONEPION, PTYCHOON Pilsbry, 1919

PTYCHOTREMA (Cont'd.)
MIRELLIA Thiele, 1933
Thaumatogulella Haas, 1951
NSENDWEA Dupuis & Putzeys, 1923
SPHINCTOSTREMA Girard, 1894
Sphinctotrema Thiele, 1931
PUPIGINELLA Pilsbry, 1919
SINOENNEA Kobelt, 1904
INDOENNEA Kobelt, 1904
STEREOSTELE Pilsbry, 1919
STREPTOSTELE H. Dohrn, 1866
Campylaxis Ancey, 1888
GRAPTOSTELE Pilsbry, 1919
RAFFRAYA Bourguignat, 1883
Ischnostele Boettger, 1915
TEXTOSTELE Venmans, 1959
TOMOSTELE Ancey, 1885
Eustreptostele Germain, 1915
VARICOSTELE Pilsbry, 1919

STROPHOCHEILOIDEA (30)
STROPHOCHEILIDAE
STROPHOCHEILUS Spix, 1827
Otis Férussac, 1821
(non Linnaeus, 1758)
Strophochilus Agassiz, 1846 (err.)
Coniclus Albers, 1850
Stropocheilus Ihering, 1929 (err.)
ANTHINUS Albers, 1850
AUSTROBORUS Parodiz, 1949
pro Microborus Pilsbry, 1926
(non Blanford, 1897)
CHILIBORUS Pilsbry, 1926
GONYOSTOMUS Beck, 1837
Gonyostoma, Goniostoma, Goniostomus,
Gonyotomus auctt. (err.)
MIRINABA Morretes, 1952
Metara Morretes, 1952
SPEIRONEPION Bequaert, 1948

MEGALOBULIMIDAE (30)
MEGALOBULIMUS Miller, 1878
Bulimus Scopoli, 1786 (non 1777)
Borus Albers, 1850
(non Agassiz, 1846)
Corus Jousseaume, 1877; Strand, 1928
Megabulimus Haas, 1935
Streptocheilus Souza Lopez, 1940
Lophocheilus Daniel, 1942
Phaiopharus, Psiloicus Morretes, 1952

DORCASIIDAE
DORCASIA Gray, 1838
TRIGONEPHRUS Pilsbry, 1905
Galaxias Beck, 1837
(non Cuvier, 1817)
TULBAGHINA Melvill & Ponsonby, 1898
Tulbaghina Gude, 1921 (err.)

ACAVOIDEA (45h - 36b)
ACAVIDAE
ACAVINAE
ACAVUS Montfort, 1810
AMPELITA Beck, 1837
EURYSTYLA Ancey, 1887
Poecilostylus Pilsbry, 1890
HELICOPHANTA Férussac, 1821
LEUCOTAENIUS von Martens in Albers, 1860
Pseudoclavator Germain, 1914
OLIGOSPIRA Ancey, 1887
STYLODON Beck, 1837
Columplica Hartmann, 1843
Pachys Albers, 1850

CLAVATORINAE
CLAVATOR von Martens in Albers, 1860

CARYODIDAE
CARYODES Albers, 1850
ANOGLYPTA von Martens in Albers, 1860
HEDLEYELLA Iredale, 1914
Liparus Albers, 1850
(non Olivier, 1807)
Panda Albers, 1860
(non Heyden, 1826)
BRAZIERESTA, PANDOFELLA, PYGMIPANDA
all Iredale, 1933
PEDINOGYRA Albers, 1860

MACROCYCLIDAE
MACROCYCLIS Beck, 1837

MEGOMPHICIDAE (Ammonitellidae)
MEGOMPHIX H. B. Baker, 1930
AMMONITELLA J. G. Cooper, 1868
Gonostoma Binney, 1878
(non Held, 1837)
POLYGYRELLA W. G. Binney, 1863
Adelodonta Ancey, 1880
POLYGYROIDEA Pilsbry, 1930 (37)

RHYTIDOIDEA
RHYTIDIDAE (Paryphantidae)
RHYTIDA Albers, 1860
Eurhytida Möllendorff, 1903
Echotrida, Murphitella, Strangesta,
Tasmaphena all Iredale, 1933
Namoitena, Occirhenea, Tasmadelos,
Montedelos all Iredale, 1943
PTYCHORHYTIDA Möllendorff, 1903
WAINUIA Powell, 1930
DELOS Hutton, 1904
pro Elaea Hutton, 1884
(non Ziegler, 1833)
Rhenea Hutton, 1893
(non Saalmüller, 1884)
Prolesophanta, Saladelos Iredale, 1933
HEBRIDELOS Solem, 1959
DELOUGAPIA Powell, 1952
DIPLOMPHALUS Crosse & Fischer, 1873
MICROPHYURA Ancey, 1882
Microphiura Möllendorff, 1903 (err.)
PSEUDOMPHALUS Ancey, 1882
MACROCYCLOIDES von Martens, 1867
MELAVITRINA Iredale, 1933
NATA Watson, 1934 (53)
NATELLA Watson, 1934
NATALINA Pilsbry, 1893
Aerope Albers, 1860
(non Leach in Desmarest, 1823)
AFRORHYTIDA Möllendorff, 1903
CAPITANA Watson, 1934
HYPERRHYTIDA Watson, 1934
OUGAPIA Crosse, 1894
Torresiropa Iredale, 1933
PARYPHANTA Albers, 1850
POWELLIPHANTA O'Connor, 1945
PRIODISCUS von Martens, 1898
RHYTIDAREX Powell, 1948
AMBORHYTIDA Climo, 1974
SCHIZOGLOSSA Hedley, 1893
VICTAPHANTA Iredale, 1933

SYSTROPHIIDAE
SYSTROPHIA L. Pfeiffer, 1855
ENTODINA Ancey, 1887
SCOLODONTA Döring, 1875

SYSTROPHIA (Cont'd.)
SYSTROPHIELLA H. B. Baker, 1925
WAYAMPIA Tillier, 1980
GUESTIERIA Crosse, 1872
HAPPIA Bourguignat, 1889
Ammonoceras L. Pfeiffer, 1855
(non Lamarck, 1822)
DREPANOSTOMELLA Bourguignat, 1889
HAPPIELLA H. B. Baker, 1925
Occulator Pilsbry, 1926
PAYENIA Rochebrune & Mabille, 1889
PROHAPPIA Thiele, 1927
MIRADISCOPS H. B. Baker, 1925
PUNCTODISCOPS H. B. Baker, 1925
Microhappia Thiele, 1927
TAMAYOA H. B. Baker, 1925
TAMAYOPS H. B. Baker, 1928
ZILCHISTROPHIA Weyrauch, 1960 (?)

CHLAMYDEPHORIDAE
CHLAMYDEPHORUS Binney, 1879
Chlamydophorus Dall, 1880
(non Wagler, 1830)
Apera Heynemann, 1885

HAPLOTREMATIDAE
HAPLOTREMA Ancey, 1881
Macrocyclis Binney & Bland, 1869
(non Beck, 1837)
Mesomphyx Gray, 1841
(non Rafinesque)
Selenites Fischer, 1877
(non Hope, 1840)
Hoplobiena, Haptotrema,
Hoptotrema all Binney, 1885 (err.)
Circinaria Beck, Pilsbry, Auctt. to 1920
(non Beck, 1837)
ANCOMENA, ANCOTREMA, GREGGIELLA,
HAPLOMENA all H. B. Baker, 1931
GEOMENE Pilsbry, 1927
Proselenites Thiele, 1927
ZOPHOS Gude, 1911
Morchia von Martens in Albers, 1860
(non A. Adams, 1860)
Baudonia Binney, 1885
(non Mabille, 1868)
AUSTROSELENITES Kobelt, 1905

PLECTOPYLIDOIDEA
SCULPTARIIDAE
SCULPTARIA L. Pfeiffer, 1855

PLECTOPYLIDIDAE (Corillidae)
PLECTOPYLIS Benson, 1860
CHERSAECIA, ENDOPLON, SINICOLA all Gude, 1899
ENDOTHYRELLA Zilch, 1960
pro Endothyra Gude, 1899
(non Phillips, 1845)
CORILLA H. & A. Adams, 1855
Atopa Albers, 1850 (part)
(non Fabricius, 1799)
AMPHICOELINA Haas, 1933

PUNCTOIDEA (45h)
ENDODONTIDAE (45b)
ENDODONTA Albers, 1850
AAADONTA Solem, 1976
ACANTHOPTYX Ancey, 1888
ANCEYODONTA Solem, 1976
ANDREFRANCIA Solem, 1960
AUSTRALDONTA Solem, 1976
COOKECONCHA Solem, 1976
GAMBIODONTA Solem, 1978

ENDODONTIDAE (Cont'd.)
GLYPTAULAX Gude, 1914 ?
HIRASEA Pilsbry, 1902 ?
FAMETESTA Pilsbry, 1902 ?
HIRASIELLA Pilsbry, 1902
HIRTUDISCUS Hylton-Scott, 1973
JANULUS Lowe, 1852
KLEOKYPHUS Solem, 1976
KONDOCONCHA Solem, 1976
LIBERA Garrett, 1881 (45b)
Garretia Cossmann, 1900
(non Pease, 1873)
Garrettina Thiele, 1931
MAUTODONTHA Solem, 1976
GARRETTOCONCHA Solem, 1976
MINIDONTA Solem, 1976
NESODISCUS Thiele, 1931
NESOPHILA Pilsbry, 1893
OPANARA Solem, 1976
ORANGIA Solem, 1976
PHILALANKA Godwin-Austen, 1898 ?
Actinaria L. Pfeiffer, 1855
PROSITALA Germain, 1915
PLANUDONTA Solem, 1976
PRICECONCHA Solem, 1973
PROTOENDODONTA Solem, 1977
PSEUDOLIBERA Solem, 1976
PSICHION Gude, 1911
RHYSOCONCHA Solem, 1978
RIKITEA Solem, 1976
RUATARA Solem, 1976
STEPHADISCUS Hylton-Scott, 1982
TAIPIDON Solem, 1976
THAUMATODON Pilsbry, 1893
ZYZZYXDONTA Solem, 1976

Placement ? (Endodontidae) (53)
ANAGLYPHULA Rensch, 1932
AFROPUNCTUM Haas, 1934
FOXIDONTA Clench, 1950
PTYCHODON Germain, 1932 ?
(non Ancey, 1888)
PSEUDOHELENOCONCHA Germain, 1932
THESKELOMENSOR Iredale, 1933

PUNCTIDAE
PUNCTUM Morse, 1864
Patulastra L. Pfeiffer, 1879
PSEUDOPUNCTUM Morrison, 1935
TOLTECIA Pilsbry, 1926
Patulastra Lindholm, 1927
(non L. Pfeiffer, 1879
Pullastra L. Pfeiffer in
Westerlund, 1889 (error)
HYDROPHREA Climo, 1974
LAOMA Gray, 1850
PHRIXGNATHUS Hutton, 1883
TAGUAHELIX Powell, 1955
LAOMAREX Powell, 1948
OBANELLA Dell, 1952
PARALAOMA Iredale, 1913 ?
PASMADITTA Kershaw & Smith, 1986 (28)
PEDICAMISTA Kershaw & Smith, 1986
PLEUROPUNCTUM Germain, 1929 ? (53)

HELICODISCIDAE
HELICODISCUS Morse, 1864
HEBETODISCUS H. B. Baker, 1929
PSEUDISCUS Morrison, 1942
CHANOMPHALUS Strebel & Pfeffer, 1880 ?
LILLOICONCHA Weyrauch, 1965
POLYGRYISCUS Pilsbry, 1948 (45a)
RADIODISCUS Pilsbry & Ferriss, 1906 ?

RADIODISCUS (Cont'd.)
RADIOCONUS H. B. Baker, 1927
RADIODOMUS H. B. Baker, 1930
STENOPYLIS Fulton, 1914
Coarctatio Haas, 1945
ZILCHOGYRA Weyrauch, 1965
TROCHOGYRA Weyrauch, 1965

CHAROPIDAE
CHAROPINAE (45b)
CHAROPA Albers, 1860
Simplicaria Suter, 1891
AFRODONTA Melvill & Ponsonby, 1908 ?
ANNOSELIX Iredale, 1939
BA Solem, 1983
BEILANIA Preston, 1913 ? (16b)
CHAUREOPA Climo, 1985
CRYPTOCHAROPA Preston, 1913
DAMONITA Climo, 1981
DIPNELIX Iredale, 1937
DISCOCHAROPA Iredale, 1913
DUPUCHAROPA Iredale, 1937
FECTOLA Iredale, 1915 (16a)
Cavellia Iredale, 1915
FLAMMOCHAROPA Climo, 1970 ?
AESCHRODOMUS Pilsbry, 1892
pro Thera Hutton, 1884
(non Stephens, 1831)
EGESTULA Iredale, 1915
GEMINOROPA Iredale, 1933
Carelliopa Dell, 1952
MOCELLA Iredale, 1915
Subfectola Powell, 1939 ?
GERONTIA Hutton, 1883
GRAEFFEDON Solem, 1983
GYROCOCHLEA Hedley, 1924
JOHANNESOCONCHA Preston, 1913 ?
HELENOCONCHA Pilsbry, 1892
HELENODISCUS Solem, 1977
HUONODON Iredale, 1945
Mylesia Climo, 1978 (16a)
LAGIVALA Solem, 1983
LAUOPA Solem, 1983
MAAFU Solem, 1983
MICROCHAROPA Solem, 1983
MONTAROPA Climo, 1984
NORFOLCIOCONCHA Preston, 1913
NOTODISCUS Thiele, 1931 (45b)
PARACHAROPIA Climo, 1983 (16b)
PARVICHAROPA Solem, 1958
PHORTION Preston, 1910
Phasis Albers, 1850
(non Hübner, 1818)
PHENACOCHAROPA Pilsbry, 1893
Tesseraria O. Boettger, 1880
(non Haeckel, 1880)
PILSBRYCHAROPA Solem, 1978
Corinomala Iredale, 1939 (nom.nud.)
PILULA Martens, 1898
PROPILULA Germain, 1918 ? (53)
PSEUDEGESTULA Dell, 1954
PTYCHODON Ancey, 1888 (16a)
Strobila Hutton, 1883
(non Morse, 1864)
Huttonella Suter, 1890
(non L. Pfeiffer, 1856)
Maoriana Suter, 1891
ALSOLEMIA Climo, 1981
pro Solemia Climo, 1978
(non Loomis, 1964)
UNILAMELLATUS Weyrauch, 1965
RETICHAROPA Solem, 1959
RHOPHODON Hedley, 1924

CHAROPINAE (Cont'd.)
RUTHVENIA Gude, 1911
Austenia Gude, 1897
(non Nevill, 1878)
Sykesia Gude, 1897
(non Pomel, 1883)
SINPLOYEA Solem, 1983
STEPHANODA Albers, 1860
SUTERIA Pilsbry, 1892
Patulopsis Suter, 1892
(non Strebel, 1879)
THERASIELLA Powell, 1848
THYSANOTA Albers, 1860
TRACHYCYSTIS Pilsbry, 1892 (Trachycystidae ?)
Pella Albers, 1860 (part)
(non Stephens, 1832)
CHALCOCYSTIS, CHILOCYSTIS Watson, 1934
CYCLOCYSTIS Watson, 1934 ? (53)
LYROCYSTIS Watson, 1934
LIPAROCYSTIS Watson, 1934
PHAULOCYSTIS Watson, 1934
XEROCYSTIS Watson, 1934
TROPIDOTROPIS Ancey, 1888
TUIMALILA Solem, 1983
VATUSILA Solem, 1983

TRUKCHAROPINAE
TRUKCHAROPA Solem, 1983
KUBARYIELLUS Solem, 1983
JOKAJDON Solem, 1983
PALIKIRUS Solem, 1983
PALLINE Solem, 1983
ROIMONTIS Solem, 1983
RUSSATUS Solem, 1983

SEMPERDONINAE
SEMPERDON Solem, 1983
HIMEROCONCHA Solem, 1983
LADRONELLUM Solem, 1983

AMPHIDOXINAE (53)
AMPHIDOXA Albers, 1850
ALLODISCUS Pilsbry, 1892
Psyra Hutton, 1884
(non Walker, 1860)
Allodisens Gabriel, 1929 (err.)
HIRSUTODISCUS Climo, 1971
PLATYRHYTIDA Pilsbry, 1895 (53)
Platystoma Ancey, 1882
(non Hörnes, 1855)
Saissetia Bayle in Ancey, 1888
(non Deplanches, 1865)
FLAMMOCONCHA Dell, 1952
FLAMMULINA von Martens, 1873
Calymna Hutton, 1884
(non Hübner, 1816)
HEDLEYOCONCHA Pilsbry, 1893
MONOMPHALUS Ancey, 1882
MUSSONULA Iredale, 1937
OREOKERA Stanisic, 1987
(Oreokera Iredale, 1933 - nom.nud.)
PARARHYTIDA Ancey, 1882
MICROMPHALIA Ancey, 1882
PLESIOPSIS Ancey, 1888
PARYPHANTOPSIS Thiele, 1928
Illonesta, Gallodema Iredale, 1944
PHACUSSA Hutton, 1883
Phelussa Iredale, 1915 ?
PROTOFLAMMULINA Climo, 1971
PSEUDALLODISCUS Climo, 1971
PSEUDOCHAROPA Peile, 1929
Howeinsulea Peile, 1929
RANFURLYA Suter, 1903

AMPHIDOXINAE (Cont'd.)
RHYTIDOPSIS Ancey, 1882
THERASIA Hutton, 1883
PHENACOHELIX Suter, 1892
Neophenacohelix Cumber, 1961
Delinitesta, Mulathena, Stenocapha,
Thyrasona all Iredale, 1933 (nom.nud.
Flammulops, Queridomus Iredale, 1937
SERPHO Hutton, 1904
Carthaea Hutton, 1884
(non Walker, 1858)
THALASSOHELIX Pilsbry, 1892
THERMIA Hutton, 1904
Pyrrha Hutton, 1884
(non Cabanis, 1849)

ROTADISCINAE (53)
ROTADISCUS Pilsbry, 1926

OTOCONCHIDAE (45h) (36b)
OTOCONCHA Hutton, 1884 (53)
MAORICONCHA Dell, 1952

DISCIDAE
DISCUS Fitzinger, 1833
Gonyodiscus Fitzinger, 1833
Diomphalus, Delomphalus Charpentier,1837
Patula Held, 1837
Euryomphala Beck, 1837 =
Euryomphala Hermannsen, 1847
Patularia Clessin, 1876
(non Swainson, 1840)
Cratera Megerle in Scudder, 1882
Goniosiscus Beck in Kobelt, 1879
Allerya Bourguignat,1878
(non Mörch, 1877)
ANTEDISCUS Baker in Pilsbry, 1948
ATLANTICA Ancey, 1887
KERAEA Gude, 1911
Julus Wollaston, 1878 (Iulus)
(non Linnaeus, 1758)
MEXICODISCUS Pilsbry, 1926
NEMATODISCUS Pilsbry, 1948
ANGUISPIRA Morse, 1864
Helix, Patula, Pyramidula auctt.(err.)
ZONODISCUS Pilsbry, 1948

OREOHELICIDAE (41c) (36b)
OREOHELIX Pilsbry, 1904
Patula Held, 1837 (auctt.)
Pyramidata Fitzinger, 1833 (auctt.)
RADIOCENTRUM Pilsbry, 1905

OOPELTIDAE
OOPELTA Mörch, 1867

Suborder Elasmognatha

SUCCINEOIDEA (45h)
SUCCINEIDAE
SUCCINEINAE
SUCCINEA Draparnaud, 1801
Lucena Oken, 1815
Tapada Studer, 1820
Amphibina, Amphibulina Hartmann, 1821
Cochlohydra Férussac, 1821
Succinia Gray in Turton, 1840
Neritostoma Mörch, 1864
(non H. & A. Adams, 1855)
Lucerna Hall, 1867 (err.)
(non Swainson, 1840)
Succinastrum Mabille, 1870
Austrosuccinea, Arborcinea Iredale, 1937

SUCCINEA (Cont'd.)
Cerinasota Iredale, 1939
Hydrophyga Lindholm, 1927
CALISUCCINEA Pilsbry, 1948
DESMOSUCCINEA Webb, 1954
NOVISUCCINEA Pilsbry, 1948
PAPUSUCCINELLA Iredale, 1941
SUCCINELLA Mabille, 1870
TRUELLA Pease, 1871 ?
BONINOSUCCINEA Habe, 1956
CAMPTONYX Benson, 1858
LITHOTIS Blanford, 1863
HELISIGA Lesson, 1831
Helisigna Gray, 1850 (err.)
HYALIMAX H. & A. Adams, 1855
JARAVA Godwin-Austen, 1882
LAXISUCCINEA Cooke, 1921
OMALONYX Orbigny, 1841 (Homalonyx)
Pelta Beck, 1837 (nom.nud.)
NEOHYALIMAX Simroth, 1896
OXYLOMA Westerlund, 1885
Hydrotropa Lindholm, 1927
NEOXYLOMA Pilsbry, 1948

CATINELLINAE
CATINELLA Pease, 1870
Mediappendix Pilsbry, 1948 (invalid)
INDOSUCCINEA Rao, 1924
NEOSUCCINEA Matekin, 1956
QUICKELLA C. Boettger, 1939
QUICKIA Odhner, 1950
BURCHELLA Patterson, 1970

ATHORACOPHOROIDEA (36b)
ATHORACOPHORIDAE (Land slugs) (45h)
ATHORACOPHORUS Gould, 1852
Janella Gray, 1850
(non Grateloup, 1838)
Neojanella Cockerell, 1891
AMPHICONOPHORA Suter, 1897
Amphikonophora auctt. (err.)
CONOPHORA Hutton, 1879 (Konophora)
Vomanus Iredale, 1915
PSEUDANEITEA Cockerell, 1891
Reflectopallium Burton, 1963
ANEITEA Gray, 1860
Aneityopsis Grimpe & Hoffmann, 1925
TRIBONIOPHORUS Humbert, 1863
NEOMECKLENBURGIA Simroth, 1918 ?
OTTONIA Simroth, 1918 ?
ANEITELLA Cockerell, 1891
PALLIOPODEX Burton, 1963

(Helicid Sigmurethra)

SAGDOIDEA
SAGDIDAE
SAGDINAE
SAGDA Beck, 1837
Epistyla, Epistylia Swainson, 1840
Parahelix Ihering, 1892
HOJEDA H. B. Baker, 1926 ?
Microphysa von Martens, 1860
(non Westwood, 1834)
HYALOSAGDA Albers, 1860
MICROSAGDA H. B. Baker, 1935
LACTEOLUNA Pilsbry, 1926
AEROTROCHUS Pilsbry, 1926
STRIOLUNA Pilsbry, 1926
MEIOPHYSEMA H. B. Baker, 1935
ODONTOSAGDA von Martens, 1860 ?
PROSERPINULA Albers in von Martens, 1860
Proserpinella Pilsbry, 1949 (em.)

SAGDINAE (Cont'd.)
STAUROGLYPTA H. B. Baker, 1935
TRIFAUX H. B. Baker, 1935
VOLVIDENS J. B. Henderson, 1914
XENODISCULA Pilsbry, 1919 ?
ZAPHYSEMA Pilsbry, 1894

AQUIBANINAE
AQUEBANA Pilsbry, 1926
EXSUAVITAS H. B. Baker, 1939
SUAVITAS Pilsbry, 1926
SETIPELLIS Pilsbry, 1926
EUCLASTARIA Pilsbry, 1926
Euclasta von Martens, 1877
(non Lederer, 1855)
YUNQUEA H. B. Baker, 1940 ?

PLATYSUCCINEINAE
PLATYSUCCINEA Ancey, 1881

GASTRODONTOIDEA
GASTRODONTIDAE
GASTRODONTA Albers, 1850
CLAPPIELLA H. B. Baker, 1929
POECILOZONITES Boettger, 1884
Bermudia Ancey, 1887
Juno Mazyck, 1889
DISCOZONITES Pilsbry, 1924
GASTRELASMUS Pilsbry, 1924
STRIATURA Morse, 1864
PSEUDOHYALINA Morse, 1864
STRIATUROPS H. B. Baker, 1928
VENTRIDENS Binney & Bland, 1869
Mesomphix Tryon, 1866 (37)
(non Rafinesque, 1819)
ELLIOTTIUS Pilsbry, 1946
VENTRICALLUS Pilsbry, 1946
ZONITOIDES Lehmann, 1862
Alienitor Iredale, 1937 (4)
ZONITELLUS H. B. Baker, 1928
PSEUDOHYALUS H. B. Baker, 1929

HELIXARIONOIDEA (36a,b)(Helicarionoidea)
EUCONULIDAE
EUCONULUS Reinhardt, 1883
Conulus Fitzinger, 1833
(non Leske, 1778)
Petasia Beck, 1837 (part)
(non Stephens, 1829)
Trochulus Westerlund, 1886 (37)
(non Gray, 1847)
Arnouldia Bourguignat,1890 (pro Conulus)
Petasina Gude & Woodward, 1921
(non Beck, 1847)
Euconulops H. B. Baker, 1928
CHETOSYNA H. B. Baker, 1941
MONOCONULUS H. B. Baker, 1941
NESOCONULUS H. B. Baker, 1941
PELLUCIDOMUS H. B. Baker, 1941
AFROCONULUS Van Moll & Van Bryggen, 1971
CANCELLOCONUS Rensch, 1932
CONEUPLECTA Möllendorff, 1893
Conobycus Thiele, 1928
Durgellina Thiele, 1928
Parasitala Thiele, 1931
Sitalina Thiele, 1931
DISCOCONULUS Reinhardt, 1883
DRYACHLOA Thompson & Lee, 1980
EURYCHLAMYS Godwin-Austen, 1899
GUPPYA Mörch, 1867
Pelia Albers, 1860 ?
(non Bell, 1836)
Spiroconulus von Martens, 1892

EUCONULIDAE (Cont'd.)
HABROCONUS Fischer & Crosse, 1872
Stenopus Guilding, 1828
(non Desmarest, 1825)
Trochoconulus Reinhardt, 1883
COCOSCONUS H. B. Baker, 1941
COCCOSLENS H. B. Baker, 1941
ERNSTIA Jousseaume, 1889
PSEUDOGUPPYA H. B. Baker, 1925
KORORIA H. B. Baker, 1941
LUCHUCONULUS Pilsbry, 1928
MACROCERAS Semper, 1870
PALAUA H. B. Baker, 1941
VELIFERA W. G. Binney, 1879

HELIXARIONIDAE (36a,b) (Helicarionidae)
HELIXARIONINAE
HELIXARION Férussac, 1821
Helicarion Férussac, 1822 (em.)
Helicarium Agassiz, 1846
Laconia Gray, 1855
Fastosarion, Luinarion Iredale, 1933
Verculerion Iredale, 1933
Parmavitrina, Peloparion Iredale, 1937
Desidarion, Mysticarion Iredale, 1941
SITALARION H. B. Baker, 1941
AMPHIBLEMA Gude, 1911
BATHIA Robson, 1914
CALDWELLIA H. Adams, 1873
EPIGLYPTA Pilsbry, 1893
INOZONITES Pfeffer, 1883
METHVENIA Robson, 1914
NESAECIA Gude, 1911
Rotularia Mörch, 1872
(non Defrance, 1827)
PARMACOCHLEA E. A. Smith, 1884
PETALOCHLAMYS Godwin-Austen, 1907
PSEUDAUSTENIA Cockerell, 1891
WESTRALCYSTIS Iredale, 1939 (45g)
WILHELMINAIA Preston, 1913 (45g)

SESARINAE
SESARA Albers, 1860
Malandena Iredale, 1933 (10a)
ADVENA Gude, 1913
ANTINOUS Robson, 1914
BELLOCONCHA Preston, 1913 ?
DENDROLAMELLARIA Preston, 1913
Lamellaria Preston, 1913 ?
(non Montagu, 1815)
DENDROTROCHUS Pilsbry, 1894
PONAPEA H. B. Baker, 1941
SANTOTROCHUS Solem, 1959
GEOTROCHUS Hasselt, 1823
Chiroktisma Gude, 1913
EURYBASIS Gude, 1913
TEGUMEN Gude, 1913
HEMIGLYPTA Möllendorff, 1893
HEMIGLYPTOPSIS Thiele, 1931
KALIELLA Blanford, 1863
Gastrodontella Möllendorff, 1901
STRIOKALLIELLA Thiele, 1931 ?
LEPIDOTRICHIA Bartsch, 1942
ATRICHOCONCHA Bartsch, 1942
HEMITRICHIELLA Zilch, 1956
Hemitrichia Möllendorff, 1888
(non Rostafinski, 1873)
MATHEWSOCONCHA Preston, 1913 ?
NESONANINA C. Boettger, 1916
NITOR Gude, 1911 (10a)
Thalassia Albers, 1860
(non Bonaparte, 1856)
Pravinitor, Modonitor Iredale, 1937

SESARINAE (Cont'd.)
ORPIELLA Gray, 1855
Eurypus Semper, 1870
(non Kirby, 1819)
Fretum Sykes, 1900
EUFRETUM H. B. Baker, 1941
FIJIA Gude, 1913
KALENDYMA Gude, 1913
HALOZONITES Pilsbry & Cooke, 1941
IRENELLA Gude, 1913
Aulacopus L. Pfeiffer, 1878
(non Serville, 1832)
OWARAHA H. B. Baker, 1941
PAPUARION Van Mol, 1973
PARMELLA H. Adams, 1867
PSEUDOKALIELLA Godwin-Austen, 1910
QUIROSELLA Clench, 1958 ?
RAHULA Godwin-Austen, 1907
RASAMA Laidlaw, 1932
Sarama Godwin-Austen, 1908
(non Moore, 1887)
Saramina Wenz, 1947
RHYSOTINA Ancey, 1887
Thomeonanina Germain, 1909
ROYBELLIA Preston, 1913
RYSSOTA Albers, 1850
Rhyssota L. Pfeiffer, 1855
Rhysota Albers, 1860
LAMARCKIELLA Möllendorff, 1898
PARARYSSOTA Bartsch, 1938
TRUKRHYSA H. B. Baker, 1941
SITALINOPSIS Thiele, 1931
SIVELLA Blanford, 1863
TADUNIA Godwin-Austen, 1918

EREPTINAE
EREPTA Albers, 1850
Microstylodonta Germain, 1921
PSEUDOPHASIS Germain, 1918
ARIOCAELATURA Germain, 1821
CTENOGLYPTA Ancey, 1904
CTENOPHILA Ancey, 1882
Tachyphasis Germain, 1918
DANCEA Zilch, 1960
pro Pseudocaelatura Wenz, 1947
(non Germain, 1921)
Ariocaelatura "Germain" Thiele, 1931
(non Germain, 1921)
DUPONTIA Godwin-Austen, 1908
HARMOGENANINA Germain, 1919
Rotulla Albers, 1850
(non Agassiz, 1841)
Cycliscus Gude, 1911
(non Schönherr, 1843)
LOUISIA Godwin-Austen, 1908
Pseudocaldwellia Germain, 1908
PACHYSTYLA Mörch, 1852
PLEGMA Gude, 1911
Coelatura L. Pfeiffer, 1877
(non Conrad, 1853)
Caelatura Fischer, 1883
(non Conrad, 1865)
PSEUDOCAELATURA Germain, 1921 (?)

MICROCYSTINAE
MICROCYSTIS Beck, 1837
Platycloster Beck, 1837
pro Helicopsis Beck, 1837
(non Fitzinger, 1833)
ALLENOCONCHA, DIEPENHEIMIA, GREENWOODOCONCHA,
IREDALEOCONCHA all Preston, 1913
CNESTICYSTIS, FACORHINA, LEUROCYSTIS
all H. B. Baker, 1933

MICROCYSTINAE (Cont'd.)
COOKEANA H. B. Baker, 1938
DIASTOLE Gude, 1913
EUANANA, LAUA H. B. Baker, 1938
FANULUM Iredale, 1913
TROCHONANITA H. B. Baker, 1938
Trochononanina Garrett, 1884
(non Mousson, 1869)
HIONA Cooke in Baker, 1940
AUKENA H. B. Baker, 1940 ?
INSULORBIS, HIONELLA, HIONARION, MINORORBIS, NESOCYCLUS, NEUTRA, OPARA, UAPUNEUTRA all H. B. Baker, 1940
KAALA Cooke in Baker, 1940
KUSAIEA H. B. Baker, 1938
LAMPROCYSTIS Pfeffer, 1883
AVARUA, GUAMIA, KERAKYSTIS, MANUREVA, MOALA, NAIAUA, RAIATEA, TONGACYSTIS all H. B. Baker, 1940
LIARDETIA Gude, 1913
BELOPYGMEUS, BELONESIA, DASYCONUS, NESOREUS, OCEANESIA all H. B. Baker, 1938
KIECONCHA Iredale, 1913 ?
MENDANA H. B. Baker, 1938
FATUOA H. B. Baker, 1938
MACRORBIS H. B. Baker, 1938
TAHUATOA H. B. Baker, 1938
UANUKA H. B. Baker, 1938
PHILONESIA Sykes, 1900
AA, HALEAKALA, HILOAA, KIPUA, MAUKA, NESARION, NUKUPIENA, OAFATUA, PIENA, PITCAIRNIA, RAPAFILA, UAFATUA all H. B. Baker, 1949
WAIHOUA H. B. Baker, 1940 ?
PITTOCONCHA Preston, 1913 ?
PUKALOA H. B. Baker, 1938
QUINTALIA Preston, 1913
SUTERELLA Iredale, 1915 ?
TENGCHIENA H. B. Baker, 1942

Synonyms ? (nom. nud.)
Dendronitor, Echonitor, Eclipsena, Expocystis all Iredale, 1933
Melocystis, Nevelasta, Periclocystis, Sodaleta, Tarocystis, Turrisitala all Iredale, 1933

ARIOPHANTIDAE
ARIOPHANTA Desmoulins, 1829
COLPARION Laidlaw, 1938 (49)
CRYPTOZONA Mörch, 1872
Xestina Pfeffer, 1878
Nilgiria Godwin-Austen, 1888
EUPLECTA Semper, 1870
HEMIPLECTA Albers, 1850
KORATIA Godwin-Austen, 1919
INDRELLA Godwin-Austen, 1901
KALIDOS Gude, 1911
PROPEBLOYETIA Germain, 1913 (19)
MALAGARION Tillier, 1979 (49)
NANINIA Sowerby, 1842
XESTA Albers, 1850
Nanina Gray, 1834
(non Risso, 1826)
Zagamena Iredale, 1941 ?
PLATYMMA Tomlin, 1938
RATNADVIPIA Godwin-Austen, 1899
RAVANA Godwin-Austen, 1901
RHYSSOTOPSIS Ancey, 1887
Haughtonia Godwin-Austen, 1899
(non Kinahan, 1859)

DYAKIINAE
DYAKIA Godwin-Austen, 1891
Semperia Godwin-Austen, 1898
(non Crosse, 1867)
ASPERITAS Gude, 1911
ELAPHROCONCHA Gude, 1911
EVERETTIA Godwin-Austen, 1891
KALAMANTANIA Laidlaw, 1931
PSEUDOPLECTA Laidlaw, 1932 ?
QUANTULA H. B. Baker, 1941
RHINOCHILIS Thiele, 1931
SASAKINA Rensch, 1930
Sasakia Rensch, 1930 (non Moore, 1896)
STAFFORDIA Godwin-Austen, 1907

PARMARIONINAE
PARMARION P. Fischer, 1855
Rigasia Gray in H. & A. Adams, 1858 ?
DAMAYANTIA Issel, 1874 (49)
DAMAYANTIELLA Hoffmann, 1940
ISSELENTIA Collinge, 1901
MICROPARMARION Simroth, 1893
COLLINGEA Simroth, 1898
PARMUNCULUS Collinge, 1899
PHILIPPINELLA Möllendorff, 1899
WIEGMANNIA Collinge, 1901

GIRASIINAE
GIRASIA Gray, 1855
Hoplites Theobald, 1864
AUSTENIA Nevill, 1878
Cryptibycus Cockerell, 1898
CERATACONTA Cockerell, 1930
CRYPTAUSTENIA Cockerell, 1898
CRYPTOGIRASIA Godwin-Austen, 1908 ?
DIHANGIA Godwin-Austen, 1916
GALONGIA Godwin-Austen, 1916
MARIAELLA Gray, 1855
Tennentia Humbert, 1862
Vega Westerlund, 1883
Dekhania Godwin-Austen, 1888

MACROCHLAMYDINAE
MACROCHLAMYS Benson, 1832 in Gray, 1847
Tanychlamys Benson, 1834
Orobia Albers, 1860
EUAUSTENIA Cockerell, 1898
PARVATELLA Blanford & Godwin-Austen, 1908
PSEUDHELICARION Möllendorff, 1894
RHADELLA Godwin-Austen, 1914 ?
APOPARMARION Collinge, 1902
BAIAPLECTA Laidlaw, 1956
Pangania Laidlaw, 1932
(non Poppius, 1914)
BAPUIA Godwin-Austen, 1918
BENSONIES H. B. Baker, 1938
Bensonia L. Pfeiffer, 1855
(non Gray, 1847)
CRYPTOSEMELUS Collinge, 1902
DALINGIA Godwin-Austen, 1907
GLYPTOBENSONIA Möllendorff, 1894
TRICHOBENSONIA Möllendorff, 1902
HOLKEION Godwin-Austen, 1908
KHASIELLA Godwin-Austen, 1899
MEGAUSTENIA Cockerell, 1912
Cryptosoma Theobald, 1857
(non Milne-Edwards, 1837)
MICROCYSTINA Mörch, 1872
MUANGNUA Solem, 1966
OXYTESTA Zilch, 1956
pro Oxytes L. Pfeiffer, 1855
(non Giebel, 1848)
PARAPARMARION Collinge, 1902

MACROCHLAMYDINAE (Cont'd.)
ROTUNGIA Godwin-Austen, 1918
SARIKA Godwin-Austen, 1907
SYAMA Godwin-Austen, 1908
TAPHROSPIRA Blanford, 1905
TERAIA Solem, 1966
VITRINULA Gray in Carpenter, 1857
Vitrinella Gray, 1855
(non C. B. Adams, 1850)
Otesia H. & A. Adams, 1858
Medyla Albers, 1860
Pareuplecta Möllendorff, 1890
OTESIOPSIS Habe, 1973
PLIOTROPIS Möllendorff, 1899

CHRONINAE (53)
CHRONOS Robson, 1914

DURGELLINAE (53)
DURGELLA Blanford, 1863
IBYCUS Heynemann, 1862
Leptodontarion P. & F. Sarasin, 1899
MINYONGIA Godwin-Austen, 1916
MYOTESTA Collinge, 1901
Ostracolethe Simroth, 1901
SAKIELLA Godwin-Austen, 1908
SATIELLA Godwin-Austen, 1908
SITALA H. Adams, 1865
Conulema Stoliczka, 1871
Moaria Chaper, 1885 ?
SOPHINA Benson, 1859

TROCHOMORPHIDAE
TROCHOMORPHA Albers, 1850
Nigritella von Martens, 1860
BENTHEMIA Forcart, 1964
COTITROCHUS, LAUHALA, LENTITROCHUS,
LUDIFICATOR all H. B. Baker, 1941
HARTMANITROCHUS Solem, 1959
BERTIA Ancey, 1887
EXRHYSOTA H. B. Baker, 1941 ?
BRAZIERIA Ancey, 1887
BRAZIERELLA, ENTOMOSTOMA, PROBRAZIERIA
all H. B. Baker, 1941
CALOSTROPHIA Ancey, 1887 ?
COXIA Ancey, 1887 ?
HOGOLUA H. B. Baker, 1941
KONDOA H. B. Baker, 1941
VIDENA H. & A. Adams, 1855
Discus Albers,1850
(non Fitzinger,1833)
Disculus Schaufuss, 1869
LIRAVIDENA Solem, 1959
PELELIUA, PERIRYUA H. B. Baker, 1941
VIDENOIDA Habe, 1955
VITRINOCONUS Semper, 1873
VITRINOPSIS Semper, 1873 ?
SUNDAVITRINA Rensch, 1932
VITRINOIDEA Semper, 1873

UROCYCLIDAE (50)
UROCYCLINAE
UROCYCLUS Gray, 1864
Elisa Heynemann, 1833
(non Reichenbach, 1854)
Elisolimax Cockerell, 1893
COMORINA Simroth, 1910
MESOCYCLUS Pollonera, 1906
MICROCYCLUS Simroth, 1896
ACANTHARION Binder & Tillier, 1983
AMATARION Van Mol, 1970
ANISOTOXON Van Goethem, 1975
ASPIDELUS Morelet, 1883

UROCYCLINAE (Cont'd.)
ASPIDOTOMIUM Degner, 1932
ATOXON Simroth, 1889
ATOXONOIDES Van Goethem, 1975
BUKOBIA Simroth, 1896
Stuhlmannia Simroth, 1895
(non Michaelsen, 1890)
BUETTNERELLA Simroth, 1910
Buettneria Simroth, 1889
(non Karsch, 1888)
CAMERUNARION Van Mol, 1970
CENTRAFICARION Van Mol, 1970
CHLAMYDARION Van Mol, 1970
DENDROLIMAX Heynemann, 1868
Verania Simroth, 1903 ?
ESTRIA Poirier, 1887
ELISOLIMAX Cockerell, 1893
KIRKIA Pollonera, 1909 ?
LEPTICHNOIDES Van Goethem, 1975
LEPTICHNUS Simroth, 1896
NUPNUS Van Goethem, 1975
PEMBATOXON Van Goethem, 1978
PHANEROPORUS Simroth, 1889
PSEUDATOXON Van Goethem, 1975
RHOPALOGONIUM Degner, 1932
SYLVARION Van Mol, 1970
SENEGALARION Van Mol, 1970
THIELARION Van Mol, 1970
TRESIA Van Goethem, 1975
TRICHOTOXON Simroth, 1889
Diplotoxon Simroth, 1897
ATRICHTOXON Simroth, 1910
POLYTOXON, SPIROTOXON Simroth, 1897
VERDCOURTIA Van Mol, 1970
VERRUCARION Van Mol, 1970

TROCHOZONITINAE (53)
TROCHOZONITES Pfeffer, 1883
TELEOZONITES Pilsbry, 1919
ZONITOTROCHUS Pilsbry, 1919
AFRICARION Godwin-Austen, 1883
GYMNARION Pilsbry, 1919
LACRIMARION Connolly, 1939
MESAFRICARION Pilsbry, 1919
BELONARION Pilsbry, 1919
SHELDONIA Ancey, 1887
Peltatus Godwin-Austen,1908
ANDRARION, KERKOPHORUS, MICROKERKUS,
all Godwin-Austen, 1912
THAPSIA Albers, 1860
Gudeella Preston, 1913
pro Thapsiella Gude, 1911
(non P. Fischer, 1885)
MEGATHAPSIA, THAPSINELLA Reemakers, 1959
TROCHONANINA Mousson, 1869
Martensia Semper, 1870
(non Agassiz,1860)
Ledoulxia Bourguignat, 1885
Eduardia P. Hesse, 1916
BLOYETIA Bourguignat, 1889
Guillainia Bourguignat, 1885
(non Crosse, 1884)
CRENATINANINA Germain, 1920
FALLOONELLA Preston, 1914
MONTANOBLOYETIA Verdcourt, 1961 ?
PERCIVALIA Preston, 1914
PLICATONANINA Verdcourt, 1961 ?
SJOSTEDTINA Verdcourt, 1961 ?
ZINGIS von Martens, 1878
ZONITARION Pfeffer, 1883
ANGUSTIVESTIS, ENTAGARICUS Pilsbry, 1919
DEGNERIA Verdcourt, 1956
GRANULARION Germain, 1912

CYSTOPELTIDAE [10a] [36b]
CYSTOPELTA Tate, 1881

VITRINOIDEA
VITRINIDAE
VITRININAE
VITRINA Draparnaud, 1801
Helicolimax Férussac, 1801
Vitrinus Montfort, 1810
Cobresia Hübner, 1810
Hyalina Studer, 1820
(non Schumacher, 1817)
Pagana Gistl, 1848
HESSEMILIMAX Shileyko, 1986 [44g]
SEMILIMACELLA Soos, 1917
Targionia Hesse, 1923
(non Signoret, 1870)
Tozzettia Hesse, 1924
EUCOBRESIA H. B. Baker, 1929
INSULIVITRINA P. Hesse, 1923
Insulina Forcart, 1946 (err.)
OLIGOLIMAX Fischer, 1878
PHENACOLIMAX Stabile, 1859
ARABVITRINA Thiele, 1931
GALLANDIA Bourguignat, 1880 ?
Trochovitrina Boettger, 1880
GUERRINA Odhner, 1954
MADEIROVITRINA Groh & Hemmen, 1986 [22]
SEMILIMAX Agassiz, 1845 [53]
Chlamydea Westerlund, 1886 ?
Vitrinopugio Ihering, 1892
CALIDIVITRINA Pilsbry, 1919
Caldivitrina Connolly, 1930 (err.)
VITRINOBRACHIUM Kunkel, 1929

PLUTONIINAE
PLUTONIA Stabile, 1864
Vitriplutonia Collinge in Cockerell, 1893

VITREINAE
VITREA Fitzinger, 1833
Crystallus Lowe, 1854
Diaphanella Clessin, 1880
Anomphala Westerlund, 1886
(non Jonas, 1846)
MONADEA Westerlund, 1902
COREOVITREA Riedel, 1967
HAWAIIA Gude, 1911
Pseudohyalina Morse, 1864 (part)
Macgillivrayella Preston, 1913 [37]
(non Ashmead, 1899)
Pseudovitrea H. B. Baker, 1928
LINDBERGIA Riedel, 1959
ECHINOPHALLUS Riedel, 1960
SPINOPHALLUS Riedel, 1962
PRISTILOMA Ancey, 1887
pro Pristina Ancey, 1886
(non Ehrenberg, 1831)
Anceyia Pilsbry, 1887
(non Mabille, 1886)
Ogaridiscus Chamberlin & Jonas, 1929
PRISTINOIDES, PRISTINOPSIS, PRISCOVITREOPS, PRISCOVITREA all H. B. Baker, 1931
PYCNOGYRA Strebel & Pfeffer, 1880
TROGLOVITREA Negrea & Riedel, 1968

ZONITIDAE
ZONITINAE
ZONITES Montfort, 1810
Tragomma Held, 1837
Helicode Dumas, 1847 =
Helicodes Herrmannsen, 1852
(non Boisduval, 1844)

ZONITES (Cont'd.)
Verticillus Moquin-Tandon, 1848
AEGOPHTHALMUS Hesse, 1910
TURCOZONITES Riedel, 1987
AEGOPINELLA Lindholm, 1927
AEGOPIS Fitzinger, 1833
BALCANODISCUS Riedel & Urbanski, 1964
ALLAEGOPIS Riedel, 1979
ARABOXYCHILUS Riedel, 1977
BIRULANA Lindholm, 1922
DORAEGOPIS Riedel, 1979
GLYPHYALINIA von Martens, 1892
Glyphyalina Pilsbry, 1898
Glyphialinia Haas, 1929
GLYPHOGNOMON, GLYPHYALOIDES, GLYPHYALUS, GLYPHYALOPS all H. B. Baker, 1928
HYALINIA Agassiz, 1837
Hyalina Férussac in Gray, 1847
(non Schumacher, 1817)
GASTRANODON O. Boettger, 1889 [42]
MESOMPHIX Rafinesque, 1819
MICROMPHIX Pilsbry, 1911
OMPHIX Pilsbry, 1911
NESOVITREA C. M. Cooke, 1921
PERPOLITA H. B. Baker, 1928
OMPHALINA Rafinesque, 1831
MORELETIA Gray, 1855
Edusa Albers, 1860
(non Gistl, 1848)
ZONYALINA von Martens, 1865
OXYCHILOPS C. Boettger, 1930
PONTOXYCHILUS Riedel, 1970
OXYCHILUS Fitzinger, 1833
Polita Held, 1837
Helicella Gray, 1847
(non Férussac, 1821)
Lucilla Lowe, 1852
Aplostoma Férussac in Moquin-Tandon, 1855
Euhyalina Albers, 1857
Omalota Megerle in Scudder, 1882
Euhyalinia Taylor, 1907
ALLOGENES Gude, 1911 ?
DROUETIA Gude, 1911
CALLORETINELLA Haas, 1934 ?
CONULOPOLITA O. Boettger, 1879
EOPOLITA Pollonera, 1916
HELICOPHANA Westerlund, 1886
Cretozonites Kobelt, 1890
HIRAMIA Pallary, 1939 [42]
LINDHOLMELLA C. Boettger, 1930
Retinella Westerlund, 1902
(non Fischer, 1877)
LONGIPHALUS Riedel, 1958
MEDITERRANEA Clessin, 1880
Hydatina Westerlund, 1886
(non Schumacher, 1817)
Diaphanella Hesse, 1916
(non Clessin, 1880)
Geodiaphana Thiele, 1917
MORLINA A. J. Wagner, 1914
Gemma Hazay, 1884
(non Deshayes, 1853)
ORTIZIUS Forcart, 1957
RIEDELIUS Hüdec, 1961
PARAEGOPIS Hesse, 1910
PARAVITREA Pilsbry, 1898
Taxeodonta Pilsbry, 1898
PARAVITREOPS, PARMAVITREA, PECTOVITREA, PETROVITREA all H. B. Baker, 1931
PATULOPSIS Strebel & Pfeffer, 1880
OMPHALINELLA H. B. Baker, 1930
PILSBRYNA H. B. Baker, 1929

ZONITINAE (Cont'd.)
RETINELLA Fischer in Shuttleworth, 1877
Mesomphix Albers, 1857
(non Rafinesque, 1819)
Aegopina Kobelt, 1878
Aegopsina Westerlund, 1886
Nesovirea Cooke, 1921
Hyalina, Hyalinia, Polites,
Vitrea, Zonites authors (err.)
GYRALINA Andreae, 1902 ?
LYRODISCUS Pilsbry, 1893
Lyra Mousson, 1872
(non Cumberland, 1816)
MELEDELLA Sturany, 1908 ?
PSEUDOPOLITA Germain, 1908 ?
SPELAEOPATULA Wagner, 1922 ?
SCHISTOPHALLUS Wagner, 1914 ?
CELLARIOPSIS Wagner, 1914
STENORHACHIODON Lindholm, 1927
TROGLAEGOPIS Riedel, 1983
VITRIZONITES W. G. Binney, 1879

GODWINIINAE (53)
GODWINIA Sykes, 1900
OMPHALOPS H. B. Baker, 1941

DAUDEBARDIIDAE
DAUDEBARDIA Hartmann, 1821
Helicophanta Férussac, 1822 (non 1821)
Daudebartia Beck, 1837
Rufina Clessin, 1878
Sieversia Kobelt, 1880
Eudaudebardia Westerlund, 1886 (part)
Dudichia Wagner, 1941
Szuchumiella Wagner, 1945
Suhumiella, Suchumiella Wagner, 1952
LIBANIA Bourguignat, 1870
pro Moussonia Bourguignat, 1866
(non O. Semper, 1865)
Cibania, Cibinia Wagner (err.)
Banatoconcha Wagner, 1941
BILANIA Shileyko, 1986
CARPATHICA A. J. Wagner, 1895 ?
Illyrica Wagner, 1895
DECEBALLIA Grossu, 1969
INGURIA Shileyko, 1986
LOTHARIA Shileyko, 1986

PARMACELLIDAE
PARMACELLA Couvier, 1804
Parmacellus Férussac, 1819
Cryptella Webb & Berthelot, 1833
Phosphorax Webb & Berthelot, 1833 ?
Drusia Gray, 1855
Clethropodium Westerlund, 1897
Euparmacella, Proparmacella Simroth, 1912
KANDAHARIA Godwin-Austen, 1914
Candaharia Godwin-Austen, 1888 (n.nud.)
? LEVANDERIA Likharev, 1980
(ncn Sturany, 1905)

MILACIDAE
MILAX Gray, 1855 (May)
Amalia Moquin-Tandon, 1855 (Sept.)
Palizzolia, Sansania Bourguignat, 1877
Lallemantia Mabille, 1868
Pirainea Lesson & Pollonera, 1882
MACROTHYLACUS H. Wagner, 1930 (47)
MICROMILAX P. Hesse, 1926
Cypria Simroth, 1810 (non Zenker, 1854)
PROMILAX H. Wagner, 1930
SUBAMALIA Pollonera, 1887
Malinastrum Bourguignat, 1864 (part)

MILAX (Cont'd.)
TANDONIA Lesson & Pollonera, 1882 ?
ASPIDOPORUS Fitzinger, 1833

LIMACOIDEA
LIMACIDAE
LIMAX Linnaeus, 1758
Limacella Brard, 1815
Eulimax Moquin-Tandon, 1855
Limacus R. Lehman, 1864
Plepticolimax Malm, 1865
Chromolimax, Gestroa, Opilolimax,
Stabilea all Pini, 1876
Heynemannia, Plecticolimax Malm, 1868
Simrothia Clessin, 1884 (part)
Plecticolimax Tryon, 1884
Macroheynemannia Simroth, 1891
CASPILIMAX P. Hesse, 1926
KASPERIA Godwin-Austen, 1914
MALACOLIMAX Malm, 1868
Microheynemannia Simroth, 1891
PROLIMAX Simroth, 1906
SVANETIA P. Hesse, 1926
VITRINOIDES Simroth, 1891
Vitrinolimax Babor, 1898
BIELZIA Clessin, 1887
Frauenfeldia Hazay, 1884
(non Clessin, 1878)
Limacopsis Simroth, 1888
CAUCASOLIMAX Likharev, 1980
DEROCERAS Rafinesque, 1820 (37)
Krynickia Kaleniczenko, 1851 (nom.nud.)
Krynickillus Kaleniczenko, 1851
Krynickellus Gray, 1855 (err.)
Krynichia Hazay, 1884
Megapelta Mörch, 1857 ?
Clytropelta Heynemann, 1867 ?
Hydrolimax Malm, 1865
(non Haldemann, 1842)
Mabillia Bourguignat, 1877 ?
Arctolimax, Chorolimax Westerlund, 1894
MALINO Gray, 1855 ? (47)
EUMILAX O. Boettger, 1881
PARALIMAX O. Boettger, 1883
FURCOPENIS Castillejo, 1983
GIGANTOMILAX O. Boettger, 1883
LEHMANNIA Heynemann, 1863 ?
Melitolimax Pollonera, 1891
AMBIGOLIMAX Pollonera, 1887
LYTOPELTE Boettger, 1886
Platyoxon Simroth, 1886
Tropidolytopelte Simroth, 1901
LIOLYTOPELTE Simroth, 1901
MEGALOPELTE Lindholm, 1914
MESOLIMAX Pollonera, 1888
METALIMAX Simroth, 1896
METALIMACOIDES Simroth, 1912
DELAVIELEUSIA Hagenmuller, 1855 ?
FAUDELIA Hagenmüller, 1855
CHANCELIA Hagenmüller, 1855
MONOCHROMA Simroth, 1896
PSEUDARION Germain, 1912
TOXOLIMAX Simroth, 1899
TURCOLIMAX Simroth, 1899
TURCOMILAX Simroth, 1901
MICHAELISIA Likharev, 1980
TAULIMAX Likharev, 1980

BOETTGERILLIDAE (36b)
BOETTGERILLA Simroth, 1910

AGRIOLIMACIDAE (36b)
AGRIOLIMAX Morch, 1865

TRIGONOCHLAMYDOIDEA
TRIGONOCHLAMYDIDAE
TRIGONOCHLAMYS O. Boettger, 1881
PHRIXOLESTES Simroth, 1901
DRILOLESTES Lindholm, 1925
Pseudomilax Simroth, 1901
(non Boettger, 1881)
Chlamydolimax Wenz, 1947
HYRCANOLESTES Simroth, 1901
Chrysalidomilax Simroth, 1901
PARMACELLILLA Simroth, 1910 ?
PSEUDOMILAX O. Boettger, 1881
SELENOCHLAMYS O. Boettger, 1883

POLYGYROIDEA (19) (Mesodontoidea)
POLYGYRIDAE
POLYGYRINAE
POLYGYRA Say, 1818
Helicodon Sowerby, 1825 (part)
pro Helicodonta Férussac, 1819
Cyclodoma Swainson, 1840
Ulostoma Albers, 1850
Anchistoma H. & A. Adams, 1855
(non Herrmannsen, 1846)
DAEDALOCHILA Beck, 1837
Daedalocheila auctt. (err.)
Triodonta Agassiz, 1846
(non Bory, 1827)
Tridonta Gray, 1847
(non Schumacher, 1817)
ACUTIDENS Pilsbry, 1956
EDUARDUS Pilsbry, 1930
ERYMODON Pilsbry, 1956
LINISA Pilsbry, 1930
LOBOSCULUM Pilsby, 1930
MONOPHYSIS Pilsbry, 1956
UPSILODON Pilsbry, 1930
RAGSDALEORBIS Webb, 1954 (33b)
SOLIDENS Pilsbry, 1956
ERECTIDENS Pilsbry, 1953 ?
GIFFORDIUS Pilsbry, 1930
MESODON Rafinesque in Férussac, 1821 (37)
Odomphium Rafinesque, 1831
Odontomphalium Agassiz, 1846
APHALOGONA Webb, 1954
APPALACHINA Pilsbry, 1940
INFLECTARIUS Pilsbry, 1940
? Xolotrema Rafinesque, 1831 (non 1819)
PATERA Albers, 1850
Odotropis Rafinesque in Pilsbry, 1930
Odontotropis Agassiz, 1846 ?
Trophodon Pilsbry, 1930
(non Rafinesque, 1831)
POLYGYRISCUS Pilsbry, 1948
PRATICOLELLA von Martens, 1892
pro Praticola Strebel & Pfeffer, 1880
(non Swainson, 1837)
Dorcasia Binney, 1878
(non Gray, 1838)
FARRAGUTIA Vanatta, 1915
FILAPEX Pilsbry, 1940
STENOTREMA Rafinesque, 1819
Chimotrema, Toxostoma, Toxotrema,
Trophodon, Stenostoma Rafinesque, 1831
Caracollatus, Cohutta Archer, 1948
EUCHEMOTREMA Archer, 1939
MAXILLIFER Pilsbry, 1940
TRILOBOPSIS Pilsbry, 1939

ASHMUNELLINAE
ASHMUNELLA Pilsbry & Cockerell, 1899
CRYPTOMASTIX Pilsbry, 1939
BUPIGONA Webb, 1970

TRIODOPSINAE (19)
TRIODOPSIS Rafinesque, 1819
Menomphis Rafinesque, 1831
Tridopsis Rafinesque in Beck, 1837 (part)
Triodontopsis Agassiz, 1846
? Micranepsia Pilsbry, 1940
Haroldorbis, Shelfordorbis Webb, 1959
ALLOGONA Pilsbry, 1939
DYSMEDOMA Pilsbry, 1939
NEOHELIX Ihering, 1892
VESPERICOLA Pilsbry, 1939
WEBBHELIX Emberton, 1988
XOLOTREMA Rafinesque, 1819
Wilcoxorbis Webb, 1952

THYSANOPHORIDAE
THYSANOPHORA Strebel & Pfeffer, 1880
LYROCONUS, MIROCONUS, SETIDISCUS
all H. B. Baker, 1927
STRIALUNA Pilsbry, 1926
VILITAS Pilsbry, 1926 ?
ITZAMNA Pilsbry, 1926
MCLEANIA Bequaert & Clench, 1939
MICROCONUS Strebel & Pfeffer, 1880
PULCHRICONUS F. G. Thompson, 1958
MICROPHYSULA Cockerell in Pilsbry, 1926 ?

CAMAENOIDEA (45h)
SOLAROPSIDAE (36b)
SOLAROPSIS Beck, 1837
Solarium Spix, 1827
(non Lamarck, 1799)
Helicella Swainson, 1840 ?
(non Férussac, 1821)
Ophiodermis Agassiz, 1846
Ophiospila Ancey, 1887
PSADARA Miller, 1878
Eupsadara Pilsbry, 1926
Heliopsis Pilsbry, 1933 ?
Psadariella Weyrauch, 1956

CAMAENIDAE (41d)
CAMAENINAE
CAMAENA Albers, 1850
Camena von Martens, 1860
Eucochlias Theobald, 1878
Camaenella Pilsbry, 1893
Pseudobba Möllendorff, 1891
MIYAKOIA Minato, 1980 (34)
ALBERSIA H. Adams, 1865
AMPHIDROMUS Albers, 1850
Goniodromus C. Bulow, 1905
Syndromus Pilsbry, 1900
AMPLIRHAGADA Iredale, 1933
Tenuigada Iredale, 1939
APATETES Gude, 1914
ARIOPHANTOPSIS Rensch, 1930
ARNEMELASSA Iredale, 1933
AUSTROCHLORITIS Pilsbry, 1891
Pareulota Möllendorff, 1891
Chloritobadistes Iredale, 1933
Nannochloritis, Patrubella Iredale, 1938
BACCALENA Iredale, 1937
BADISTES Gould, 1862
Meridolum Iredale, 1933
BASEDOWENA Iredale, 1937
BAUDINELLA Thiele, 1931
Gonobaudinia Iredale, 1933
BEDDOMEA Nevill, 1878
Phengus Jousseaume, 1894
(non Albers, 1850)
BENTOSITES Iredale, 1933
BURMOCHLORITIS Godwin-Austen, 1920

CAMAENINAE (Cont'd.)

CALVIGENIA Iredale, 1938
CALYCIA H. Adams, 1865
CARACOLUS Montfort, 1810
Discodoma Swainson, 1840
Serpentulus H. & A. Adams, 1855
Volvulus Oken, 1815
CARINOTRACHIA Solem, 1985
CHLORITIS Beck, 1837
Erigone Albers, 1850
(non Savigny, 1826)
Semicornus Klein in H. & A. Adams, 1855
Sulcobasis Tapparone-Canefri, 1883
Trichochloritis Pilsbry, 1891
Ptychochloritis Möllendorff, 1902
Eustomopsis Gude, 1906
Helerigone Strand, 1910
Sphaerospira Kobelt, 1914
(non Mörch, 1867)
Trachychloritis Haas, 1934
Aleatelix, Goldielix, Parvisheba,
Sheba, Timasensus all Iredale, 1941
Nipponochloritis Habe, 1955
Verdichloritis Clench, 1957
Neochloritis Minato, 1982 ?
CHORITISANAX Iredale, 1933
CHOSENELIX Pilsbry, 1927
COLONICONCHA Pilsbry, 1933
CRATERODISCUS McMichael, 1959 ?
CRYPTAEGIS Clapp, 1923
CRISTILABRUM Solem, 1981
CUPEDORA Iredale, 1933
DAMOCHLORA Iredale, 1938
Perochlora Iredale, 1938
DIRUTRACHIA Iredale, 1937
DISCOCONCHA Rensch, 1937
Opterigone Iredale, 1941
DISCOMELON Iredale, 1938 ?
DISTEUSTOMA Iredale, 1941
DIVELLOMELON Iredale, 1933
DORCASIDEA Iredale, 1941
EXILIBADISTES Iredale, 1933
EXILIGADA Iredale, 1939
EURYCRATERA Beck, 1837
Theliodomus, Thelidonta Swainson, 1840
Thelodomus, Thelydomus, Thelodonta auct.
Leiochila, Parthena Albers, 1850
Pachystoma Albers, 1850
(non Guilding, 1828)
Liochila von Martens, 1860
Dendrochlis Pilsbry & Brown, 1910
FATULABIA Iredale, 1937
GALADISTES Iredale, 1938
GALLODERMA Iredale, 1941
GANESELLA Blanford, 1863
Liocystus Mörch, 1872
Coliolus Tapparone-Canefri, 1887
Trochomorphoides Nevill, 1878
Fruticotrochus Kobelt, 1879
Globotrochus Haas, 1935
GANESELLOIDES Yen, 1969
GEMITELIX Iredale, 1933
GLOREUGENIA Iredale, 1933
GLYPTORHAGADA Pilsbry, 1890
Eximiorhagada Iredale, 1933
Halmatorhagada Iredale, 1933
GNAROSOPHIA Iredale, 1933
GRABAUIA Yen, 1935
GRANULOMELON Iredale, 1933
HADRA Albers, 1860
ISOMERIA Albers, 1850
JACKSONENA Iredale, 1937
KANABOHELIX Pilsbry, 1927

CAMAENINAE (Cont'd.)

KENDALLENA Iredale, 1941
KENDRICKIA Solem, 1985
KIMBORAGA Iredale, 1933
LABYRINTHUS Beck, 1837
Lampadion Röding, 1798
Lyriostoma, Lyrostoma Swainson, 1840
Ambages Gude, 1912
LACUSTRELIX Iredale, 1937
LANDOURIA Godwin-Austen, 1918
MANDARINA Pilsbry, 1895
Boninia Pilsbry, 1894
MECYNTERA Iredale, 1941
MELOSTRACHIA Solem, 1979
Melostrachia Iredale, 1938 (nom.nud.)
MERACOMELON Iredale, 1933
Contramelon, Findomelon Iredale, 1937
MESODONTRACHIA Solem, 1985
MIKIRIA Godwin-Austen, 1918
MOELLENDORFFIA Ancey, 1887
Proctostoma Mabille, 1887
Trichelix Ancey, 1887
Moellendorffiella Pilsbry, 1905
MOULDINGIA Solem, 1984
MUSSONENA Iredale, 1938
NECVIDENA Iredale, 1941
NEOCEPOLIS Pilsbry, 1891
NEVERITIS Iredale, 1937
NINGBINGIA Solem, 1981
OBSTEUGENIA Iredale, 1933
OBBA Beck, 1837
Pusiodon Swainson, 1840
Gallina Hartmann, 1843
Philina Albers, 1850 (part)
Obbina Semper, 1873
Obbiberus Haas, 1935
OCCIRHENEA Iredale, 1933
OFFACHLORITIS Iredale, 1933
ORDTRACHIA Solem, 1984
OREOBBA Pilsbry, 1894
Janira Albers, 1850
(non Leach, 1813)
PALLIDELIX Iredale, 1933
PARACHLORITIS Ehrmann, 1912
PARATROCHUS Pilsbry, 1893 ?
PARGLOGENIA Iredale, 1938
PLANISPIRA Beck, 1837
Cristigibba Tapparone-Canefri, 1883
Australgibba Iredale, 1933
Pseudopapuina Haas, 1934
Minicispira, Phrenigibba, Setogibba,
Spatiolabia all Iredale, 1941
PLECTORHAGADA Iredale, 1933
Idamera Iredale, 1939
PLEURODONTE Fischer von Waldheim, 1807
Caprinus Montfort, 1810
Dentellaria Schumacher, 1817
Pleurodonta Beck, 1837
Lucerna, Lucidula, Lucernella
all Swainson, 1840
Chrysodon Ancey, 1887 (non Oken, 1815)
Gonostomopsis Pilsbry, 1889
PLEUROXIA Ancey, 1887
Angasella Angas, 1864 (non Crosse, 1864)
Angasietta, Gantomia Iredale, 1939
POLYDONTES Montfort, 1810
Leiostoma Swainson, 1840
Pachystoma Albers, 1850 (part)
(non Guilding, 1828)
Luquillia Crosse, 1892
Grandodomus, Hispaniolana Pilsbry, 1931
Lissembryon Pilsbry, 1933
Salleana Clench, 1962

CAMAENINAE (Cont'd.)
POLYGYRATIA Gray, 1847
Ophiogyra Albers, 1850
Ridleya Ancey, 1901
PROTOTRACHIA Solem, 1984
PRYMNBRIAREUS Solem, 1981
PSEUDOPARTULA L. Pfeiffer, 1855
QUISTRACHIA Iredale, 1939
RAMOGENIA Iredale, 1938
RETROTERRA Solem, 1985
RHAGADA Albers, 1860
Thetagada, Tumegada Iredale, 1939
Bellrhagada Iredale, 1938 (45g)
SATSUMA A. Adams, 1868
Coniglobus Pilsbry & Hirase, 1906
Luchuhadra Kuroda & Habe, 1949
SEMOTRACHIA Iredale, 1933
Catellotrachia Iredale, 1933
Spernachloritis Iredale, 1933
SETOBAUDINIA Iredale, 1933
SINUMELON Iredale, 1930
Notobadistes Cotton & Godfrey, 1932
SPHAEROSPIRA Mörch, 1867
Micardista Iredale, 1933
SPURLINGIA Iredale, 1933
STEGODERA von Martens, 1876
Steganodera Kobelt, 1879
TEMPORENA Iredale, 1933
THERSITES L. Pfeiffer, 1855
Annakelea Iredale, 1933
TOLGACHLORITIS Iredale, 1933
TORRESITRACHIA Iredale, 1939
Magitrachia Iredale, 1941
TRACHIA Albers, 1860
Eurystoma Albers, 1850
(non Rafinesque, 1818)
Philidora Morgan, 1885
TRACHIOPSIS Pilsbry, 1893
Melotrachia Iredale, 1933
TRADEUSTOMA Iredale, 1941
TRAUMATOPHORA Ancey, 1887
Vulnus Sykes, 1904
TROZENA Iredale, 1938
TURGENITUBULUS Solem, 1981
VAROHADRA Iredale, 1933
Figuladra Iredale, 1933
VENTOPELITA Iredale, 1933
VIDUMELON Iredale, 1933
WESTRALTRACHIA Iredale, 1933
Parrhagada, Zygotrachia Iredale, 1939
XANTHOMELON von Martens, 1860
Globorhagada Iredale, 1933
YAKUCHLORITIS Habe, 1956
ZACHRYSIA Pilsbry, 1894
Auritesta, Chrysias, Megachrysia
all Pilsbry, 1929
Torrechrysias Bonilla, 1936
ZYGHELIX Iredale, 1937

PAPUININAE (41d)
PAPUINA von Martens, 1860
Geotrochus Swainson, 1832
(non Hasselt, 1823)
Eugenia von Martens, 1860
Merope Albers, 1860 ?
(non Newman, 1838)
Cymotropis von Martens, 1860 ?
Insularia Tapparone-Canefri, 1883
Carmerope, Caroletitia, Hombronula,
Letitia, Molmerope, Tepomusa,
Saccoletitia, Zenolina,
Zetemina all Iredale, 1941
Wahgia Clench & Turner, 1959

PAPUININAE (cont'd.)
CANEFRIULA Iredale, 1941
Medistoma, Paulodorra Iredale, 1941
CLAUDETTEA Iredale, 1941
CRYSTALLOPSIS Ancey, 1887
Cristovala Clench, 1958
FORCARTIA Clench & Turner, 1962
MEGALACRON Rensch, 1934
Pileolus Lesson, 1831
(non Sowerby, 1823)
Glomerata Rensch, 1934
Emiralena, Lisprelia, Lullicola,
Solmopina, Pinnadena all Iredale, 1941
MELIOBBA Iredale, 1940
Netogobba Iredale, 1941
MUNICEPS Iredale, 1941
NOCTEPUNA Iredale, 1933
PAPUANELLA Clench & Turner, 1959
PAPUEXUL Iredale, 1933
PAPUSTYLA Pilsbry, 1893
POSORITES Iredale, 1933
QUIROSENA Iredale, 1941
RHYNCHOTROCHUS Möllendorff, 1895
Henga, Kathadena, Pompalabia,
Violenga all Iredale, 1941
RHYTIDOCONCHA Rensch, 1933
SMEATONIA Iredale, 1941
SOLMODORA Iredale, 1941
SOLMOGADA Iredale, 1941
SOLMOPESTA Iredale, 1941
SOLMOTELLA Iredale, 1941

HELICOIDEA
SPHINCTEROCHILIDAE (36b)
SPHINCTEROCHILA Ancey, 1887
Mima Westerlund, 1886 (non Meigen, 1820)
Rima Pallary, 1910 (err.)
ALBEA Pallary, 1909
Calcarina Moquin-Tandon, 1848
(non Orbigny, 1826)
Leucochroa von Martens, 1860
(non Beck, 1837)
Candidissima Pallary, 1910
CARIOSULA Pallary, 1910
ZILCHENA Forcart, 1972

HYGROMIIDAE
HYGROMIINAE
HYGROMIA Risso, 1826
Hygromanes Herrmannsen, 1847
Sciaphila Westerlund, 1902
(non Meigen, 1818)
LOZEKIA Hüdec, 1970
ALTAICOLA Lindholm, 1927
ANGIOMPHALA Shileyko, 1978
LENTIGA Shileyko, 1978
ARCHAICA Shileyko, 1970
RIEDELIA Shileyko, 1972
UGAMA Shileyko, 1978
CHILANODON Westerlund, 1897
CILIELLA Mousson, 1872
Lepinota Westerlund, 1889
CIRCASSINA P. Hesse, 1921
JASONELLA Lindholm, 1927
CRETIGENA Shileyko, 1972
DIODONTELLA Lindholm, 1929
EUARCHAICA Shileyko, 1970
EUOMPHALIA Westerlund, 1889
Euryomphala Westerlund, 1897
(non Beck, 1837)
HARMOZICA Lindholm, 1927
HESSEOLA Lindholm, 1927
OSCARBOETTGERIA Lindholm, 1927

HYGROMINAE (Cont'd.)
FRUTICOCAMPYLAEA Kobelt, 1871
SHILEYKOIA Hüdec, 1972
KARABAGHIA Lindholm, 1927
LEJEANIA Ancey, 1887
LINDHOLMOMNEME Haas, 1936
METAFRUTICICOLA Ihering, 1892
Pseudocampylaea Hesse, 1884
(non L. Pfeiffer, 1877)
Cressa Westerlund, 1889
(non Beck, 1871)
Latonia Westerlund, 1889
(non H. Meyer, 1843)
Metafruticola Westerlund, 1902 (err.)
Westerlundia Kobelt, 1904
CAUCASOCRESSA P. Hesse, 1921 ?
MONACHOIDES Gude & Woodward, 1921
Monachella Gude & Woodward, 1921
(non Salvadori, 1874)
Pseudomonacha Pfeffer, 1929
NANAJA Shileyko, 1978
PERFORATELLA Schlüter, 1838
Trochiscus Held, 1837
(non Heyden, 1826)
Petasia Beck, 1837
(non Stephens, 1829)
Dibothrion L. Pfeiffer, 1855
PORTUGALA Gittenberger, 1980
PROSTHENOMPHALA Baidashnikov, 1985
PYRENAEARIA P. Hesse, 1921
SEMIFRUTICICOLA A. J. Wagner, 1914
STENOMPHALIA Lindholm, 1927
Micromphalia Lindholm, 1927
(non Ancey, 1882)
BATUMICA Shileyko, 1978
URTICICOLA Lindholm, 1927
ZENOBIELLA Gude & Woodward, 1921
Zenobia Gray, 1821
(non Oken, 1815)
Pseudotrichia Likharev, 1949

TRICHIINAE (44b)
TRICHIA Hartmann, 1841
? Trochulus Chemnitz,1784 in Lindholm,1827
Erethismus Gistl, 1848
Hispidella Lowe, 1854
(non Lowe, 1852)
Fruticola Held,1837 in von Martens,1860
(non Herrmannsen, 1847)
Capillifera Honigmann, 1906
DIOSCURIA Lindholm, 1927
Helixflaveola Mousson, 1863 ?
(non Kaleniczenko, 1853)
EDENTIELLA Polinski, 1929
Filicinella Polinski, 1929 ?
PETASINA Beck, 1847
Rimula Lowe, 1854 (non 1852)
(non Defrance, 1827)
Petasiella Gude & Woodward, 1921
ANOPLITELLA Lindholm, 1929 ?
CAUCASIGENA Lindholm, 1927
CAUCASOFIXUS Lindholm, 1927
HYGROHELICOPSIS Shileyko, 1978
KOKOTSCHASHVILIA Hüdec & Lezhawa, 1969 (44b)
LEUCOZONELLA Lindholm, 1927
NARINULA Shileyko, 1978
pro Narina Shileyko, 1978
(non Distant, 1906)
ODONTOTREMA Lindholm, 1927
PLICUTERIA Shileyko, 1978
PONENTINA P. Hesse, 1921
TEBERDINIA Shileyko, 1978
XEROCAMPYLEA Kobelt, 1871

HALOLIMNOHELICINAE (36b)
HALOLIMNOHELIX Germain, 1913
Blayneyella, Burungaella, Elgonella,
Larogiella, Mikenoella,
Nakurella all Preston, 1914
Congohelix Pfeffer, 1931
MASSAIHELIX Germain, 1913 ?
Nassaihelix Fischer-Piette, 1947

HELICELLIDAE ?
HELICELLINAE
HELICELLA Férussac, 1821
Planatella Clessin, 1876
Pseudoxerophila Westerlund, 1879
Xerobulla, Xerofriga, Xerolaxa,
Xerolenta, Xerolincta, Xeromicra
all Monterosato,1892
JABALCONIA Ortiz de Zarate, 1952
Pseudoxerotricha Ortiz de Zarate,1950
(non Boettger, 1911)
XEROTHRACIA Schütt, 1962
XEROTRICHA Monterosato, 1892
Polytrichia Pallary, 1922
Cyclomonilearia Pfeffer, 1929 ?
CANDIDULA Kobelt, 1871
Xeroalbina, Xerocodia, Xerogyra,
Xerovaga, Xerotringa, Xeroplexa,
Xerofalsa, Xeroamanda Monterosato, 1892
Xeramanda Kobelt, 1904 (err.)
CERNUELLA Schlüter, 1838
Heliomanes Brown, 1844
(non Newman, 1840)
Xerolauta, Xerolena, Xerolutea,
Xerovaria all Monterosato, 1892
Xeroampula, Xeromunda Monterosato,1892 ?
ALTENIELLA Clerx & Gittenberger, 1977
MICROXEROMAGNA Ortiz de Zarate, 1950 (53)
XEROCINCTA Monterosato, 1892
XEROMOESTA Monterosato, 1892
COCHLICELLA Férussac, 1820 (37) (Risso,1826)(53)
Longaeva Megerle in Menke, 1828
Elisma Leach in Turton, 1831 (Elismia)
Cochlicellus Beck, 1837
Xeroacuta Monterosato, 1892
HELICOPSIS Fitzinger, 1833 (Trichiinae - 36b)
Striatinella Clessin, 1876 (part)
Striatella Westerlund, 1876
(non Brot, 1870)
Costatella Kimakowicz, 1890
(non Dall, 1870)
Xeroalbina Monterosato, 1892 (part)
Martha A. J. Wagner, 1914
XEROLEUCA Kobelt, 1877
LEUCHROCHROA Beck, 1837
Jacosta Gray, 1821 (ICZN 431) (53)
Numidica Issel, 1885 ?
(non Cocco, 1832)
Tropidocochlis Locard, 1893 (part)
XEROMAGNA Monterosato, 1892
Xeroambigua, Xerofusca, Xerolissa,
Xeroplana, Xerovera Monterosato, 1892
MONACHA Fitzinger, 1833 (53) (Monacheae - err.)
Carthusiana Kobelt, 1871
Theba Risso in authors (non Risso)
ASHFORDIA Taylor, 1917
BOEMICA Shileyko, 1978
CYRNOTHEBA Germain, 1929
METATHEBA P. Hesse, 1914
PARATHEBA P. Hesse, 1914
PLATHYTHEBA Pilsbry, 1895
Nummulina Kobelt,1871
(non Orbigny,1826)
SZENTGALYA Pinter, 1977

HELICELLINAE (Cont'd.)
MONILEARIA Mousson, 1872
PAEDHOPLITA Lindholm, 1927 (44c)
TROCHOIDEA Brown, 1827
Turricula Beck, 1837
(non Schumacher, 1817)
Trochula Schlüter, 1838
Obelus Hartmann, 1843 (53)
Crenea Albers, 1850 (part)
(non Risso, 1826)
Xeroclivia Monterosato, 1892
Xerocochlea Monterosato, 1892
Tropidocochlis Locard, 1893 (part)
XEROCLAUSA Monterosatoe, 1892
XEROCRASSA Monterosato, 1892
Xerocauta Monterosato, 1892
Xerolaeta Monterosato, 1892 ?
Xeroptychia Westerlund, 1902
Ptycodia Kobelt, 1904
pro Xeroptyca Monterosato, 1892 ?
Xeropthychia Lamy, 1911
Ereminella Pallary, 1919
Xeroptycha P. Hesse, 1920
Heldia P. Hesse, 1926 ?
Xerobarcana, Xeroregima Brandt, 1959
XEROPICTA Monterosato, 1892
XEROSECTA Monterosato, 1892
KALITINIA Hüdec & Lezhava, 1967

HELICIDAE
HELICINAE
HELIX Linnaeus, 1758
Cochlea Da Costa, 1778
Glischrus Stüder, 1820
Helicogena Férussac, 1819 (part)
Pomatia Leach in Turton, 1831
Coenatoria Held, 1837
Callunea Megerle in Scudder, 1882
Megastoma Megerle in Scudder, 1882
Pomatiella Pallary, 1909
(non Clessin, 1889)
Cunula Pallary, 1936
Tammouzia Pallary, 1939
CANTAREUS Risso, 1826
Canthareus, Cantarelus auctt. (err.)
Tapada Gray in Turton, 1840
(non Stüder, 1820)
CRYPTOMPHALUS Agassiz in Charpentier, 1837
Cornu Born, 1778 ?
Erctella Monterosato, 1894 ?
IDIOMELA Cockerell, 1921 ?
MALTZANELLA P. Hesse, 1917
Maltzania P. Hesse, 1908
(non O. Boettger, 1881)
NAEGELEA Hesse, 1918
Naegelia Gude, 1921
PACHYPHALLUS P. Hesse, 1918
PHYSOSPIRA C. Boettger, 1914
RHODODERMA P. Hesse, 1918
TYRRHENARIA P. Hesse, 1918
ABCHASOHELA Hudec & Lezhawa, 1971 (Aciculidae ?)
pro Hela Hüdec & Lezhawa, 1970
(non Münster, 1830)
ALABASTRINA Kobelt, 1904
Alabastra Kobelt, 1904
ATLASICA Pallary, 1917
GUILIA Pallary, 1926
LECHATELIERIA Pallary, 1926
LOXANA Bourguignat, 1898
Beaumieriana Pallary, 1926
MICHAUDIA Pallary, 1926
SIRETIA Pallary, 1926
TINGITANA Pallary, 1919

HELICINAE (Cont'd.)
ALLOGNATHUS Pilsbry, 1888
IBERELLUS P. Hesse, 1908
Balearica Kobelt, 1904
(non Brisson, 1760)
CAUCASOTACHEA C. Boettger, 1909
LINDHOLMIA P. Hesse, 1918
CAUSA Shileyko, 1971 ?
CEPAEA Held, 1837
Tachea Leach in Turton, 1831
(non Fleming, 1822)
Hystrionica Megerle in Scudder, 1882
Palaeotachea Joos, 1912
Austrotachea, Holcotachea, Hyalotachea,
Megalotachea, Mesotachea, Paratachea,
Pleurotachea all Pfeffer, 1929
OCTADENA Shileyko, 1978
CODRINGTONIA Kobelt, 1898
EOBANIA P. Hesse, 1913
EREMINA L. Pfeiffer, 1855
Eremophila Kobelt, 1871
Erinna Mörch, 1865
(non H. & A. Adams, 1858)
Eremiopsis C. Boettger, 1909
Nomma Pallary, 1924
Exiliberus Iredale, 1942
EUPARYPHA Hartmann, 1843 ?
HELIXENA Backhuys, 1977
HEMICYCLA Swainson, 1840
Mycena Albers, 1850
Trachytachea Pfeffer, 1929
HESSEA O. Boettger, 1911
IBERUS Montfort, 1810
Carocolla Schumacher, 1817
Euiberus Westerlund, 1889
Transiberus Monterosato, 1892
MASSYLAEA Möllendorff, 1898
Joleaudiella Pallary, 1926
Platyotala Pfeffer, 1929
NESIBERUS Haas, 1934
ISAURICA Kobelt, 1901
LEVANTINA Kobelt, 1871
ASSYRIELLA P. Hesse, 1908
GYROSTOMELLA P. Hesse, 1911
Gyrostoma Hesse, 1908
(non Kirby & Spence, 1828)
MACULARIA Albers, 1850
MAUROHELIX P. Hesse, 1917
Gaetulia Kobelt, 1898
(non Stål, 1864)
Wiegmannia Hesse, 1916
(non Collinge, 1901)
MURELLA L. Pfeiffer, 1877
AMBIGUA Westerlund, 1902
Opica Kobelt, 1904
MARMORANA Hartmann, 1844
TYRRHENIBERUS Kobelt, 1904
OTALA Schumacher, 1817
Archelix Albers, 1850
DESERTICOLA P. Hesse, 1911
DUPOTETIA Kobelt, 1904
Pachydupotetia Pfeffer, 1929
PELASGA P. Hesse, 1908
Pseudofigulina P. Hesse, 1917
PSEUDOTACHEA C. Boettger, 1909
ROSSMAESSLERIA Hesse, 1907
STEKLOVIELLA Shileyko, 1979
pro Steklovia Shileyko, 1978
(non Schlickum & Strauch, 1972)
TACHEOCAMPYLAEA L. Pfeiffer, 1877
TACHEOPSIS C. Boettger, 1909
THEBA Risso, 1826
Xerphila Held, 1837

GEOMITRINAE (53)
GEOMITRA Swainson, 1840
Coronaria Lowe, 1852
CRASPEDARIA Lowe, 1852
SERRATOROTULA Groh & Hemmen, 1986 (22a)
ACTINELLA Lowe, 1852
FAUSTELLA Mandahl-Barth, 1950
Rimula Lowe, 1852
(non Defrance, 1827)
HISPIDELLA Lowe, 1852
PLEBECULA Lowe, 1852
Helicoma Lowe, 1854 ?
DISCULA Lowe, 1852
Turritella Wollaston, 1878
(non Lamarck, 1799)
CALLINA Lowe, 1854
HYSTRICELLA Lowe, 1854
HETEROSTOMA Hartmann, 1844
LEMNISCIA Lowe, 1854
OCHTHEPHILA Beck, 1837
Placentula Lowe, 1852
(non Lamarck, 1822)
Disculella Pilsbry, 1895
CASEOLUS Lowe, 1852
HELICOMELA Lowe, 1854
Wenzia Pfeffer, 1929
LEPTOSTICTEA Mandahl-Barth, 1950
PSEUDOCAMPYLAEA L. Pfeiffer, 1877
SPIRORBULA Lowe, 1852
Irus Lowe, 1852
(non Oken, 1821)
STEENBERGIA Mandahl-Barth, 1950
TECTULA Lowe, 1852
MANDAHLIA Forcart, 1965 ?

HELICIGONINAE (Campylaeinae)
HELICOGONA Férussac in Risso, 1826
Caracolla, Chilotrema Turton, 1831
Latomus Fitzinger, 1833
Lenticula Held, 1837
JOSEPHINELLA Haas, 1936
ARIANTA Leach in Turton, 1831
Arionta von Martens in Albers, 1860
CAMPYLAEA Beck, 1837
Eucampylaea L. Pfeiffer, 1879 (part)
ARIANTOPSIS Wagner, 1927
DELPHINATIA Hesse, 1931
DINARICA Kobelt, 1902
Sabliaria Brusina, 1904
Joossia Pfeffer, 1929
FAUSTINA Kobelt, 1904
Cattania Brusina, 1904
LIBURNICA Kobelt, 1904
Botteria Brusina, 1904
WLADISLAWIA Wagner, 1927
CHILOSTOMA Fitzinger, 1833 ?
Corneola Held, 1837
CAMPYLAEOPSIS Sturany & Wagner, 1914
CINGULIFERA Held, 1837
DROBACIA Brusina, 1904
Partschia Boettger, 1911
Dobracia Ehrmann, 1933
KOSICIA Brusina, 1904
THIESSEA Kobelt, 1904
CYLINDRUS Fitzinger, 1833
ELONA H. & A. Adams, 1855 (Elonidae - 21)
Sterna Albers, 1850
(non Linnaeus, 1758)
ISOGNOMOSTOMA Fitzinger, 1833
Plicostoma Schlüter, 1838
Isognomonostoma Agassiz, 1846
VIDOVICIA Brusina, 1904
Hazaya Soos, 1909

HELICODONTINAE
HELICODONTA Férussac in Risso, 1826 (non 1819)
Chilodon, Helicodon Ehrenberg, 1831
Trigonostoma Fitzinger, 1833
(non Blainville, 1825)
Vortex Beck, 1837
Gonostoma Held, 1837
(non Rafinesque, 1810)
Euphemia Leach in Beck, 1847
CANARIELLA Hesse, 1918
Paracanariella Pfeffer, 1929
Pachycarocollina Pfeffer, 1929
Praecanariella Pfeffer, 1929
CARACOLLINA Beck, 1837
Caesarella Pfeffer, 1929
Praeoestophorella Pfeffer, 1929
DREPANOSTOMA Porro, 1836
Contorta Megerle in Villa, 1841
GASULLIELLA Gittenberger, 1980
LINDHOLMIOLA Hesse, 1931
MASTIGOPHALLUS Hesse, 1918
OESTOPHORA Hesse, 1907
OESTOPHORELLA Pfeffer, 1929 ?
SOOSIA Hesse, 1918
TRISSEXODON Pilsbry, 1895
Protodrepanostoma Germain, 1929

LEPTAXINAE
LEPTAXIS Lowe, 1852
Cryptaxis, Katostoma Lowe, 1854
LAMPADIA Lowe, 1854
Mitra Albers, 1850
(non Martyn, 1784)

BRADYBAENIDAE (41b)
BRADYBAENINAE
BRADYBAENA Beck, 1837
Fruticicola Held, 1837
Eulota Hartmann, 1842
Eulotella von Martens, 1891
Manchurohelix Taki, 1936
Ezohelix Kuroda & Emura, 1938
Koreanohadra Kuroda & Habe, 1949
Phaeohelix Kuroda & Habe, 1949
Virginihelix Kuroda & Habe, 1949
ACUSTA von Martens in Albers, 1860
AINOHELIX Kuroda & Taki, 1933
KARAFTOHELIX Pilsbry, 1927
MASTIGEULOTA Pilsbry, 1895
TOROBAENA Haas, 1935
AEGISTA Albers, 1850
Neoaegista Azuma, 1955
Lepidopisum Kuroda & Habe, 1958
COELORUS Pilsbry, 1900
PLECTOTROPIS von Martens, 1860
Thea Albers, 1850
(non Mulsant, 1846)
ARMANDIELLA Ancey, 1901 ?
Armandia Ancey, 1901
(non Philippi, 1881)
CATHAICA Möllendorff, 1884
Eucathaica Andreae, 1900
CAMPYLOCATHAICA Andreae, 1900
KAZNAKOVIELLA Lindholm, 1922
PLIOCATHAICA Andreae, 1900
XEROCATHAICA Andreae, 1900
CHALEPOTAXIS Ancey, 1887
COCCOGLYPTA Pilsbry, 1895
DOLICHEULOTA Pilsbry, 1901
EUHADRA Pilsbry, 1890
HAPLOHELIX Pilsbry, 1919
LAEOCATHAICA Möllendorff, 1899
TRICHOCATHAICA Gude, 1919 ?

BRADYBAENINAE (Cont'd.)
METODONTIA Möllendorff, 1886 (Metodonta - err.)
Tetrodontina Ancey, 1887
NESEULOTA Ehrmann, 1912
NESIOHELIX Kuroda & Emura, 1943
PLECTEULOTA Möllendorff, 1892 ?
PONSADENIA Shileyko, 1978
MESASIATA Shileyko, 1978
TARBAGATAJA Shileyko, 1978
PSEUDASPASITA Möllendorff, 1902
PSEUDIBERUS Ancey, 1887
PLATYPETASUS Pilsbry, 1894
SEMIBULIMINUS Möllendorff, 1899
STILPNODISCUS Möllendorff, 1899
URGUESSELLA Preston, 1914
Debeauxhelix Bacci, 1943
PSEUDOBULIMINUS Gredler, 1886 (41b)
Buliminopsis Heude, 1890
BULIMINIDIUS Heude, 1890
Funiculus Heude, 1890
(non Scudder, 1882)
GIARDIA Ancey, 1907 (Girardius - err.)
RUDENS Heude, 1890
SECUSANUS Gredler, 1894
STENOGYROPSIS Möllendorff, 1899
TRISHOPLITA Jacobi, 1898
VICARIIHELIX Pilsbry, 1919

HELICOSTYLINAE (41b)
HELICOSTYLA Férussac, 1821
Cochlostyla Férussac, 1821
Cochlostylus Fischer, 1883
Orthostylus Beck, 1837
Pithohelix, Pythohelix Swainson, 1840
COCHLODRYAS von Martens, 1860
Poecilus Pfeiffer, 1879
(non Bonelli, 1813)
DRYOCOCHLIAS Möllendorff, 1898
HELICOBULINUS Broderip, 1841
Chromocochlea Hartmann, 1843 =
Chromatocochlias Agassiz, 1846
Coenobita Gistl, 1848
(non Latreille, 1829)
Helicobulimus Möllendorff, 1890
Rhymbocochlias Möllendorff, 1895
HYPSELOSTYLA von Martens, 1868
OPALLIOSTYLA Pilsbry, 1896
Eudoxus Albers, 1850
(non Kirby in Agassiz, 1846)
ORUSTIA Mörch, 1853
PACHYSPHAERA Pilsbry, 1892
CALOCOCHLIA Hartmann, 1840
Calocochlea Hartmann, 1843
Callicochlias Agassiz, 1847
Callocochlias Mörch, 1857 ?
Halocochlea Bartsch, 1932
ANIXA Pilsbry, 1894
Axina Albers, 1850
(non Kirby, 1817)
PYROCHILUS Pilsbry, 1892
Phania Albers, 1860
(non Meigen, 1824)
TRACHYSTYLA Pilsbry, 1892
CANISTRUM Mörch, 1852 (? non Fabricius, 1823)
CHLORAEA Albers, 1850
CHROMATOSPHAERA Pilsbry, 1892
CORASIA Albers, 1850
LEYTIA Pilsbry, 1892
PFEIFFERIA Gray, 1853
CHRYSALLIS Albers, 1850
DOLICHOSTYLA Pilsbry, 1896
Prochilus Albers, 1860
(non Illiger,1811)

HELICOSTYLINAE (Cont'd.)
MESANELLA Clench & Turner, 1952
PHENGUS Albers, 1850
PHOENICOBIUS Mörch, 1852
STEATODRYAS Pilsbry, 1932
Hypoptychus Pilsbry, 1893
(non Steindachner, 1880)
Ptychostylus Möllendorff, 1888
(non Sandberger, 1870)
TRICHEULOTA Pilsbry, 1895 (36c)
Chloritis Semper, 1877
(non Beck, 1837)

HELMINTHOGLYPTIDAE (41a;33b)
Xanthonychidae (36b)
HELMINTHOGLYPTINAE
HELMINTHOGLYPTA Ancey, 1887
Helix, Arianta, Arionta authors (part)
(non Arianta Leach, 1831)
Charodotes Pilsbry, 1939
COYOTE Reeder & Roth, 1988 (40x)
ROTHELIX Miller, 1985
DINOTROPIS Pilsbry & Cockerell, 1937 (53)
EPIPHRAGMOPHORA Döring, 1873
Doeringia Ihering, 1929
ANGRANDIELLA Ancey, 1887 ?
KARLSCHMIDTIA Haas, 1955 (53)
EREMARIONTA Pilsbry, 1913
EREMARIONTOIDES Miller, 1981
GLIABATES Webb, 1959
GLYPTOSTOMA Bland & Binney, 1873
GREGGELIX W. B. Miller, 1972
MARTIRELIX W. B. Miller, 1982
HUMBOLDTIANA Ihering, 1892
LYSINOE H. & A. Adams, 1855
Aglaja Albers, 1850
(non Renier, 1804)
Odontura Fischer & Crosse, 1870
(non Rambur, 1838)
Priodontura H. Fischer, 1899
MICRARIONTA Ancey, 1880
CHAMAEARIONTA Berry, 1930
MOHAVELIX Berry, 1943
MONADENIA Pilsbry, 1895 (37) (Monadeniidae - 36c)
Aglaia von Martens, 1860 (part)
Aglaja W. G. Binney, 1890
(non Albers, 1850)
CORYNADENIA Berry, 1940
SHASTELIX Roth, 1981
PLESARIONTA Pilsbry, 1939
POLYMITA Beck, 1837
SONORELIX Berry, 1943
HERPETEROS Berry, 1947
SONORELLA Pilsbry, 1900
MASCULUS Pilsbry, 1939
MYOTOPHALLUS Pilsbry, 1939
SONORANAX Pilsbry, 1939
TRYONIGENS Pilsbry, 1927
XERARIONTA Pilsbry, 1913

XANTHONYCHINAE
XANTHONYX Crosse & Fischer, 1867
AVERELLIA Ancey, 1887 ?
Coelospira Ancey, 1886 (non Hall, 1858)
MIRAVERELLIA H. B. Baker, 1922
DISCOLEPIS Ancey, 1904
TRICHODISCINA von Martens, 1892
Trichodiscus Strebel, 1880
(non Ehrenberg, 1830)
BUNNYA H. B. Baker, 1941 (53)
CRYPTOSTRACON W.G. Binney,1879 (Cryptostrakon,err.)
LEPTARIONTA Crosse & Fischer, 1872
METOSTRACON Pilsbry, 1900 ?

CEPOLINAE (37)
CEPOLIS Montfort, 1810
Cepolum Montfort, 1810
BELLACEPOLIS Pilsbry, 1943
CORYDA Albers, 1850
Histrio L. Pfeiffer, 1855
CYSTICOPSIS Mörch, 1852
EURYCAMPTA Von Martens, 1860
GULADENTIA Clench & Aguayo, 1951
HEMITROCHUS Swainson, 1840
Phaedra Albers, 1850
(non Horsfield, 1829)
Polytaenia von Martens, 1860
JEANNERETIA Pfeiffer, 1877
LEVICEPOLIS H. B. Baker, 1943
PLAGIOPTYCHA L. Pfeiffer, 1856
Monodonta Bartsch, 1932
(non Lamarck, 1801)
TAENIORAPHE Pilsbry, 1933
DIALEUCA Albers, 1850
Leptoloma von Martens, 1860

ARIONOIDEA (36b)
ARIONIDAE
ARIONINAE
ARION Férussac, 1819
Lochea Moquin-Tandon, 1855
Macroarion P. Hesse, 1926
(non Johnston, 1833)
CARINARION P. Hesse, 1926
Carinella Mabille, 1870
(non Johnston, 1833)
KOBELTIA Seibert, 1873
Baudonia Mabille, 1868
MESARION P. Hesse, 1926
Prolepis Moquin-Tandon, 1855 ?
Tetraspis Hagenmüller, 1855
MICROARION P. Hesse, 1926
ARIUNCULUS Lesson, 1881
Ischnusarion Pollonera, 1890

ARIONINAE (Cont'd.)
LETOURNEUXIA Bourguignat, 1886
GEOMALICUS Allman, 1843
Eugeomalacus Mabille, 1870 (part)
ARRUDIA Pollonera, 1890

ANADENINAE
ANADENUS Heynemann, 1863
Anadeninus Simroth, 1912
ANADENULUS Cockerell, 1890
NIPPONARION Yamaguchi & Habe, 1955
PROPHYSAON Bland & Binney, 1873 (37)
Limacarion J. G. Cooper, 1879
Phenacarion Cockerell, 1890
MIMETARION Pilsbry, 1948

ARIOLIMACINAE
ARIOLIMAX Mörch, 1859
Aphallarion Pilsbry & Vanatta, 1896
MEADARION Pilsbry, 1948
HESPERARION Simroth, 1891
MAGNIPELTA Pilsbry, 1953 ?
UDOSARX Webb, 1954 (Zacoleinae)
ZACOLEUS Pilsbry, 1903

BINNEYINAE
BINNEYA J. G. Cooper, 1863
Bineeia J. G. Cooper in
Binney & Bland, 1869 (err.)
ALLOTHYRA Pilsbry, 1948
HEMPHILLIA Bland & Binney, 1872

PHILOMYCIDAE
PHILOMYCUS Rafinesque, 1820
Limacella Blainville, 1817 (37)
(non Brard, 1815)
Eumelus Rafinesque, 1820
Tebenophorus A. Binney, 1842
MEGHIMATIUM Hasselt, 1824
Incillaria Benson, 1842
PALLIFERA Morse, 1864 (Pallifer auctt. - em.)

References:

Babrakzai, W. B. Miller & O. G. Ward "Cytotaxonomy: Ariz. Oreohelicidae" Bull.AMU 40:4-11:1975 (1)

Bishop, M. J. "Zonitoides: identity of Alienitor" J.Malac.Soc.Aust.4(1-2):7-8 - 1978 (4)

Britton, K. M. "Onchidiacea: worldwide review-" J.Moll.Stud.50:179-191 - 1984 (6)

Breure, A. S. H. "Catalog of Bulimulidae"
I. Amphibuliminae:Basteria 37:51-56 - 1973 (7)
II.Odontostominae:Ibid.38:109-127 - 1974 (7a)
---- "Bulimulidae" Malacologia 18:107-114 -1979 (8)

Burch, J. B. "Freshwater gastropods: No.America (N. of Mexico)"Journ.Conch.115(1-2):3-9.1978(10a)
---- "Limpet genus Brondelia"Bull.AMU.2:88.1984(10-b)
---- & C.M. Patterson "The systematic position of the Athoracophoridae" Malacologia 9(1):259-260 - 1969 (12)

Climo, F.M."New higher classification: New Zealand Rhytidae" Jour.Roy.Soc.N.Z.7(1):59-65 1977-78
---- "Classification of New Zealand Arionacea" Nat.Mus.N.Z. Rec. 1(12):177-201 - 1978 (16a)
---- "Paracharopa" Ibid 2(14):151-161 - 1983 (16b)

Cooke, C. M. & Y. Kondo "Tornatellidae and Achatinellidae" B. P. Bishop Mus.Bull.222 -1960 (17)

Dance, S. P. "Bruggenea, new genus: from Borneo" Arch.Moll.102(1-3):131-132 - 1972 (18)

Emberton, K.C. "E. North American Triodopsinae" Malacologia 28(1-2):159-273 - 1988 (19)

Forcart,L."Systematische: gattung Sphincterochila" Arch.Moll.102(4-6):147-164 - 1972 (20)

Gascoigne, T. "Provis. classification: Ascoglossa" Jour.Moll.Stud.51(1):8-22 - 1985 (20x)

Gittenberger, E. "On Elona: Elonidae, Family nov." Malacologia 18:139-145 - 1979 (21)
---- "Speleodentorcula Gen.nov.: Griechenland" Zool.Medid.Leiden 59(19):221-228 - 1985 (21a)

Gomez, B.J. & E. Angulo "Systematic position of Cryptazeca"Arch.Moll.118(1-3):57-62.1987(22x)

Groh, K. & J.Hemmen"Vitriniden: Madiera-Archipels" Arch.Moll.116(4-6):183-217 - 1986 (22)
---- "Serratorotula" Ibid.117(1-3):33-38.1986 (22a)

Gundrizer, V. A. & Y.I. Starabogatov "New species: Yenisei Basin" ZZ 58(8):1130-1135 (ES) -1979 (Veliger review 27(3):342 - 1985) (23)

Harry, H. W. "Catalogue: Neotropical Planorbidae" Malacologia 1(1):33-53 - 1962 (24)

References (Pulmonata, Cont'd.)
Heller, J. "Enidae of the Aegean Islands"
J.Moll.Stud.42:371-393 - 1976 (25)

Hubendick, B. "Phylogenic & Tiergeoprapia der (26)
Siphonariacea"Zool.Bidr.Uppsala.24:1-216-1945

Jaume,M.L. & A.de la Torre "Catalogo: fauna Cuba"
Circular Mus.Bibliot.Zool.Habana.(1-8) 1972
---- "Los Urocoptidae de Cuba" (reprint of 1972)
Ciencias Biologicas,U.de Habana 4(53)1976(27)

Kershaw, R.C. & B.J. Smith "- two punctid genera:
Tasmania" Vict.Nat.103(1):16-18 - 1986 (28)

Kruglov, N. D. & Y. Starobogatov "Subgenus Galba:
other Lymnaea" Zool.Zh.64(1):24-35 -1985 (29)

Leme, J.L.M."Neotropical Strophocheiloidea"1973(30)
Arq.Zool.Mus.Zool.U.Sao Paulo 23(5):320-333

Mead, A.R."Transfer Leucotaenius: Achatinidae to
Acavidae"Arch.Moll.116(4-6):137-155-1986 (31)

Meier-Brook, C. "Taxonomic status of Gyraulus"
Malacologia 24(1-2):1-113 - 1983 (32)

Miller, W. B. "New genus & species of Greggelix"
Veliger 24(4):345-348 - 1982 (33)
---- "A new subgenus of Helminthoglypta"
Veliger 28(1):94-98 - 1985 (33a)

---- & Reeder, Babraksai, Fairbanks "New, revised
taxa in Northa American terrestrial mollusca
(N.of Mexico) published since 19 March,1948"
I. Tryonia 11:1-14 - 1984 (33b)

Minato, H. "Land shells of Miyako Islands"
Venus 39(2):77-97 (1982 ?) (34)
---- & T. Habe "--Ujigunto and Kusakakigunto
islets" Venus 41(2):124-140 - 1982 (34a)
---- & T. Habe "New genus-Japanese Clausiliidae"
Venus 41(4):240-246 - 1983 (34b)

Minichev, Y.S. "Systematic placement: Soleolifera"
Biol.Morya 1:31-38 (ES) - 1975 (35)

Nordsieck, H."Anatomie und systematic: Clausilien"
"System: rezenten europaischen Clausilien"(36)
Arch.Moll.109(4-6):249-275;1979(1963 - 1980)
---- "System: Stylommatophora, with special regard
to the systematic position of Clausiliidae"
I. Arch.Moll.116(1-3):1-24 - 1985 (36a)
II. Ibid. 117(1-3):93-116 - 1986 (36b)
---- "Revision des systems der Helicoida"
Arch.Moll.118(1-3):9-50 - 1987 (36c)

Pilsbry, H. A. "Land Mollusca of North America"
Acad.Nat.Sci.Philad.Vol.I,II -1939-1948 (37)

Pinter, L. "Helicidae s.l." (Comments on Shileyko's
revision) Malacologia 16(1):223-225.1977 (38)

Ponder, W.F. "Review of the Rissoidae"
Rec.Aust.Mus.Suppl.4:1-219 - 1985 (39)
---- "Glacidorbacea: operculate freshw. gastropod"
Zool.Jour.Linn.Soc.87(1):152-182 -1986 (39a)

Radoman, P. "Ein Beitrag zur kenntnis: Bosniens"
Arch.Moll.104(1-3):81-84 - 1974 (40)

Reeder, R.L. & B. Roth "New subgenus: Helmintho-
glypta" Veliger (3-4):252-257 - 1988 (40x)

Richardson, L. "Catalog of species" (41)
"Helminthoglyptidae" Tryonia 6; 1982 ? (41a)
"Bradybaenidae" Ibid.9(1-2):1-253;1983 (41b)
"Oreohelicidae" Ibid.10:1-30; - 1984 (41c)
"Camaenidae" Ibid.12(1-2):1-479; -1985 (41d)

Riedel, A. & H. Pieper "Die systematisch stellung:
Gastranodon"Arch.Moll.110(4-6)185-189 1980(42)
---- "Uper einige Zonitidae aus dem Iran"
Arch.Moll.111(4-6):181-189 - 1981 (42a)
---- "Revision der gattung Zonites, Montfort"
Annales Zool.Warsz.41(1):1-42 - 1987 (42b)

Shileyko, A.A. "The mollusks of - Lauriinae in the
USSR" (Review in Veliger, 1975 (44)
---- "Organization and systematics: Orculidae"
NDVS(4):47-58 - 1976 (Veliger review) (44a)
---- "On the systematics of Trichia, s. l."
Malacologia 17(1):1-56 - 1978 (44b)
---- "New names for two taxa of Helicoidea"
Arch.Moll.110(1-3):107 - 1979 (44c)
---- "Type species: Crimea, Caucases" I,II Enidae
III.Buliminidae.Veliger 22(4):395 1980 (44d)
---- "System of the order Geophila: Helicidae" in
"Morphology, systematics, phylogeny:Mollusks"
Trans.Zool.Inst.Acad.Sci.USSR Vol.80 - 1979
Sp.Occ.Pub.6 Harvard U. - 1985 (44e)
---- "Some Data on taxonomy of Daudebardiidae"
Proc.Zoo.Inst.Leningr.148:97-123 -1986 (44f)
---- "System and Philogeny of Vitrinellidae"
Ibid.148:124-157 - 1986 (44g)

Solem, A."Endodontoid land snails:Pacific Islands"
I. Endodontidae:1-508 - 1976
II. Punctidae, Charopidae:1-336 - 1978
Field Mus.Nat.Hist.Chicago (45b)
---- "Camaenids: Western and Central Australia"
I. Taxa of Trans-Australian distribution
Rec.West.Aust.Mus.Suppl.10:1-142 1979 (45cd)
---- II. Taxa from Kimberly: Amphlirhagada
Ibid. Supp.11:143-320 - 1981 (45c)
---- III. Taxa from the Ningbing ranges
Ibid. Suppl.11:321-425 - 1981b (45d)
---- IV. Taxa :Kimberly: Westraltrachia
Ibid. Suppl.17:427-705 - 1984 (45e)
---- V. Remaining Kimberly genera and Addenda
Ibid Suppl.20:707-981 - 1985 (45f)
---- "Small land snails: North.Australia" - 1981
I.Gyliotrachela:J.Mal.Soc.Aust.5(1-2):87-100
II. Westralcystis. 5(3-4):175-193 -1981 (45g)
---- "Classification of Land Mollusca"in Fretter
& Peake (Eds.) Pulmonata, Vol.2A-1978 (45h)

Stanisic, J. "Studies on Charopidae: Oreokera"
J.Malac.Soc.Aust.8:1-21 - 1987 (46)

Te, G. A. "New classification: family Physidae"
Arch.Moll.110(4-6):179-184 - 1980 (47)

Thompson, F.G. & H.G. Lea "Helicarionids from
SE. U.S." Malac.Rev.13:37-44 - 1980 (48)

Tillier, S. "Malagarion paenelimax, gen. nov.-
Madagascar" Veliger 21(3):361-368 1979(49)

Van Goethem "La Systematique des Urocyclinae"
Malacologia 16(1):133-138 - 1977 (50)

Walker, J.C."Comp. anatomy: Bullinus, Isidorella"
J.Mal.Soc.Aust.4(4):233 - 1984 (51)

Zilch, A."Euthyneura" (Tiel 2) in Schindewolf,ed.
"Handbuch der Paleozoologie" 1959-60 (53)

Class Cephalopoda (4a)
Subclass Nautiloidea (Tetrabranchiata)
Order Nautilida

NAUTILIDAE
NAUTILUS Linnaeus, 1758

Subclass Coleoidea (Dibranchiata)
Order Sepiida
Suborder Spirulina
SPIRULIDAE
SPIRULA Lamarck, 1801
Lituina Link, 1806
Lituus Gray, 1849

Suborder Sepiina
SEPIIDAE
SEPIA Linnaeus, 1758
Eusepia Naef, 1923
HEMISEPIUS Steenstrup, 1875
Hemisepion Rochebrune, 1884
METASEPIA Hoyle, 1885
SEPIELLA Gray, 1849
Diphtherosepion Rochebrune, 1884

SEPIIDAE: Placement ? (6)
ACANTHOSEPION, LOPHOSEPION, SPATHIDOSEPION, all Rochebrune, 1884
Acanthosepia, Lophosepia, Spathidosepia all Naef, 1923 (em.)
ASCAROSEPION Rochebrune, 1884 (Ascarosepia - em.)
Amplisepia Iredale, 1926
CRUMENASEPIA, DECORISEPIA, GLYPTOSEPIA, SOLITOSEPIA, MESEMBRISEPIA, PONDERISEPIA all Iredale, 1926
DORATOSEPION Rochebrune, 1884 (Doratosepia - em.)
Andreaesepia Grimpe, 1922
Arctosepia Iredale, 1926
PLATYSEPIA Naef, 1923
RHOMBOSEPION Rochebrune, 1884 (Rhombosepia - em.)
Parasepia Naef, 1923
TENUISEPIA Cotton, 1932

SEPIOLIDAE
SEPIOLINAE
SEPIOLA Leach, 1817
Eusepiola Grimpe, 1922
HEMISEPIOLA, HETEROSEPIOLA Grimpe, 1922 ?
EUPRYMNA Steenstrup, 1887
Fidenas Gray, 1849
INIOTEUTHIS Verrill, 1881
RONDELETIOLA Naef, 1921
Rondeletia Naef, 1916 (non Goode & Bean, 1894)
SEPIETTA Naef, 1912
Sepidium Levy, 1912

ROSSIINAE
ROSSIA Owen, 1834 (10)
Epistychina Gistl, 1848
Allorossia Grimpe, 1922
? Franklinia Norman,1890 (non Jerdon,1863)
AUSTROROSSIA Berry, 1918
NEOROSSIA Boletzky, 1971 (14)
SEMIROSSIA Steenstrup, 1881

HETEROTEUTHINAE
HETEROTEUTHIS Gray, 1849
STEPHANOTEUTHIS Berry, 1909
IRIDOTEUTHIS Naef, 1912
NECTOTEUTHIS Verrill, 1883
SEPIOLINA Naef, 1912
STOLOTEUTHIS Verrill, 1881

SEPIADARIIDAE
SEPIADARIUM Steenstrup, 1881
SEPIOLOIDEA Orbigny, 1845

IDIOSEPIIDAE
IDIOSEPIUS Steenstrup, 1881
Idiosepion Fischer, 1882 (em.)
Microteuthis Ortmann, 1888
Naefidium Grimpe, 1920
Berryidium Grimpe, 1931

Order Teuthida (Decapoda)
Suborder Myopsida

LOLIGINIDAE (6)
LOLIGO Schneider, 1784 (1)
DORYTEUTHIS Naef, 1912
HETEROLOLIGO Natsukari, 1984
NIPPONOLOLIGO Natsukari, 1983
PHOTOLOLIGO Natsukari, 1984
ALLOTEUTHIS Wülker, 1920
Teuthis Gray, 1849 (non Linnaeus, 1758)
Acroteuthis Berry, 1913 (non Stolley, 1911)
Acruroteuthis Berry, 1920 ?
Acrololigo Grimpe, 1921
LOLIOLOPSIS Berry, 1929
LOLIOLUS Steenstrup, 1856
LOLLIGUNCULA Steenstrup, 1881
SEPIOTEUTHIS Blainville, 1824
Chondrosepia Rüppell & Leuckart, 1828
UROTEUTHIS Rehder, 1945

PICKFORDIATEUTHIDAE (10)
PICKFORDIATEUTHIS Voss, 1953

Suborder Oegopsida
LYCOTEUTHIDAE
LYCOTEUTHINAE
LYCOTEUTHIS Pfeffer, 1900
Enoploteuthis Chun, 1900 (non Orbigny, 1839)
Thaumatolampas Chun, 1903
Asthenoteuthion Pfeffer, 1912
Leptodontoteuthis Robson, 1926
Oregoniateuthis Voss, 1956
NEMATOLAMPAS Berry, 1913
SELENOTEUTHIS Voss, 1958

LAMPADIOTEUTHINAE
LAMPADIOTEUTHIS Berry, 1916

ENOPLOTEUTHIDAE
ENOPLOTEUTHINAE
ENOPLOTEUTHIS Orbigny, 1839
ABRALIA Gray, 1849
Enoploion Pfeffer, 1912
ASTEROTEUTHIS Pfeffer, 1900
STENOBRALIA Grimpe, 1931
ABRALIOPSIS Joubin, 1896 (12)
Compsoteuthis Pfeffer, 1900
Nepioteuthion Pfeffer, 1912
Prodromoteuthis Pfeffer, 1912
MICRABRALIA Pfeffer, 1900
ENIGMOTEUTHIS Adam, 1973
WATASENIA Ishikawa, 1913
pro Watasea Ishikawa, 1913 (non Jordan & Snyder, 1901)

ANCISTROCHEIRINAE
ANCISTROCHEIRUS Gray, 1849
Thelidioteuthis Pfeffer, 1900

PYROTEUTHINAE
PYROTEUTHIS Hoyle, 1904
Charybditeuthis Viviani, 1914
PTERYGIOTEUTHIS H. Fischer, 1896
Pterygonepion, Iotethion Pfeffer, 1912

OCTOPOTEUTHIDAE (Octopodoteuthidae)
OCTOPOTEUTHIS Rüppell, 1844
Octopodoteuthis Krohn, 1845
Verania Krohn, 1847 (auctt.)
Octopodoteuthopsis Pfeffer, 1912
TANINGIA Joubin, 1931
Cucioteuthis Steenstrup, 1882 (auctt.)

ONYCHOTEUTHIDAE
ONYCHOTEUTHIS Lichtenstein, 1818
Teleonychoteuthis Pfeffer, 1900
ANCISTROTEUTHIS Gray, 1849
CHAUNOTEUTHIS Appellöf, 1890
KONDAKOVIA Filippova, 1972 (2)
MOROTEUTHIS Verrill, 1881
Moroteuthopsis Pfeffer, 1908
ONYKIA Lesueur, 1821 (Onychia auctt. - err.)
Steenstrupiola Pfeffer, 1884
Teleoteuthis Verrill, 1885

GONATIDAE
GONATUS Gray, 1849
Cheiloteuthis, Lestoteuthis Verrill,1880
EOGONATUS Nesis, 1972
BERRYTEUTHIS Naef, 1921 (Berryiteuthis - error)
Pfefferiopsis M. Ishikawa, 1924
GONATOPSIS Sasaki, 1920
BOREOTEUTHIS Nesis, 1971

CTENOPTERYGIDAE
CTENOPTERYX Appellöf, 1880 (14)

BATHYTEUTHIDAE
BATHYTEUTHIS Hoyle, 1885
Benthoteuthis Verrill, 1885
Chunoteuthis Robson, 1921 ?
Indoteuthis Grimpe, 1922 ?

HISTIOTEUTHIDAE
HISTIOTEUTHIS Orbigny, 1939
Cranchia Férussac, 1835 (part)
(non Leach, 1817)
Lolidona Risso, 1854
Calliteuthis Verrill, 1880
Loligopsis Owen, 1881 (part)
Histiopsis Hoyle, 1885 (juv.)
Meleagroteuthis Pfeffer, 1900
Stigmatoteuthis Pfeffer, 1900
Histiothauma Robson, 1948

PSYCHROTEUTHIDAE
PSYCHROTEUTHIS Thiele, 1920

ARCHITEUTHIDAE
ARCHITEUTHIS Steenstrup, 1857 (Architeuthus - err.)
Megaloteuthis Kent, 1874
Dinoteuthis More, 1875
? Mouchezia Velain, 1877 (Mouchezis, err.)
Megateuthis Hilgendorf, 1880
Plectoteuthis Owen, 1881
Steenstrupia Kirk,1882 (non Forbes, 1846)
Dubioteuthis Joubin, 1899 ?

NEOTEUTHIDAE (Alluroteuthidae)
NEOTEUTHIS Naef, 1921
ALLUROTEUTHIS Odhner, 1923
NOTOTEUTHIS Nesis & Nikitina, 1986 (4c)

WALVISTEUTHIDAE (4b)
WALVISTEUTHIS Nesis & Nikitina, 1986

BRACHIOTEUTHIDAE
BRACHIOTEUTHIS Verrill, 1881
Tracheloteuthis Steenstrup, 1882
Verrilliola Pfeffer, 1884
Entomopsis Rochebrune, 1884
? HISTIOCHROMIUS Pfeffer, 1912

OMMASTREPHIDAE
ILLICINAE
ILLEX Steenstrup, 1880
TODAROPSIS Girard, 1890

TODARODINAE
TODARODES Steenstrup, 1880
Sagittatus Risso, 1854
MARTIALIA Rochebrune & Mabille, 1889
NOTOTODARUS Pfeffer, 1912

ORNITHOTEUTHINAE
ORNITHOTEUTHIS Okada, 1927

OMMASTREPHINAE
OMMASTREPHES Orbigny, 1835 (14)
Cycria Leach in Gray, 1849
Lolimnites Risso, 1854
DOSIDISCUS Steenstrup, 1857
EUCLEOTEUTHIS Berry, 1916
HYALOTEUTHIS Gray, 1849
STHENOTEUTHIS Verrill, 1880
Symplectoteuthis Pfeffer, 1900

THYSANOTEUTHIDAE
THYSANOTEUTHIS Troschel, 1857
CIRROBRACHIUM Hoyle, 1904 (?)

LEPIDOTEUTHIDAE (Pholidoteuthidae)
LEPIDOTEUTHIS Joubin, 1895
PHOLIDOTEUTHIS Adam, 1950
TETRONYCHOTEUTHIS Pfeffer, 1900

BATOTEUTHIDAE (16)
BATOTEUTHIS Young & Roper, 1968

CYCLOTEUTHIDAE (17)
CYCLOTEUTHIS Joubin, 1919
? Lycoteuthis Joubin, 1900
(non Pfeffer, 1900)
DISCOTEUTHIS Young & Roper, 1969

CHIROTEUTHIDAE
CHIROTEUTHIS Orbigny, 1839 (4)
Doratopsis Rochebrune, 1884 (larvae)
Hyaloteuthis Pfeffer, 1884
(non Gray, 1849)
Leptoteuthis Verrill, 1884
(non Meyer, 1834)
Chirothauma Chun, 1910
Chiridioteuthis Pfeffer, 1912
Diaphanoteuthis, Toroteuthis Tomlin, 1931
Bigelowia Macdonald & Clench, 1934
? CHIROSOMA Joubin, 1912
ASPEROTEUTHIS Nesis, 1980
CHIROPSIS Joubin, 1933 (4a)
CHIROTEUTHOIDES Berry, 1920
TANKAIA Sasaki, 1929
VALBYTEUTHIS Joubin, 1931
Planktoteuthis Pfeffer, 1912 (larvae)

MASTIGOTEUTHIDAE
MASTIGOTEUTHIS Verrill, 1881

MASTIGOTEUTHIS (Cont'd.)
Chiroteuthopsis Pfeffer, 1900
Idioteuthis Sasaki, 1916 =
Iridioteuthis Sasaki, 1929
ECHINOTEUTHIS Joubin, 1933

JOUBINITEUTHIDAE
JOUBINITEUTHIS Berry, 1920
Valdemaria Joubin, 1931

PROMACHOTEUTHIDAE
PROMACHOTEUTHIS Hoyle, 1885

GRIMALDITEUTHIDAE
GRIMALDITEUTHIS Joubin, 1898
Enoptroteuthis Berry, 1920 (Juv.)

Family incertae sedis:
PARATEUTHIS Thiele, 1921

CRANCHIIDAE
CRANCHIINAE (2)
CRANCHIA Leach, 1817
DRECHSELIA Joubin, 1931
LEACHIA Lesueur, 1821
Perothis Rathke, 1835
Dyctidiopsis, Pyrgopsis Rochebrune,1884
LIOCRANCHIA Pfeffer, 1884
Fuscocranchia Joubin, 1920

TAONIINAE
TAONIUS Lesueur, 1861 (Steenstrup ?)
Desmoteuthis Verrill, 1881
Phasmatopsis Rochebrune, 1884 ?
BATHOTHAUMA Chun, 1906
Leucocranchia Joubin, 1912
BELONELLA Lane, 1957
Toxeuma Chun, 1906
EGEA Joubin, 1933
GALITEUTHIS Joubin, 1898
? Zygaenopsis Rochebrune, 1884
(non Felder, 1874)
? Zygocranchia Hoyle, 1909
? Euzygaena Chun, 1910
Crystalloteuthis Chun, 1906
Taonidium Pfeffer, 1900
Phasmatoteuthion Pfeffer, 1912
HELICOCRANCHIA Massy, 1907
Hensenioteuthis Pfeffer, 1900
Ascocranchia Voss, 1962
LIGURIELLA Issel, 1908
Vossoteuthis Nesis, 1974
MEGALOCRANCHIA Pfeffer, 1912
Corynomma Chun, 1906
Ascoteuthis Berry, 1920
Carynoteuthis Voss, 1960
MESONYCHOTEUTHIS Robson, 1925
SANDALOPS Chun, 1906
Uranoteuthis Lu & Clarke, 1974
TEUTHOWENIA Chun, 1910
Owenia Prosch, 1847
(non Chiaje, 1844)
Verrilliteuthis Berry, 1916
Anomalocranchia Robson, 1924

Order Vampyromorpha
VAMPYROTEUTHIDAE
VAMPYROTEUTHIS Chun, 1903
Melanoteuthis Joubin, 1912
Hymenoteuthis Berry, 1916
? Watasella Sasaki, 1920
? Danateuthis, Retroteuthis,
Hensenoteuthis all Joubin, 1929

Order Octopoda
Suborder Cirrata
CIRROTEUTHIDAE (Stauroteuthidae)
CIRROTEUTHIS Eschricht, 1838
? Epulo Gistl, 1848
CIRROTEUTHOPSIS Grimpe, 1920 ?
CIRROCTOPUS Naef, 1923 ?
CHUNIOTEUTHIS Grimpe, 1916
CIRROTHAUMA Chun, 1911
FROEKENIA Hoyle, 1904
GRIMPOTEUTHIS Robson, 1932
STAUROTEUTHIS Verrill, 1879

OPISTHOTEUTHIDAE
OPISTHOTEUTHIS Verrill, 1883
Teuthidiscus Berry, 1918 ?

Suborder Incirrata
BOLITAENOIDEA
BOLITAENIDAE
BOLITAENA Steenstrup, 1859
DORSOPSIS Thore, 1949
ELEDONELLA Verrill, 1884
JAPETELLA Hoyle, 1885
Chunella Sasaki,1920 (non Kukenthal,1902)
Bolitaenella Grimpe, 1922

AMPHITRETIDAE
AMPHITRETUS Hoyle, 1885

IDIOCTOPODIDAE
IDIOCTOPUS Taki, 1962

OCTOPODOIDEA (Heteroglossa)
OCTOPODIDAE
OCTOPODINAE
OCTOPUS Lamarck, 1798
Polypus Schneider, 1784 in Hoyle, 1901
Eledonenta Rochebrune, 1884
Macrotritopus Grimpe, 1922
AMPHIOCTOPUS Fischer, 1882
CALLISTOCTOPUS Taki, 1964
ENTEROCTOPUS Rochebrune & Mabille, 1889
MACROCTOPUS Robson, 1928
PARACTOPUS Naef, 1923
Pseudoctopus Grimpe, 1925
PINNOCTOPUS Orbigny, 1845
ROBSONELLA Adam, 1938
Joubinia Robson, 1929 (preoccupied)
TRITAXEOPUS Owen, 1881
CISTOPUS Gray, 1840
DANOCTOPUS Joubin, 1933
Berrya Hoyle, 1939 (Adam ?)
Hapaloctopus Taki, 1962
EUAXOCTOPUS Voss, 1971
HAPALOCHLAENA Robson, 1929
MACROCHLAENA Robson, 1929 ?
PTEROCTOPUS Fisher, 1882
SASAKINELLA Taki, 1964
SCAEURGUS Troschel, 1857

ELEDONINAE (Ozaeninae)
ELEDONE Leach,1817 (Heledone Agassiz,1846 -err.)
Moschites Schneider, 1785
Ozaena Rafinesque, 1814 (Ozoena - err.)
Epistrophea Gistl, 1848
Hoylea, Hallia Rochebrune, 1886
Acantheledone Fort, 1941
MEGALELEDONE Taki, 1961
PARELEDONE Robson, 1932
TETRACHELEDONE Voss, 1955
VELODONA Chun, 1915
VOSSELEDONE Palacio, 1978

BATHYPOLYPODINAE
BATHYPOLYPUS Grimpe, 1921
BENTHELEDONE Robson, 1932
BENTHOCTOPUS Grimpe, 1921
Atlantoctopus Grimpe, 1921 ?
GRANELEDONE Joubin, 1918
GRIMPELLA Robson, 1928
HAPTOCHLAENA Grimpe, 1922 ?
TERETOCTOPUS Robson, 1929
THAUMELEDONE Robson, 1930

VITRELEDONELLIDAE
VITRELEDONELLA Joubin, 1918

LAETMOTEUTHIDAE (Placement ?) (6)
LAETMOTEUTHIS Berry, 1913

ARGONAUTOIDEA
ALLOPOSIDAE
ALLOPOSUS Verrill, 1880
? Haliphron Steenstrup, 1859
Alloposina Grimpe, 1922
Heptapus Joubin, 1929

OCYTHOIDAE
OCYTHOE Rafinesque, 1814
Parasira Steenstrup, 1861

TREMOCTOPODIDAE (14)
TREMOCTOPUS Delle Chiaje, 1830
Philonexis Orbigny, 1835

ARGONAUTIDAE
ARGONAUTA Linnaeus, 1758

References:
Cohen, A.C. "Systematics, distribution: Loligo; N.Atl." Malacologia 15(2):299-367 - 1976 (1)

Filippova, J.A. "New data: squids from Scotia Sea: (Antarctic)" Malacologia 11(2):391-406.1977(2)

Klumov S.K. & V.L. Yukhov "Oegopsidae: Mesonychoteuthis"(1975) Veliger review 19(4):442.1977(3)

Nesis, K. N. "Cephalopods: Interior American seas: Cheiroteuthidae" TRO 100:259-288(ES) 1977 (4)
---- "Cephalopoda of the World" (Russian, 1982) TFH.Publications Ltd.:1-351 - 1987 (4a)
---- & I.V. Nikitina "New family: abyssal squids" Zool.Zhur.65(1):47-54 - 1986 (4b)
---- "New genus of squid: Southeast Pacific" Ibid.65(2):290-299 - 1986 (4c)

Roper,C.F.E., M.J.Sweeney & C.E.Nauen "Cephalopods: World" FAO Fisheries Synopsis.125 #3 -1984 (6)

---- & R. E. Young "Review of the Valbyteuthidae" Proc.U.S.Nat.Mus.Vol.123(3612):1-9 - 1967 (7)

Voss, G.L. "New family, genus; myopsid squid" (10) Bull.Mar.Sci.Gulf Carib.3(4):602-609 -1953
---- "The family Lycoteuthidae" Bull.Mar.Sci. 12:264-305 - 1962 (11)
---- "Cephalopods of the Philippine Islands" U.S.Nat.Mus.Bull.234 - 1963 (12)

Voss, N.A. "Cephalopoda, N. Atl.: Histioteuthidae" Bull.Mar.Sci.19(4):713-867 - 1969 (13)
---- "Status, trends: Cephalopod systematics" Symp.Zool.Soc.Lond.38:49-60 - 1977 (14)

Young, R. E. & C. F. E. Roper "Bathyteuthidae" Antarct. Res.Ser. 11:185-202 - 1968 (16)
---- "Monograph of the Cephalopoda of the North Atlantic: the family Cycloteuthidae" Smith.Contr.Zool.5:1-24 - 1969 (17)

Class Bivalvia
Subclass Protobranchia
Order Solemyoida (1)

SOLEMYOIDEA (7)
SOLEMYIDAE (1)
SOLEMYA Lamarck, 1818
Solenimya Bowdich, 1822 (nom.van.)
Solenomya Children, 1823 (nom.van.)
Stephanopus Scacchi, 1833
Solenymia Swainson, 1840 (nom.van.)
Solenymya Schaufuss, 1869 (nom.van.)
PETRASMA Dall, 1908
SOLEMYARINA Iredale, 1939
Zesolemya Iredale, 1939
Solemyaria Habe, 1951 (nom.null.)
ACHARAX Dall, 1908

NUCINELLOIDEA (7)
NUCINELLIDAE
NUCINELLA Wood, 1851
Pleurodon Wood, 1840
(non Harlan, 1834)
Nuculina Orbigny, 1844
(non Porro, 1837)
Cyrillona Iredale, 1929
Neopleurodon Hertlein & Strong, 1940
HUXLEYIA A. Adams, 1860 (Apr.)
Cyrilla A. Adams, 1860 (June)
(unnec. nom.nov. pro Huxleyia)

HUXLEYIA (Cont'd.)
Diabolica Jousseaume, 1897
Cyrillista Iredale, 1929

Order Nuculoida

NUCULOIDEA (1)
NUCULIDAE
NUCULA Lamarck, 1799
Polyodonta Megerle, 1811
Nuclea Turton, 1822 (err.)
Lembulus Leach in Sowerby, 1842 (61)
(non Risso, 1826)
Lamellinucula Schenck, 1944
PRONUCULA Hedley, 1902
BREVINUCULA Thiele, 1934
NUCULOMA Cossmann, 1907
Nuculopsis Woodring, 1925
(non Girty, 1911)
Leionucula Quenstedt, 1930
Lionucula Thiele, 1934 (err.)
Ennucula Iredale, 1931
Eunucula Cotton, 1947 (err.)
Austronucula Powell, 1939
ACILA H. & A. Adams, 1858
TRUNCACILA Schenck in Grant & Gale, 1931
CONDYLONUCULA Moore, 1977 (61)
LINUCULA Marwick, 1931 (? fossil)
RUMPTONUCULA Bergmans, 1978 (61)
SINONUCULA Xu, 1985

PRISTIGLOMIDAE (53)
PRISTIGLOMA Dall, 1900
Pro Glomus Jeffreys, 1876
(non Gistl, 1848)
MICROGLOMA Saunders & Allen, 1973
PSEUDOGLOMUS Dall, 1898

NUCULANOIDEA
TINDARIIDAE (55)
TINDARIA Bellardi, 1875
Tyndaria Fischer, 1886 (em.)
Deminucula Iredale, 1931

LAMETILIDAE (54, 1)
LAMETILA Allen & Saunders, 1973
PRELAMETILA Allen & Saunders, 1973
PHASEOLUS Monterosato, 1875

RADIIDENTIDAE (61)
RADIIDENS Egoreva & Starobogatov, 1975
Pro Silicula Thiele, 1935 (1934) (61)
(non Jeffreys, 1879)

NUCULANIDAE
LEDELLINAE
LEDELLA Verrill & Bush, 1897
pro Junonia Seguenza, 1877
(non Hubner, 1819)
Comitileda Iredale, 1924
Magaleda Iredale, 1929
TINDARIOPSIS Verrill & Bush, 1897

SPINULINAE
SPINULA Dall, 1908 (7)
BATHYSPINULA Filatova, 1958

NUCULANINAE
NUCULANA Link, 1807
Leda Schumacher, 1817 (Lida auctt.-err.)
Scaeoleda, Thestyleda Iredale, 1929
Eptoleda, Kamaleda, Zygonoleda,
Exocholeda all Iredale,1939 (n.nd.)
Costelloleda Hertlein & Strong, 1940
Politoleda Hertlein & Strong, 1940
Robaia Habe, 1958
COSTANUCULANA Habe, 1951
Costanucula Edwards & Hopwood,1966 (err.)
JUPITERIA Bellardi, 1875
Sacella Woodring, 1925
pro Ledina Sacco, 1898 (Dec.)
(non Dall, 1898; Apr.)
Teretileda Iredale, 1929
LEMBULUS Risso, 1826
SINOLEDA Xu, 1985
ADRANA H. & A. Adams, 1858
POROLEDA Tate, 1893 ?
PROPELEDA Iredale, 1924 ?
SCISSULADRANA Petuch, 1987 ?
TENUILEDA Habe, 1977 ?
ZEALEDA Marwick, 1924 ?

YOLDIIDAE
YOLDIINAE
YOLDIA Moller, 1842
Tepidoleda Iredale, 1939
AEQUIYOLDIA Soot-Ryen, 1951
KALAYOLDIA Grant & Gale, 1931
CNESTERIUM Dall, 1898
Scissula Dall, 1908
(non Dall, 1900)
MEGAYOLDIA Verrill & Bush, 1897
Multidentata Krishtofovich, 1964
ORTHOYOLDIA Verrill & Bush, 1897

YOLDIELLINAE
YOLDIELLA Verrill & Bush, 1897
ADRANELLA Verrill & Bush, 1898
MICROYOLDIA Verrill & Bush, 1897
OVALEDA Iredale, 1925
PORTLANDIA Morch, 1857
Pseudoportlandia Woodring, 1925
Parayoldiella Filatova, 1971
SAREPTA Adams, 1860

SILICULIDAE (1) (?)
SILICULA Jeffreys, 1879
(unnec. nom.nov. pro Phaseolus
Monterosato, 1875) (61)
LAMELLILEDA Cotton, 1930

MALLETIIDAE (1)
MALLETIA Des Moulins, 1832
Solenella Sowerby, 1833
Ctenoconcha Gray, 1840
Mallettia, Mellitia auctt. (err.)
Solenarca Herrmannsen, 1848 (err.)
Pseudomalletia Fischer, 1886
Katadesmia Dall, 1908
Clencharia Clarke, 1961
MALLETIELLA Soot-Ryen, 1957
Bathymalletia Kuroda & Habe, 1971
MINORMALLETIA Dall, 1908

NEILONELLIDAE (Saturniidae - 1)
NEILONELLA Dall,1881 (61)
Saturnia Seguenza, 1877
(non Schrank, 1802) (61)
Austrotindaria Fleming, 1948
NEILO Adams, 1854
Carinineilo Kuroda & Habe, 1971
PROTONUCULA Cotton, 1930
PSEUDOTINDARIA Saunders & Allen, 1977 (?)

Subclass Pteriomorphia
Order Arcoida

ARCOIDEA
ARCIDAE
ARCINAE
ARCA Linnaeus, 1758
Daphne Poli, 1791 =
Daphnoderma Poli, 1795
Arcaria Rafinesque, 1815 (em.)
Cyphoxis Rafinesque, 1818
Arcacites Schlotheim, 1820 (invalid)
Archa, Arcula, Aroa auctt. (err.)
Navicula Blainville, 1825
Byssoarca Swainson, 1833
Cibota Morch, 1853
(non Herrmannsen, 1852)
Arcoptera Heilprin, 1887
Mesocibota Iredale, 1939
EONAVICULA Arkell, 1929
BARBATIA Gray, 1842
Modioliformia Deshayes, 1860
Granoarca Conrad, 1862
Plagiarca Conrad, 1875
Soldania Stephani & Pantanelli, 1878 ?
(non Orbigny, 1826)
Savignyarca Jousseaume, 1891
Asperarca Sacco, 1898
Obliquarca Sacco, 1898
Pugilarca Marwick, 1928
Destacar Iredale, 1936
Abarbatia Dall, Bartsch & Rehder, 1938
Barbarca Dall, Bartsch & Rehder, 1938
Hawaiarca Dall, Bartsch & Rehder, 1938

BARBATIA (Cont'd.)
Barbatirus, Mimarcaria, Miratacar, Opularca, Thronacar, Ustularca all Iredale, 1939
Jacksonarca Harris & Palmer, 1946
Nipponarca Habe, 1951 ?
Tucetonella Habe, 1961
ACAR Gray, 1857
Vitracar Iredale, 1939
CALLOARCA Gray, 1857
CUCULLAEARCA Conrad, 1865
Polynema Conrad, 1873
CORALIARCA Habe & Okutani, 1968 ?
FUGLERIA Reinhart, 1937 (7)
LITHARCA Gray, 1842
Litharea Gray, 1840 (nom.nud.)
Lithoarca Paetel, 1875 (err.)
TETRARCA F. Nordsieck, 1969 ?
TRISIDOS Röding, 1798
Trisis Oken, 1815 (invalid)
Parallelepipedum Mörch, 1850
Epitrisis Iredale, 1939

ANADARINAE
ANADARA Gray, 1847
Rhomboides Anton, 1838 ? (61) (non Goldfuss, 1820)
Anomalocardia Mörch, 1853; Adams, 1857 (non Schumacher, 1817)
Cara Gray, 1857
Anadarca, Anadaria, Anadura auctt.(err.)
Nemoarca Conrad, 1869
Pectinatarca Sacco, 1898
Diluvarca Woodring, 1925
Tegilarca Iredale, 1939
Esmerarca Olsson, 1961
CUNEARCA Dall, 1898
Imparilarca Iredale, 1929
Potiarca Iredale, 1939
GRANDIARCA Olsson, 1961
HATAIARCA Nodo, 1966
LARKINIA Reinhart, 1935
RASIA Gray, 1857
SCAPHARCA Gray, 1847
Sectiarca Olsson,1961
SCAPHARCOPSIS F. Nordsieck, 1969 ? (61)
THYAS Gray, 1857 (?)
BATHYARCA Kobelt, 1891
Microcucullaea Iredale, 1929
Indacar Iredale, 1939
BENTHARCA Verrill & Bush, 1898
LUNARCA Gray, 1857
Argina Gray, 1840 (nom.nud.) (non Hübner, 1816)
Lunaria Gray, 1847 (err.)
Lunularca Paetel, 1875 (err.)
Mabellarca Iredale, 1939
Arginarca McLean, 1951
SAMACAR Iredale, 1936
SCAPHULA Benson, 1834 ?
SENILIA Gray, 1842

NOETIIDAE
NOETIINAE
NOETIA Gray, 1857
Noetiella Thiele & Jaeckel, 1931
BARBATIELLA Jousseaume in Lamy, 1917
Paranoetia Thiele, 1934
EONTIA MacNeil, 1938
SHELDONELLA Maury, 1917
Didimacar Iredale, 1939

STRIARCINAE (43a)
STRIARCA Conrad, 1862
Galactella Cossmann & Peyrot, 1912
ARCOPSIS Koenen, 1885
Gabinarca, Mulinarca, Spinearca, Verilarca all Iredale, 1939
ESTELLACAR Iredale, 1939
Estellarca Habe, 1977 (err.)
RECTANGULARCA Eames, 1957
SCELIDIONARCA Oliver, 1987

PARALLELODONTIDAE
PARAGRAMMATODON Nomura & Niino, 1940
PORTERIUS Clark, 1925
Pseudogrammatodon Arkell, 1930

CUCULAEIDAE
CUCULLAEA Lamarck, 1801
Cucullana Lichtenstein, 1818
Cucullastis Finlay & Marwick, 1937
Cucullona Finlay & Marwick, 1937

LIMOPSOIDEA
LIMOPSIDAE
LIMOPSIS Sassi, 1827
Trigonocoelia Nyst & Galeotti, 1835
Cnisma Mayer, 1868
Limopsilla, Lissarcula Thiele, 1923
Loringella, Phrynelima Iredale, 1929
Glycilima Iredale, 1931
FELICIA Mabille & Rochebrune, 1889 ?
PECTUNCULINA Orbigny, 1843
Cosmetopsis Rovereto, 1898
Aspalima Iredale, 1929
Senectidens, Versipella Iredale, 1931
Circlimopa, Oblimopa Iredale, 1939
CRENULILIMOPSIS Kuroda & Habe, 1971
EMPLECONIA Dall, 1908
LISSARCA Smith, 1877 ?
Austrosarepta Hedley, 1899
NIPPONOLIMOPSIS Habe, 1951

GLYCYMERIDIDAE
GLYCYMERIS Da Costa, 1778 (nom.cons.ICZN 1414)
Axinaea Poli, 1791
Axinaeoderma Poli, 1795
Tuceta Röding, 1798
Pectunculus Lamarck, 1799 (non Da Costa, 1778)
Pectunculites Krueger, 1823 (invalid)
Pectunculopsis Ebert, 1890
Pseudaxinea Monterosato, 1892
Axinia Dautzenberg, 1893 (err.)
Axinoderma Herrmannsen, 1945 (err.)
Grandaxinaea, Veletuceta Iredale, 1931
Glycymerula Finlay & Marwick, 1937
Manaia Finlay & Marwick,1937 (fossil ?)
AXINACTIS Mörch, 1861
GLYCYMERELLA Woodring, 1925
Pectunpennaceus Lamarck, 1819
AXINOLA Hertlein & Grant, 1972
MELAXINAEA Iredale, 1930
Melanaxinaea Habe, 1951 (err.)
TUCETONA Iredale, 1931
Tucetilla, Tucetopsis Iredale, 1939
BELLAXINAEA Nicol, 1984

PHILOBRYIDAE
PHILOBRYA Carpenter, 1872 (7) (? 24)
pro Bryophila Carpenter, 1864 (non Treitschke, 1825)
Briophila Carpenter, 1864 (err.)
Philippiella Pfeffer, 1886 (61)

PHYLOBRYIDAE
PHILOBRYA (Cont'd.)
Stempelleria Clasing, 1918
Stempellia Odhner, 1921
(non Leger & Hesse, 1910)
ADACNARCA Pelseneer, 1903
HOCHSTETTERIA Velain, 1877
Cosatova Iredale, 1939
HOCHSTETTERINA Thiele, 1934
MICROMYTILUS Cotton, 1931
NEOCARDIA Sowerby, 1892
NOTOMYTILUS Hedley, 1916
AUPOURIA Powell, 1937
COSA Finlay, 1927
Denticosa Iredale, 1930 ?
CRATIS Hedley, 1915
DELTAODON Barnard, 1962 ?

Order Mytiloida

MYTILOIDEA (41)
MYTILIDAE
MYTILINAE
MYTILUS Linnaeus, 1758
Callitriche Poli, 1791 ?
Callitricoderma Poli, 1795
Mytilarius Froriep, 1806
Mytilites Wahlenberg, 1819 (inv.)
Eumytilus Ihering, 1900
CRASSIMYTILUS Skarlato & Starobogatov, 1979
CRENOMYTILUS Soot-Ryen, 1955
Mytiloconcha auctt. (non Conrad, 1862)
SINOMYTILUS Thiele, 1934
AULACOMYA Mörch, 1853
BRACHIDONTES Swainson, 1840
Brachydontes, Brachyodon, Brachyodontes,
Branchydontes, Branchydonta auctt.(err.)
AUSTROMYTILUS Laseron, 1956
CHOROMYTILUS Soot-Ryen, 1952
HORMOMYA Mörch, 1853
ISCHADIUM Jukes-Browne, 1905
LIMNOPERNA Rochebrune, 1882 (?)
MYTELLA Soot-Ryen, 1955
MYTILASTER Monterosato, 1883
Mytilidonta Coen, 1935
PEREGRINAMOR Shoji, 1938 ?
PERNA Philipsson, 1788
Perna Oken, 1815 (inv.)
Pernites Krueger, 1823 (inv.)
Chloromya Mörch, 1853
Mytiloconcha Conrad, 1862
PERUMYTILUS Olsson, 1961
SCOLIMYTILUS Olsson, 1961
SEMIMYTILUS Soot-Ryen, 1955
SEPTIFER Récluz, 1848
MYTILISEPTA Habe, 1951
STAVELIA Gray, 1858
TRICHOMYA Ihering, 1900
Dentimodiolus Iredale, 1939

MODIOLINAE
MODIOLUS Lamarck, 1799 (nom. conserv.)
pro Volsella Scopoli, 1777 (suppr.)
Modiola Lamarck, 1801
Modiolites Krueger, 1823 (inv.)
Modiolina Giebel, 1852 (nom.nud.)
Perna H. & A. Adams, 1858
(non Philipsson, 1788)
Eumodiolus Ihering, 1900
Nudiola Monterosato, 1917
FULGIDA Jousseaume in Lamy, 1919
GIBBOMODIOLA Sacco, 1898
JOLYA Bourguignat, 1876

MODIOLUS (Cont'd.)
MODIOLATUS Jousseaume, 1893
MODIOLUSIA Yamamoto & Habe, 1958
ADIPICOLA Dautzenberg, 1927
pro Myrina H. & A. Adams, 1854
(non Fabricius, 1808)
Terua Dall, Bartsch & Rehder, 1938 (26)
AMYGDALUM Megerle, 1811
Modiella Monterosato, 1884
BENTHOMODIOLUS Dell, 1987 (26)
DACRYDIUM Torell, 1859 ?
QUENDREDA Iredale, 1936
GEUKENSIA Van der Poel, 1959
pro Arcuatula Soot-Ryen, 1955
(non Jousseaume in Lamy, 1919)
IDASOLA Iredale, 1915 (41) (unnec. n.n.) (61)
pro Idas Jeffreys, 1876 (26)
(non Mulsant & Verreaux, 1876)
Miridas Iredale, 1939
Myrinopsis F. Nordsieck, 1969
Habepegris Bernard, 1978 (7)
LIOBERUS Dall, 1898
MODIOLULA Sacco, 1898

Placement ?
AEDIMYTILUS Olsson, 1961
BATHYMODIOLUS Kenk & Wilson,1985 (nom.nud.?)(35)
(Priority ? in Laubier & Desbryeres, 1984)
CIBOTICOLA Iredale, 1939
FUNGIACAVA Soot-Ryen, 1969
STENOLENA Dall, Bartsch & Rehder, 1938
XENOSTROBUS Wilson, 1967

CRENELLINAE
CRENELLA Brown, 1827
Stalagmium Conrad, 1833
Hippagus, Myoparo Lea, 1833
Nuculocardia Orbigny, (? 1853)
Crenellodon Edwards in Newton, 1891 ?
ARCOPERNA Conrad, 1865 (fossil only ?)
ARCUATULA Jousseaume in Lamy, 1919
Lamya Soot-Ryen, 1958
ARVELLA Bartsch in Scarlato, 1960
BOTULA Mörch, 1853 ?
Botulopa Iredale, 1939
GREGARIELLA Monterosato, 1883
Botulina Dall, 1889
Tibialectus Iredale, 1939
MEGACRENELLA Habe & Ito, 1965
MUSCULISTA Yamamoto & Habe, 1958
MUSCULUS Röding, 1798
Modularia Beck, 1838
Lanistes Swainson, 1840
(non Montfort, 1810)
Modiolaria Lovén, 1846
Lanistina Gray, 1847
Planimodiola Cossmann, 1887
PROPETILUS Iredale, 1937
RYENELLA Fleming, 1959
Modiolarca Gray, 1843
Lanistina Soot-Ryen, 1955
(non Gray, 1847)
VILASINA Bartsch in Scarlato, 1960
SOLAMEN Iredale, 1924
EXOSIPERNA Iredale, 1929
RHOMBOIDELLA Monterosato, 1884
TRICHOMUSCULUS Iredale, 1924
VIGNADULA Kuroda & Habe, 1971

LITHOPHAGINAE
LITHOPHAGA Röding, 1798
Lithophagus Megerle, 1811
Lithodomus Cuvier, 1817

LITHOPHAGINAE
LITHOPHAGA (Cont'd.)
Lithoglyphus Sturm, 1821
(non Hartmann, 1821)
Lithodoma Blainville, 1823
Lithotormus Nitzch, 1825 (err.)
Tamarindiformis Herrmannsen, 1849
Dactylus Mörch, 1861; Jousseume, 1898
(non Schumacher, 1817)
RUPIPHAGA Olsson, 1961
ADULA H. & A. Adams, 1857
DIBERUS Dall, 1898
Exodiberus, Salebrolabis Iredale, 1939
LABIS Dall, 1916
Doliolabis Iredale, 1939
LEIOSOLENUS Carpenter, 1856 (61)
MYAPALMULA Iredale, 1939
MYOFORCEPS Fischer, 1886
STUMPIELLA Soot-Ryen, 1955
ZELITHOPHAGA Finlay, 1927

Order Pterioida
Suborder Pteriina

PTERIOIDEA
PTERIIDAE
PTERIA Scopoli, 1777
Avicula Bruguière, 1791
Glaucus Poli, 1795
(non Gmelin, 1791)
Glaucoderma Poli, 1795
Anonica Oken, 1815 (Invalid)
Anomia Deshayes, 1815 (err. pro Anonica)
(non Linnaeus, 1758)
Austropteria Iredale, 1931
ELECTROMA Stoliczka, 1871
Electrina von Martens, 1872
(? non Gray, 1850)
PTERELECTROMA Iredale, 1939
MAGNAVICULA Iredale, 1939 (?)
PINCTADA Röding, 1798
Unionium Link, 1807
Margaritiphora Megerle, 1811
Margarita Leach, 1814
Perlamater Schumacher, 1817
Meleagrina Lamarck, 1819
Pintadina Blainville, 1826
Meleagris Fischer, 1835 (err.)
Wolfataria Gray, 1840 ?
Margaritites Agassiz, 1845 (n.null.)

MALLEIDAE (Vulsellidae) (7)
MALLEUS Lamarck, 1799
Malleigenus Renier, 1807 (invalid)
Malleolus Rafinesque, 1815 (em.)
Tudes Oken, 1815 (invalid)
Himantopoda Schumacher, 1817
Himanthopoda Latreille, 1825 (err.)
MALVUFUNDUS de Gregorio, 1885
Parimalleus Iredale, 1931
Parvimalleus Salisbury, 1932 (nom.null.)
Brevimalleus McLean, 1947
pro Fundella de Gregorio, 1884
(non Zeller, 1848)
Malvifundus Habe, 1977
VULSELLA Röding, 1798
Vulselligenus Renier, 1807 (invalid)
Reniella Swainson, 1840
Vulcella Orbigny, 1846 (nom.null.)
Baphia Gray, 1847
Pulsella Weyenbergh, 1875 (nom.null.)
Abisa, Madrela de Gregorio, 1884
Madrila Fuchs, 1887 (nom.null.)

ISOGNOMONIDAE
ISOGNOMON Solander in Lightfoot, 1786
Isognoma Lightfoot, 1786
Melina Philipsson, 1788
Perna Bruguière, 1789
(non Philipsson, 1788)
Pernigenus Renier, 1807 (invalid)
Sutura Megerle, 1811
Pernaria Rafinesque, 1815 (emend.)
Hippochaeta Sangiovanni, 1884 ?
Aviculoperna Pchelintsev, 1931
(non Cossmann, 1887)
Anisoperna, Parviperna Iredale, 1939
ISOGONUM Röding, 1798
Pedalion Dillwyn, 1817
Malleoperna Iredale, 1939
CRENATULA Lamarck, 1803
Crenatuligenus Renier, 1807 (invalid)
Cranatula Children, 1823 (nom.null.)
Dalacia Gray, 1825

PULVINITIDAE
FORAMELINA Hedley, 1914

Suborder Pinnina

PINNOIDEA
PINNIDAE (49)
PINNA Linnaeus, 1758
Chimaera Poli, 1791
(non Linnaeus, 1758)
Chimaeroderma Poli, 1795
Pinnarius Froriep, 1806
Pinnigenus Renier, 1807 (invalid)
Pinnula Rafinesque, 1815 (em.)
Pennaria Mörch, 1853
(non Blainville, 1818)
Steptopinna von Martens, 1880
Pina Koch, 1922 (err.)
Exitopinna, Quantulopinna,
Subitopinna all Iredale, 1939
Conulariopsis Sugiyama, 1942 ?
CYRTOPINNA Mörch, 1853
OXISMA Rafinesque, 1819 ?
Oxysma Herrmannsen, 1847 (nom.van.)
Otysma Paetel, 1875 (err.)
ATRINA Gray, 1842
Servatrina Iredale, 1939
CONULARIOPSIS Sugiyama, 1942 ?
STREPTOPINNA von Martens, 1880 (61)

Order Limoida

LIMOIDEA
LIMIDAE
LIMA Bruguière, 1797
Lima Cuvier, 1798 (invalid)
Mantellum Röding, 1798
Limaria Rafinesque, 1815
(non Link, 1807)
Glaucion Oken, 1815 (invalid)
Radula Mörch, 1853
(non Gray, 1842)
Austrolima Iredale, 1929
Australima Cotton, 1931 (nom.null.)
ACESTA H. & A. Adams, 1858
Callolima Bartsch, 1913
ANTARCTOLIMA Habe, 1977
Squamilima Fleming, 1978
CTENOIDES Mörch, 1853
Divaricolima Rovereto, 1898 (fossil)
Ptenoides Smith, 1904 (nom.null.)
DIVARILIMA Powell, 1958

LIMIDAE (Cont'd.)
LIMARIA Link, 1807
Mantellum Mörch, 1853
(non Röding, 1798)
Submantellum Olsson & Harbison, 1953
LIMATULELLA Sacco, 1898
STABILIMA Iredale, 1939
LIMATULA Wood, 1839
Cystigenus Renier, 1807 (invalid)
Cystia Renier, 1807 (invalid)
Limacula Gray, 1840 (invalid)
Limula Orbigny, 1846 (nom.null.)
Linatulina Herrick, 1888 (error)
LIMATULETITA Fleming, 1978
LIMEA Bronn, 1831
Limaea Gray, 1847 (nom.van.)
NOTOLIMEA Iredale, 1924
ESCALIMA Iredale, 1929
GEMELLIMA Iredale, 1929
ISOLIMEA Iredale, 1929
Isolima Masahito & Habe, 1976 (err.)
PLATILIMARIA Habe, 1977
PLICACESTA Vokes, 1963
PROMANTELLUM Iredale, 1939
WINCKWORTHIA Glibert & Van de Poel, 1965

Order Ostreoida
Suborder Ostreina

OSTREOIDEA
OSTREIDAE (27)
OSTREINAE
Tribe Ostreini
OSTREA Linnaeus, 1758
Ostreites Herrmannsen, 1847
pro Ostracites Gesner, 1758 (non binom.)
Ostreum da Costa, 1776 (nom.van.)
Ostracites Gmelin, 1793 (rej.)
Peloridoderma Agassiz, 1846
pro Peloriderma Poli, 1795
pro Peloris Poli, 1791
Ostracarius Duméril, 1806 (nom.van.)
Ostreigenus Renier, 1807 (rej.)
Ostraea Sowerby, 1839 (nom.van.)
Monoeciostrea Orton, 1928 (nom.van.)
Ostroea Tolmer, 1928 (nom.null.)
Osrea Tzankov, 1932 (nom.null.)
Monoeciostraea Thiele, 1934 (nom.van.)
? Anomiostrea Habe & Kosuge, 1966 (inv.)
EOSTREA Ihering, 1907
Anodontostrea Suter, 1917
Tiostrea Chanley & Dinamani, 1980
AGEROSTREA Vyalov, 1936
NANOSTREA Harry, 1985
OSTREOLA Monterosato, 1884
PLANOSTREA Harry, 1985
Tribe Cryptostreini
CRYPTOSTREA Harry, 1985
BOONEOSTREA Harry, 1985
TESKEYOSTREA Harry, 1985
Tribe Undulostreini
UNDULOSTREA Harry, 1985
Tribe Pustulostreini
PUSTULOSTREA Harry, 1985

CRASSOSTREINAE
Tribe Crassostreini
CRASSOSTREA Sacco, 1897
Gryphaea Fischer, 1886
(non Lamarck, 1801)
Dioeciostrea Orton, 1928
Dioeciostraea Thiele, 1934 (nom.null.)
Angustostrea Vyalov, 1936

CRASSOSTREA (Cont'd.)
Grassostrea Vyalov, 1948 (nom.null.)
Somalidacna Azzaroli, 1958
Crassotrea Miyake & Noda, 1962 (n.nl.)
Tribe Striostreini
SACCOSTREA Dollfus & Dautzenberg, 1920
Saxostrea Iredale, 1936
Sanostrea Miyake & Noda, 1962 (n.null.)
STRIOSTREA Vyalov, 1936
PARASTRIOSTREA Harry, 1985

LOPHINAE
Tribe Lophini
LOPHA Roding, 1798
Alectryonia Fischer, 1807
Alectrionia, Actryonia, Alcetryonia,
Alectronia, Alectryossia,
Alectryouia, Alektryonia auctt.(err.)
Pretostrea Iredale, 1939
ALECTRYONELLA Sacco in Bellardi & Sacco, 1897
DENDOSTREA Swainson, 1835
Tribe Myrakeenini
MYRAKEENA Harry, 1985
ANOMIOSTREA Habe & Kosuge, 1966

GRYPHAEIDAE
PYCNODONTEINAE
Tribe Hyotissin
HYOTISSA Stenzel, 1971
PARAHYOTISSA Harry, 1985
PLIOHYATISSA Harry, 1985
NUMISMOIDA Harry, 1985
Tribe Neopycnodontini
NEOPYCNODONTE Stenzel, 1971

PLICATULOIDEA
PLICATULIDAE
PLICATULA Lamarck, 1801
Plicatuligenus Renier, 1807 (inv.)
Micatula Carpenter, 1859 (nom.null.)
Plycatula Monterosato, 1877 (n.null.)
SPINIPLICATULA Habe, 1977

DIMYIDAE
DIMYA Rouault, 1850 (Dimia - err.)
Margariona Dall in Kobelt, 1882
DIMYARINA Iredale, 1936 ?
DIMYELLA Moore, 1970
BASILIOMYA Bayer, 1971

Suborder Pectinina

PECTINOIDEA
PECTINIDAE
CHLAMYDINAE
CHLAMYS Röding, 1798
Clamys Leach, 1815 (nom.null.)
Chalmys Dollfus & Dautzenberg, 1886 (nom.null.)
Chlamydina Cossmann, 1909
pro Myochlamys Ihering, 1907
(non Fairman, 1876)
Belchlamys, Mimachlamys, Scaeochlamys,
Talochlamys, Veprichlamys,
Coralichlamys all Iredale,1939
Porosichlamys Lucas, 1979 (?)
ARGOPECTEN Monterosato, 1889
Plagioctenium Dall, 1898
Plagiopecten G. D. Hanna, 1924 (n.null.)
Plagiopectenium Stewart, 1930 (n.null.)
Haumea Dall, Bartsch & Rehder, 1938
Corymbichlamys, Edentiplica,
Volachlamys all Iredale,1939

CHLAMYDINAE
CHLAMYS (Cont'd.)
AZUMAPECTEN Habe, 1977 ?
CRYPTOPECTEN Dall, Bartsch & Rehder, 1938
EQUICHLAMYS Iredale, 1929
LEOCHLAMYS MacNeil, 1967
MANUPECTEN Monterosato, 1889 (Apr.)
Felipes Locard in Carus,1889 (Nov/Dec.)
OCHOTOCHLAMYS Milova, 1978
SWIFTOPECTEN Hertlein, 1935
AEQUIPECTEN Fischer, 1886 (61x)
Aequivalvis Jekelius, 1915 (nom.null.)
Equipecten B. L. Clark, 1929 (nom.null.)
PERAPECTEN Wagner, 1985
KARNEKAMPIA Wagner, 1988 (61xa)

PECTININAE
Tribe Hinnitini
HINNITES Defrance, 1821 (16)
Hinnita Férussac, 1822 (nom.van.)
Hinnus S. Wood in Gray, 1826
Hynnites Herrmannsen, 1847 (nom.van.)
LEPTOPECTEN Verrill, 1897
PACIPECTEN Olsson, 1961
LYROPECTEN Conrad, 1862
NODIPECTEN Dall, 1898
Nodopecten Olsson, 1914 (nom.null)
PEDUM Bruguière, 1791 (Lamarck, 1799) ?
Pedinus Rafinesque, 1815 (nom.nud.)
(non Latreille, 1796)
PLACOPECTEN Verrill, 1897
Nipponopecten Masudo, 1962 ?

Tribe Decatopectenini (62)
DECATOPECTEN Rüppel in Sowerby, 1839
Pallium Martini, 1774 (non-binom.)
Decadopecten Swainson, 1840 (err.)
Dentipecten Gray, 1847
Comptopallium Iredale, 1939
ANGUIPECTEN Dall, Bartsch & Rehder, 1938
ANNACHLAMYS Iredale, 1939
BRACTECHLAMYS Iredale, 1939
EXCELLICHLAMYS Iredale, 1939
FLEXOPECTEN Sacco, 1897
Glabropecten Sacco, 1897
Lissopecten Verrill, 1897
Proteopecten Monterosato, 1899
Flexipecten Oppenheim, 1903 (nom.null.)
GLORIPALLIUM Iredale, 1939
Gloriopallium Habe, 1977 (err.)
JUXTAMUSIUM Iredale, 1939
MIRAPECTEN Dall, Bartsch & Rehder, 1938
SOMALIPECTEN Waller, 1986
SEMIPALLIUM Lamy, 1928
Pera Chemnitz, 1784 (non-binom.)
Complicachlamys Iredale, 1939
MESOPEPLUM Iredale, 1929
Notochlamys Cotton, 1930
Dendopecten Hertlein, 1936

Tribe Eburneopectenini
CYCLOPECTEN Verrill, 1897
Cycopecten de Gregorio,1898 (nom.null.)
Cyclochlamys Finlay, 1926 (nom.null.)
Chlamydella Iredale, 1929
Arctinula Thiele, 1934
Aretinula Neave, 1939 (nom.null.)
Catillopecten Iredale, 1939
DELECTOPECTEN Stewart, 1930
Deletopecten Salibury, 1930 (nom.null.)
Similipecten Winckworth, 1932
HEMIPECTEN Adams & Reeve, 1848
Semipecten Fischer, 1886

Tribe Eburneopectenini (Cont'd.)
HYALOPECTEN Verrill, 1897
Hylopecten Stewart, 1930 (nom.nud.)
PALLIOLUM Monterosato, 1884
Pseudamusium Verrill, 1897 (em.)
Pseudomuseum B. L. Clark, 1918 (n.nl.)
Palliorum Oyama, 1944 (nom.null.)
PSEUDAMUSSIUM Mörch, 1853
Pseudamussium Klein, 1753 (pre-Linnaean)
Peplum Bucquoy, Dautzenberg
& Dollfus, 1889

Tribe Pectenini
PECTEN Müller, 1776
Vola Klein, 1753 (pre-Lin.)
Argus Poli, 1791
(non Scopoli, 1763)
Argoderma Poli, 1795
Pallium Schröter, 1802
Pectinium Link, 1807
Pandora Megerle, 1811
(non Lamarck, 1799)
Pectenus Rafinesque, 1815 (nom.nud.)
Janira Schumacher, 1817 (nom.null.)
(non Leach, 1813)
Pectinites Schlotheim, 1813
(non Gesner, 1758)
Vola Mörch, 1853
(non Mulsant, 1850)
Argodesma Paetel, 1875 (nom.null.)
Notovola Finlay, 1926
Serratovola Habe, 1951
EUVOLA Dall, 1898
FLABELLIPECTEN Sacco, 1897
OPPENHEIMOPECTEN Teppner, 1922
BATHYPECTEN Schein-Fatton, 1985 ?
CRASSODOMA Bernard, 1986 ?
YABEPECTEN Masuda, 1963 (fossil ?)

PATINOPECTININAE
PATINOPECTEN Dall, 1898
Platinopecten Dall, 1914 (nom.null.)
Mizuhopecten Masuda, 1963 (fossil ?)
MINNIVOLA Iredale, 1939

SYNCYCLONEMIDAE
SYNCYCLONEMA Meek, 1864
Syndonema, Sinsyclonema, Synciclonema,
Syncyclomena, Synclonema,
Syndiclomena authors (n.null.)
Pectinella Verrill, 1897 (16)

PROPEAMUSSIIDAE
PROPEAMUSSIUM de Gregorio, 1884
Propeamusium Dall, 1886 (nom.van.)
Paramusium Verrill, 1897
Paramussium de Gregorio,1898 (n.null.)
Propeannessium Cossmann &
Pisarro, 1906 (nom.null.)
Flavamussium, Luteamussium Oyama, 1951
PARVAMUSSIUM Sacco, 1897
Variamussium Sacco, 1897
Parviamussium Dall, 1898 (nom.null.)
Variopecten Broili in Zittel,
1924 (nom.null.)
Ctenamusium Iredale, 1929
Glyptamusium Iredale, 1939
Graptamussium, Squamamussium,
Xenamussium all Oyama, 1944
Bathyamussium Oyama, 1951 (invalid)
Micramussium Oyama, 1951
Polynemamussium Habe, 1951
ADAMUSSIUM Thiele, 1934 ?

PROPEAMUSSIDAE (Cont'd.)
AMUSIUM Röding, 1798 ?
Anusium Gray, 1826 (nom.null.)
Pleuronectes Bronn, 1831
(non Linnaeus, 1758)
Pleuronopecten Tucker-Rowland, 1938
pro Pleuronectia Swainson, 1840 ?
Amussium Herrmannsen, 1846 (nom.van.)

SPONDYLIDAE
SPONDYLUS Linnaeus, 1758
Podopsis Lamarck, 1819
Gaideropa Deshayes, 1831
Spondilus Verany, 1846 (nom.null.)
Eleutherospondylos Dunker, 1882
Spodylus Nagao, 1934 (nom.null.)
Lanilda, Sponvola Iredale, 1939
CORALLOSPONDYLUS Monterosato, 1917
ELTOPERA Iredale, 1939

ANOMIOIDEA
ANOMIIDAE
ANOMIA Linnaeus, 1758
Echion Poli, 1791
Echionoderma Poli, 1795
Fenestella Röding, 1798
Operculella Monterosato, 1915
(non Reichenbach, 1828) ?
ENIGMONIA Iredale, 1918
pro Aenigma Koch in Martini & Chemnitz, 1846
(non Newman, 1836)
HETERANOMIA Winckworth, 1922
ISOMONIA Dautzenberg & Fischer, 1897 ?
PATRO Gray, 1850 (61)
Patros Gray, 1847 (nom.null.)
PODODESMUS Philippi, 1837
MONIA Gray, 1850
Monica Coen, 1933 (non H.& A.Adams, 1855)
TEDINIA Gray, 1853 (1851)

PLACUNIDAE
PLACUNA Solander in Lightfoot, 1786
Placenta Philipsson, 1788
Ephippium Röding, 1799
Sellaria Link, 1807
Hemiplacuna Gray, 1850 ?
Placunema Stoliczka, 1871
PLACUNANOMIA Broderip, 1832
Placumomya Whitfield, 1885 (nom.null.)
Placunanomya Whitfield, 1885 (nom.null.)
Placuanomia Olsson, 1961 (nom.null.)

Subclass Paleoheterodonta
Order Trigonoida

TRIGONIOIDEA
TRIGONIIDAE
NEOTRIGONIA Cossmann, 1912

Order Unionoida (Freshwater)

UNIONOIDEA
MARGARITIFERIDAE (58) (ICZN 495)
MARGARITIFERA Schumacher, 1816 (Valid; ICZN 495)
Margaritana Schumacher, 1817 (Invalid)
Margartifera Schumacher, 1816 (Invalid)
Damaris Leach in Swainson, 1823
Damalis Leach, 1847
(non Fabricius, 1805)
Baphia Mörch, 1853
(non Gray, 1847)
Cumberlandia Ortmann, 1912 (25)
MARGARITANOPSIS Haas, 1910

MARGARITIFERA (Cont'd.)
PSEUDUNIO Haas, 1910
pro Potodoma Herrmannsen, 1847
(non Meigen, 1800)

UNIONIDAE
UNIONINAE
UNIO Philipsson, 1788 (Valid, ICZN)
Limnaea Poli, 1791 (invalid)
Limnaeoderma Poli, 1795
Uniigenus Renier, 1807 (inv.)
Lymnium Oken, 1815 (inv.)
Unionea Rafinesque, 1815
Mysca Turton, 1822
Brachydonta Schlüter, 1838 (nom.nud.)
Margarita Lea, 1838 (non 1836)
Myisca Agassiz, 1846 (err.)
Limnoderma Herrmannsen, 1847 (err.)
Chondrostea Gistl, 1848
Margaron Lea, 1852
pro Margarita Lea, 1836
(non Leach, 1814)
Nodularidia Cockerell, 1901
pro Nodularia Conrad, 1853
Pictunio Jatzko, 1962
EOLYMNIUM Prashad, 1919
ACUTICOSTA Simpson, 1900
AFRONAIA Haas, 1962
AFROPARREYSIA Haas, 1936
ARCIDOPSIS Simpson, 1900
ARCONAIA Conrad, 1865
CAFFERIA Simpson, 1900
CANTHYRIA Swainson, 1840 ?
CAUDICULATUS Simpson, 1900
CHAMBERLAINIA Simpson, 1900
Simpsonia Rochebrune, 1904
CHONDROSTES Gistl (1847) 1848 ?
CRASSUNIO Modell, 1964 ?
CTENODESMA Simpson, 1900
Cristadens, Christadens Simpson, 1914
CUNEOPSIS Simpson, 1900
DIAURORA Cockerell, 1903
pro Aurora Simpson, 1900
(non Ragonot, 1887)
ELONGARIA Haas, 1913
NANNONAIA Haas, 1913
HAASODONTA McMichael, 1956
HEUDEANA Frierson, 1922
HYRIOPSIS Conrad, 1853
Limnoscapha Lindholm, 1932
INVERSIDENS Haas, 1911
LANCEOLARIA Conrad, 1853
Pericylindrica Tomlin, 1930
pro Cylindrica Simpson, 1900
(non Clessin, 1882)
Cylindrus Modell, 1964 (err.)
LEGUMINAIA Conrad, 1865
Seguminaia Paetel, 1890 (err.)
MICROCONDYLAEA Vest, 1866
Microcondylus Drouet, 1879 (err.)
PSEUDODONTOPSIS Kobelt, 1912
Pseudodopsis Modell, 1964 (err.)
PSEUDOLEGUMINAIA Germain, 1911
MESAFRA Haas, 1936
MIDDENDORFFINAIA Moskvicheva &
Starobogatov, 1973
NEPHRONAIAS Crosse & Fischer, 1894
Caenonaias, Leptonaias, Simonaias
all Crosse & Fischer, 1894
Graphonaias Fischer & Crosse, 1894
Graptonaias Fischer & Crosse, 1894 (err.)
ELLIPTOIDEUS Frierson, 1927
MARTENSNAIAS Frierson, 1927

UNIONINAE

NEPHRONAIAS [Cont'd.]
NEPHRITICA Frierson, 1927
RETICULATUS Frierson, 1927
SPHENONAIAS Crosse & Fischer, 1894
Barynaias Crosse & Fischer, 1894
OXYNAIA Haas, 1913
PHYSUNIO Simpson, 1900
LENS Simpson, 1900
VELUNIO Haas, 1914
PRESSIDENS Haas, 1910
PRISODONTOPSIS Tomlin, 1928
pro Pseudavicula Simpson, 1900
[non Etheridge, 1892]
PROLANCEALARIA Moskvicheva, 1973
PROHYRIOPSIS Haas, 1914
PRONODULARIA Starobogatov, 1970
PROTUNIO Haas, 1913
PSEUDOPARREYSIA Woodward, 1965
PSEUDOPOTOMIDA Moskvicheva & Starobogatov, 1973
PTYCHORHYNCHUS Simpson, 1900
SCABIES Haas, 1911
SCHISTODESMUS Simpson, 1900
Schistodesma, Shistodesma auctt. [err.]
SOLENAIA Conrad, 1868
UNIONETTA Haas, 1955
pro Unionella Haas, 1913
[non Etheridge, 1888]
Unionea Haas, 1969 [err.]
[non Rafinesque, 1815]
VIRGUS Simpson, 1900
LEIOVIRGUS Haas, 1912
Nesonaia Haas, 1912

Placement [fossil ?]
ODHNERELLA Modell, 1964
PSEUDOCOELATURA Germain, 1921
RHABDOTOPHORUS Russell, 1935
RICHTOFENIA Modell, 1964
SCRIPTOLAMPROTULA Modell, 1964
TUBERUNIO Lindholm, 1932

ANODONTINAE

ANODONTA Lamarck, 1799
Glochidium Rathke, 1797 [invalid]
Anodontigenus Renier, 1807 [rej.]
Cista Hübner, 1810
Anodon Oken, 1815 [em.]
Anodontes Cuvier, 1817 [em.]
Edentula Nitzsch, 1820
Lipodonta Nitzsch, 1820
Anodonte Fischer von Waldheim,1807,1823
Onodon Partington, [1835-37] [err.]
Anodontina Schlüter, 1853
Colletopterum Bourguignat, 1880
Euanodonta Westerlund, 1880
Nayadina de Gregorio, 1914
[non Munier-Chalmas, 1864]
Anodota Petrbok, 1930
Liouvillea Bede, 1932
[non Chevreux, 1911]
Euphrata Pallary, 1933
Sinanodonta Modell, 1945
pro Pteranodon Fischer, 1886
[non Marsh, 1876]
ANEMINA Haas, 1969
pro Haasiella Lindholm, 1925
[non Pocock, 1910]
ARNOLDINA Hannibal, 1912
BRACHYANODON Crosse & Fischer, 1894
GABILLOTIA Servain, 1890
MESANODON Crosse & Fischer, 1894
PYGANODON Crosse & Fischer, 1894

ANODONTA [Cont'd.]
UTTERBACKIANA Frierson, 1927
Utterbachia Baker, 1927 [err.]
Utterbackia Baker,1927 [err.]
ALASMIDONTA Say, 1818
Monodonta Say, 1816
[non Lamarck, 1799]
Alasmodonta, Alasmisodonta, Alasmedonta,
Alasmesodonta, Anelasmodon auctt. [err.]
Amblasmodon Rafinesque, 1831
Anadontina Schlüter, 1838 ?
Uniopsis Swainson, 1840
Hemidonta Swainson, 1840
pro Hemiodon Swainson, 1840
[non Swainson, 1840, p.191] [61]
Anelasmodonta Herrmannsen, 1846 [em.]
Uniopsis Agassiz, 1852
[non Swainson, 1840]
BULLELLA Simpson, 1900
DECURAMBIS Rafinesque, 1831
Decarambis Paetel, 1875 [err.]
Rugifera Simpson, 1900
PRESSODONTA Simpson, 1900
pro Calceola Swainson, 1840
[non Lamarck, 1799]
PROLASMIDONTA Ortmann, 1914
ARCIDENS Simpson, 1900
ARKANSIA Ortmann & Walker, 1912
AMURANODONTA Moskvicheva, 1973
ANODONTOIDES Baker, 1898 [61]
pro Anodontopsis Simpson in Baker, 1898
[non McCoy, 1851]
BULDOWSKIA Moskvicheva, 1973
CRISTARIA Schumacher, 1817
Dipsas Leach, 1814
[non Laurenti, 1768]
Barbala Gray, 1825
Appius Menke, 1830
Dianisotis Rafinesque, 1831
Barbata Sowerby, 1839 [err.]
Craspedodonta Küster, 1842 ?
Cleone Gistl, 1848 [pro Dipsas Leach]
Dypsas Kobelt, 1880 [err.]
Crassitesta Simpson, 1900
Clione Simpson, 1900 [err.]
[non Pallas, 1774]
PLETHOLOPHUS Simpson, 1900
CRISTARIOPSIS Moskvicheva, 1973
LASMIGONA Rafinesque, 1831
Elasmogona Agassiz, 1846 [em.]
Elasmogena Herrmannsen, 1852 [err.]
PLATYNAIAS Walker, 1918
PTEROSYNA Rafinesque, 1831
Complanaria, Megadomus Swainson, 1840
Pterosygna Simpson, 1900 [err.]
SULCULARIA Rafinesque, 1831
Alasminota Ortmann, 1914
LEPIDODESMA Simpson, 1896
OGURANODONTA Kuroda & Habe, 1987 ?
PEGIAS Simpson, 1900 ?
Pagias Ortmann, 1921 [err.]
PILSBRYOCONCHA Simpson, 1900
PSEUDANODONTA Bourguignat, 1876
Pseudoanodonta Picaglia, 1893 [err.]
SIMPSONELLA Cockerell, 1903
pro Dalliella Simpson, 1900
[non Cossmann, 1895]
SIMPSONICONCHA Frierson, 1914 [16]
pro Simpsonaias Frierson, 1914
STROPHITUS Rafinesque, 1820
JUGOSUS Simpson, 1914
PSEUDODONTOIDEUS Frierson, 1927
Pseudodontideus Thiele, 1934 [err.]

AMBLEMINAE (Quadrulinae) (25)

Tribe Amblemini

AMBLEMA Rafinesque, 1820 (non 1819, Suppressed)
Bariosta Rafinesque, 1831
Crenodonta Schlüter, 1838
Emblema Deshayes, 1840
Baryosta Agassiz, 1846 (err.)
PSORULA Haas, 1930
BALWANTIA Prashad, 1919 (30)
CAELATURA Conrad, 1853 (30) (Coelatura ?)
Pharaonia Bourguignat, 1880
Reneus Jousseaume, 1886 ?
Renatus Rochebrune, 1904 (em.)
Horusia, Iaronia Pallary, 1924
Jaronia Thiele, 1934 (em.)
GRANDIDIERIA Bourguignat, 1885
Ruellania Bourgiugnat, 1885
(unnec. n.nov. pro Grandidieria)
KALLIPHENGA Haas, 1936
KISTINAIA Haas, 1936
LAEVIROSTRIS Simpson, 1900
MWERUELLA Haas, 1936
Miocruella Salisbury, 1949 (em.)
RHYTIDONAIA Haas, 1936
ZAIRELLA Haas, 1962
ZAIRIA Rochebrune, 1886
Zaira Simpson, 1900 (err.)
COKERIA Marshall, 1916
CONTRADENS Haas, 1913
Schizocleithrum Haas, 1913
SPRICHIA Modell, 1942 (61)
Sprickia auctt. (err.)
CYCLONAIAS Pilsbry, 1922 (25)
Rotundaria Agassiz, 1852
(non Rafinesque, 1820)
MICRONAIAS Simpson, 1900
DISCOMYA Simpson, 1900
ENSIDENS Frierson, 1911
UNIANDRA Haas, 1912
LAMELLIDENS Simpson, 1900
Lamelledens Simpson, 1900 (err.)
LAMPROTULA Simpson, 1900
Gibbosula Simpson, 1900
PARUNIO Ping, 1931
MAGNONAIAS Utterback, 1915 (Suppr.proposed)
Megalonaias Utterback, 1915 (61)
NITIA Pallary, 1924
ORTHONYMUS Agassiz, 1852
PARREYSIA Conrad, 1853
Parreyssia, Parrysia auctt. (err.)
NYASSUNIO Haas, 1936
RADIATULA Simpson, 1900
Indonaia Prashad, 1918
PLECTOMERUS Conrad, 1853
Gonamblus Rafinesque, 1831 ?
POPENAIAS Frierson, 1927
POTOMIDA Swainson, 1840
Potamida Agassiz, 1846 (err.)
Psilunio, Rytia Stefanescu, 1896
Sabbaia Cossmann, 1897
Rhombunio Germain, 1911
Migranaia Hannibal, 1912 (Migranaja)
PSEUDOBAPHIA Simpson, 1900
PSEUDODON Gould, (1844-1845)
Pseudodus Morgan, 1885 (err.)
Nasus Simpson, 1900
(non Basilewski, 1855)
Trigonodon Conrad, 1865
(non Sismonda, 1847)
BINEURUS Simpson, 1900
CHRYSOPSEUDODON Haas, 1921
COSMOPSEUDODON Haas, 1920
DIPLOPSEUDODON Haas, 1921

PSEUDODON (Cont'd.)
INDOPSEUDODON Prashad, 1922
MONODONTINA Conrad, 1853
Suborbiculus Simpson, 1900
OBOVALIS Simpson, 1900
PSORONAIAS Crosse & Fischer, 1894
QUADRULA Rafinesque, 1820
Theliderma Swainson, 1840
AMPHINAIAS Fischer & Crosse, 1894
Pustulosa Thiele, 1934
pro Bullata Frierson, 1927
(non Jousseaume, 1875)
LUTEACARNEA Thiele, 1934
pro Striata Frierson, 1927
(non Boettger, 1878)
OBLIQUATA Frierson, 1927
Scalenaria Rafinesque, 1820
Scatenaria Rafinesque, 1820 (err.)
PLEURONAIA Frierson, 1927
QUINCUNCINA Ortmann, 1922
RHOMBUNIOPSIS Haas, 1920
ROTUNDARIA Rafinesque, 1820 ? (61)
SCHEPMANIA Haas, 1912
TRAPEZOIDEUS Simpson, 1900
TRITOGONIA Agassiz, 1852
Tritogenia Mörch, 1853 (err.)

Tribe Pleurobemini

PLEUROBEMA Rafinesque, 1819 (61)
ELLIPTIO Rafinesque, 1819
Eurynia Rafinesque, 1819 ?
Cunicula Swainson, 1840
Curricula Gray, 1847 (err.)
Eurynea, Eurynaia, Eurinea, Euryma,
Eurinia, Rurynea auctt. (err.)
FUSCONAIA Simpson, 1900 (25)
Lintoxia Rafinesque, 1820 ?
Lyntoxia Agassiz, 1846 (err.)
Fusconia Frierson, 1927 (err.)
HEMISTENA Rafinesque, 1820
Stenelasma Herrmannsen, 1849 (?)
pro Lastena Rafinesque, 1820
Flexiplis Rafinesque, 1831 ?
Flexiptis Neave, 1940 (err.)
Odatelia Rafinesque, 1832
Lostena Gray, 1847
Hemilastena Rafinesque in Agassiz, 1852
Sayunio de Gregorio, 1914 ?
LEXINGTONIA Ortmann, 1914
Levingtonia Ortmann, 1925 (err.)
PLETHOBASUS Simpson, 1900
UNIOMERUS Conrad, 1853
Unionerus Utterback, 1915 (nom.null.)

Tribe Gonideini

GONIDEA Conrad, 1857
Limnobasilissa Hannibal, 1912
BRAZZAEA Bourguignat, 1885 (30)
Bruzzaea, Brazzea auctt. (err.)

Tribe Lampsilini

LAMPSILIS Rafinesque, 1820
Aeglia Swainson, 1840
DELPHINONAIAS Fischer & Crosse, 1894
DISCONAIAS Fischer & Crosse, 1894
MESONAIAS Crosse & Fischer, 1894
PHYLLONAIAS Fischer & Crosse, 1894
ORTMANNIANA Frierson, 1927
VENUSTACONCHA Thiele, 1934
pro Venusta Frierson, 1927
(non Boettger, 1877)
ACTINONAIAS Crosse & Fischer, 1893
CARUNCULINA Simpson, 1898 (30) (Corunculina -err.)
Toxolasma Rafinesque, 1831 ?
CYPROGENIA Agassiz, 1852
Crypogenia Fischer, 1886 (err.)

AMBLEMINAE (Cont'd.)
CYRTONAIAS Crosse & Fischer, 1894 (61)
Cyrtronais Frierson, 1927 (err.)
DROMUS Simpson, 1900 (30, 16)
Conchodromus Haas, 1930
DYSNOMIA Agassiz, 1852
Epioblasma Rafinesque, 1831
Disnomia Bielz, 1869 (err.)
Epilobasma Simpson, 1900 (err.)
CAPSAEFORMIS Frierson, 1927
PENITA Frierson, 1927
PILEA Simpson, 1900
SCALENILLA Ortmann & Walker, 1922
TORULOSA Frierson, 1927
TRUNCILLOPSIS Ortmann & Walker, 1922
ELLIPSARIA Rafinesque, 1820
Plagiola Rafinesque, 1819
Plagiolopsis Thiele, 1934
ENSINAIA Starobogatov, 1970
FRIERSONIA Ortmann, 1912
GLEBULA Conrad, 1853
LEMIOX Rafinesque, 1831
Conradilla Ortmann, 1921
LEPTODEA Rafinesque, 1820
Lasmonos Rafinesque, 1831
Monelasmus Agassiz, 1846
Monelagmus Paetel, 1875
Paraptera Ortmann, 1911
LIGUMIA Swainson, 1840
Sintoxia Rafinesque, 1820 ?
Ligumea Swainson, 1840 (err.)
Syntoxia Agassiz, 1847 (err.)
Legumia Agassiz, 1852 (err.)
Ligumina Philippi, 1853 (err.)
Syntonia Schaufuss, 1869 (err.) ?
Syntoina Paetel, 1875 (err.) ?
MEDIONIDUS Simpson, 1900
MICROMYA Agassiz, 1852 ?
(non Rondani, 1840) (61)
OBLIQUARIA Rafinesque, 1820
Oblicaria Orbigny, 1846 (err.)
OBOVARIA Rafinesque, 1819
PSEUDOON Simpson, 1900 (as Pseudoon)
Actionaias Ortmann, 1925
PACHYNAIAS Fischer & Crosse, 1894
Arotonaias von Martens, 1900
Ptychoderma Simpson, 1900
POTAMILUS Rafinesque, 1818 (20)
Proptera Rafinesque, 1819
Metaptera Rafinesque, 1820
Metaphera Rafinesque, 1820 (em.)
Megaptera Modell, 1964 (err.)
Symphynota Lea, 1829
Lymnadia, Naidea Swainson, 1840
Symphinota Villa, 1841 (err.)
Limnadea Agassiz, 1846 (em.)
PTYCHOBRANCHUS Simpson, 1900
SUBTENTUS Frierson, 1927
SIMPSONUNIO Starobogatov, 1970
TRUNCILLA Rafinesque, 1819
Amygdalonaias Crosse & Fischer, 1894
VILLOSA Frierson, 1927

HYRIIDAE
HYRIINAE
PRISODON Schumacher, 1817
Triquetra Klein in Mörch, 1853
(non Blainville, 1828)
Hyria Lamarck, 1819
Hyris auctt. - err.
Ilyiria Fleming, 1822
Prysodon Stefanescu, 1896
Hyriana Simpson, 1900

HYRIINAE (Cont'd.)
CALLONAIA Simpson, 1900
CASTALIA Lamarck, 1819
Tetraplodon Spix, 1827
CASTALIELLA Simpson, 1900
CASTALINA Ihering, 1891
CHEVRONAIS Olsson & Wurtz, 1951
Chevronaias Haas, 1960 (err.)
DIPLODON Spix, 1827
Iridea Swainson, 1840
Iridella Haas, 1969 (err.)
BULLOIDEUS Simpson, 1900
RHIPIDODONTA Mörch, 1853
Cyclomya Simpson, 1900
SCHLESCHIELLA Modell, 1950
DIPLODONTITES Marshall, 1922
ECHYRIDELLA McMichael & Hiscock, 1958
ECUADOREA Marshall & Bowles, 1932
HYRIDELLA Swainson, 1840 (39)
Propehyridella Cotton & Gabriel, 1932
LORTIELLA Iredale, 1934
PROTOHYRIDELLA Cotton & Gabriel, 1932
VELESUNIO Iredale, 1934
Microdontia Tapparone-Canefri, 1883
Hyridella Cotton & Gabriel, 1932
(non Swainson, 1840)
Alathyria, Centralhyria, Westralunio,
Rugoshyria, Hyridunio all Iredale,1934
MESOHYRIDELLA McMichael, 1957
PAXYODON Schumacher, 1817
Paxiodonta Deshayes, 1832
Pachyodon Agassiz,1846 (err.)
(non Meyer)
TRIPLODON Spix, 1827
Naia Swainson, 1840
Harmandia Rochebrune, 1882
TRIQUETRANA Simpson, 1900

CUCUMERUNIONINAE
CUCUMERUNIO Iredale, 1934
Cucumaria Conrad, 1853
(non Blainville, 1830)
Cucumeria Simpson, 1900 (nom.null.)
Aparcthyria Iredale, 1934 ? (61)
Quaesithyria Iredale, 1943

RECTIDENTINAE
RECTIDENS Simpson, 1900

MUTELOIDEA
MUTELIDAE
MUTELA Scopoli, 1777
Purpurina Herrmannsen, 1852 (61)
(non Orbigny, 1850)
Spatha Lea, 1838
Calliscapha Swainson, 1840
Mutelina Bourguignat, 1885
Utela Moore, 1898 (err.)
Pseudomutela Simpson, 1900
CHELIDONOPSIS Ancey, 1887
pro Chelidonura Rochebrune, 1886
(non A. Adams in Sowerby, 1850)
Chelidoneura Simpson, 1900 (err.)
MONCETIA Bourguignat, 1885
Monoetia auctt. - err.
ASPATHARIA Bourguignat, 1885
ARTHROPTERON Rochebrune, 1904 ?
CHAMBARDIA Bourguignat, 1890
Spathopsis Simpson, 1900
Leptospatha Rochebrune & Germain, 1904
pro Spathella Bourguignat, Dec., 1885
(non T. Hall, Nov., 1885)
Mitriodon Rochebrune, 1904

MUTELIDAE (Cont'd.)
IRIDINA Lamarck, 1819
Platiris Lea, 1838
Platyris Agassiz, 1848 (err.)
Eufira Gistl, 1848 (unnec. pro Iridina)
Iridella Conrad, 1853 (nom.nud.)
Euphira Paetel, 1875 (err.)
CAMERONIA Bourguignat, 1879
PLEIODON Conrad, 1834
Pleodon Swainson, 1840 (err.)
Pliodon Agassiz, 1846 (err.)
PSEUDOSPATHA Simpson, 1900
pro Burtonia Bourguignat, 1883
(non Bonaparte, 1850)
PTEROMUTELA Starobogatov, 1970 ?
BERPOLIS Leach, 1825 ? (61)

MYCETOPODIDAE (46)
MYCETOPODINAE
MYCETOPODA Orbigny, 1835
Mycetopus Orbigny, 1847 (err.)
MYCETOPODELLA Marshall, 1927

ANODONTITINAE
ANODONTITES Bruguière, 1792
Patularia Swainson, 1840
Glabaris Gray, 1847
Haplothaerus Conrad, 1874
Euryanodon Crosse & Fischer, 1894
Pachyanodon, Scolianodon, Styganodon
all von Martens, 1900
Anodotites Marshall, 1930 (err.)
Anondontites Marshall, 1930 (err.)
Glabris Prashad, 1932 (err.)
LAMPROSCAPHA Swainson, 1840
Virgula Simpson, 1900
RUGANODONTITES Marshall, 1931

LEILINAE ?
LEILA Gray, 1840
Columba Lea, 1837
(non Linnaeus, 1758)
Columbia Modell, 1964 (err.)
PSEUDOLEILA Crosse & Fischer, 1894 ? (61)

MONOCONDYLAEINAE (46)
MONOCONDYLAEA Orbigny, 1835
Monocondylus Morelet, 1866 ?
Moncondylaea, Monocondylea, Moncondyloia,
Monocondilaea auctt. (err.)
Spixoconcha Pilsbry, 1893
pro Aplodon Spix, 1827
(non Rafinesque, 1819)
FOSSULA Lea, 1870
Fossicula Marshall, 1925 (err.)
HAASICA Strand, 1932
pro Marshalliella Haas, 1931
(non Kieffer, 1914)
Iheringella Pilsbry, 1893
pro Plagiodon Lea, 1856
(non Duméril & Bibron, 1841)
Jheringella Thiele, 1934 (em.)
TAMSIELLA Haas, 1931

ETHERIIDAE
ETHERIA Lamarck, 1807 (Aetheria Oken, 1818 – em.)
ACOSTAEA Orbigny, 1851 (Acostea Gray, 1854 – err.)
Mulleria Férussac, 1824
Muelleria Philippi, 1853
Eumulleria Anthony, 1907
BARTLETTIA A. Adams, 1866 (31)
Bartellettia Paetel, 1875 (err.)
Rochanaia Morretes, 1941

ETHERIIDAE (Cont'd.)
PSEUDOMULLERIA Anthony, 1907 ?

UNIONACEA "Incertae sedis" (61)
AXIMEDIA Rafinesque, 1820
COELATURA Conrad, 1853
Corlatura Preston, 1930 (err.)
DIPLASMA Rafinesque, 1831
HEMISOLASMA Rafinesque, 1831
LAPIDOSUS Simpson, 1900
LEUCOSILLA Rafinesque, 1831
LIMNOICA Gray, 1857
MYA Röding, 1798 (non Linnaeus, 1758)
NEMRODIA Pallary, 1939
SCHIZOSTOMA Schlüter, 1838 (nom.nud.)
(non Bronn, 1834)
SECURILLA Drouet, 1855

(See Addenda B, p.144, for supraspecific "group names" not recognized, but validly proposed.)

Subclass Heterodonta
Order Veneroida

LUCINOIDEA
LUCINIDAE
LUCININAE
LUCINA Bruguière, 1797
Egraca Leach in Gray, 1852
Lepilucina Olsson, 1965
CALLUCINA Dall, 1901
Callucinopsis Chavan, 1959
Pseudolucinisca Chavan, 1959
EPILUCINA Dall, 1901
LUCINISCA Dall, 1901
Lucinesca Cossmann, 1921 (err.)
PARVILUCINA Dall, 1901
Cavalinga Chavan, 1937
Microloripes Cossmann, 1910
PLEUROLUCINA Dall, 1901
Dallucina Olsson & Harbison, 1953
RADIOLUCINA Britton, 1972 (17)
BARBIERELLA Chavan, 1938
CODAKIA Scopoli, 1777
Lentillaria Schumacher, 1817
Lenticularia Gray, 1840 (err.)
Chama Martini in Mörch, 1853
(non Linnaeus, 1758)
Anfilla de Gregorio, 1885
Pexocodakia Iredale, 1930
Codackia Lopes & Alvarenga, 1955 (err.)
CTENA Mörch, 1860
Jagonia Récluz, 1869
Jaconia Tryon, 1884 (err.)
Jugonia Jousseaume, 1894 (err.)
TALOCODAKIA Iredale, 1936
EPICODAKIA Iredale, 1930
HERE Gabb, 1866
LINGA de Gregorio, 1885
Quasilucina Stewart, 1930 (fossil ?)
BELLUCINA Dall, 1901
Cardiolucina Sacco, 1901 (fossil ?)
Vellucina Andrews, 1971 (err.)
LORIPES Poli, 1791 (43)
Loripoderma Poli, 1795
Ligula Lamarck, 1818
(non Bloch, 1782)
Loripedes Sowerby II, 1839 (err.)
Lucinida Orbigny, 1846
KELETISTES Oliver, 1986
PILLUCINA Pilsbry, 1921
Sydlorina Iredale, 1930
WALLUCINA Iredale, 1930

LUCININAE (Cont'd.)
MEGAXINUS Brugnone, 1880
Magaxinus Monterosato, 1891 (err.)
Stewartia Olsson & Harbison, 1953
NEVENULORA Iredale, 1930
PHACOIDES Blainville in Gray, 1847 (61)
Cavilucina Fischer, 1887 ?
Dentilucina Fischer, 1887
pro Lucina Lamarck, 1801
(non Lamarck, 1799)

MYRTEINAE
MYRTEA Turton, 1822
Cyrachaea Leach, 1847 (61)
Eulopia Dall, 1901
Myrtaea Dall, 1901 (err.)
Notomyrtea Iredale, 1924 ?
GONIMYRTEA Marwick, 1929
Alucinoma Habe, 1958
LUCINOMA Dall, 1901
Triodonta Gray, 1851
(non Bory, 1827)
Thiatira Leach, 1819
MONITILORA Iredale, 1930
PROPHETILORA Iredale, 1930

MILTHINAE
MILTHA H. & A. Adams, 1857
Milthea Meek, 1876 (err.)
ANODONTIA Link, 1807
Lucina Lamarck, 1799
(non Bruguière, 1797)
CAVITIDENS Iredale, 1930
LORIPINUS Monterosato, 1883
EAMSIELLA Chavan, 1951
pro Pseudolucina Chavan, 1947
(non Wilckens, 1909)
GIBBOLUCINA Cossmann, 1910
Elathia Issel, 1869 ?
EOMILTHA Cossmann, 1910 (1912)
PEGOPHYSEMA Stewart, 1930
Lissosphaira Olsson, 1961
RAWYA Strougo, 1976
SAXOLUCINA Stewart, 1930 (fossil ?)

DIVARICELLINAE
DIVARICELLA von Martens, 1880
Davaricella Woodring, 1925 (err.)
Divaricalla Aslanian, 1961 (err.)
EGRACINA Chavan, 1951
DIVALINGA Chavan, 1951
VIADERELLA Chavan, 1951
DIVALUCINA Iredale, 1936
LUCINELLA Monterosato, 1883 (61)
pro Cyclas Klein in Mörch, 1853
(non Lamarck, 1799)
POMPHOLIGINA Dall, 1901

Genera Dubia (61)
AUSTRIELLA Tennison-Woods, 1881
ELATHIA Issell, 1869
JAGONELLA Selli, 1944 (Invalid)
Tuberculina de Gregorio, 1882
(non Ebray, 1858)

FIMBRIIDAE
FIMBRIA Megerle, 1811
(non Bohadsch, 1761 -invalid)
Corbis Cuvier, 1817
Idothea Schumacher, 1817
(non Fabricius, 1796)
Idotoea Deshayes, 1835 (err.)

MACTROMYIDAE
BATHYCORBIS Iredale, 1930

THYASIRIDAE
THYASIRINAE (7)
THYASIRA Leach in Lamarck, 1818 (4)
Thyassira, Thyatira, Thiatyra, Thiatisa, Thyaseira, Thyarsira auctt. (err.)
Cryptodon Turton, 1822
Ptychina Philippi, 1836
Bequania Leach in Brown, 1844 (61)
Clausina Jeffreys, 1847
(non Brown, 1827)
Crypton Verrill, 1880 (err.)
Schizothaerus Locard, 1896
(non Conrad, 1853)
Prothyasira Iredale, 1930
CONCHOCELE Gabb, 1866
LEPTAXINUS Verrill & Bush, 1898 (16)
GENAXINUS Iredale, 1930
Vaticinaria Dall, 1901 ?
MAORITHYAS Fleming, 1950
PARATHYASIRA Iredale, 1930
PHILIS Fischer, 1861
TAURAXINUS Sacco, 1901 (16)
MENDICULA Iredale, 1924
TOMBURCHUS Harry, 1966 (61)

AXINOPSIDINAE (7)
AXINOPSIDA Keen & Chavan in Chavan, 1951
pro Axinopsis Sars, 1878
(non Tate, 1868)
ADONTORHINA Berry, 1947
AXINULUS Verrill & Bush, 1898

UNGULINIDAE
UNGULINA Roissy, 1805
Unguline Daudin in Bosc, 1801
Clotho Blainville, 1824 (61)
(non St. Fond, 1808)
CYCLADICAMA Valenciennes in Rousseau, 1854
Joannisiella Dall, 1895
pro Joannisia Dall, 1895
(non Monterosato, 1884)
Cycladichama Habe, 1964 (err.)
Cyclochama Habe & Kosuge, 1966 (err.)
TORALIMYSIA Iredale, 1936
DIPLODONTA Bronn, 1831
Diglodonta auctt. (err.)
Taras Risso, 1826 (nom.dub.)
Mysia Leach in Bronn, 1827
(non Lamarck, 1818)
Glocomene Leach in Gray, 1852
Mittrea Gray, 1854
ZEMYSINA Finlay, 1927
FELANIA Récluz, 1851
FELANIELLA Dall, 1899
ZEMYSIA Finlay, 1927
MICROSTAGON Cossmann, 1896 (1899)
pro Goodallia Deshayes, 1860
(non Turton, 1822)
MINIPISUM Yabe, 1961 (61)
NUMELLA Iredale, 1924
PHLYCTIDERMA Dall, 1899
PEGMAPEX Berry, 1960
TIMOTHYNUS Harris & Palmer, 1946
Sphaerella Conrad, 1838
(non Sommerfelt, 1824)

CYRENOIDIDAE
CYRENOIDA Joannis, 1835
Cyrenella Deshayes, 1836
Cyrenodonta H. & A. Adams, 1857

GALEOMMATOIDEA (Leptonoidea) (16)
GALEOMMATIDAE
GALEOMMA Sowerby in Turton, 1825
Galeoma Fleming, 1828 (err.)
Hiatella Costa, 1828 ?
(non Daudin, 1801)
Parthenope Scacchi, 1833
(non Weber, 1795)
Hirudinaria Scacchi, 1833 (fossil ?)
Galeomna Hanley, 1944 (err.)
Galeomnia Adams, 1862 (err.)
AMPHILEPIDA Dall, 1899
ACLISTOTHYRA McGinty, 1955
AMBUSCINTILLA Iredale, 1936
BARCLAYIA A. Adams, 1875 (Barclaya, err.)
COLEOCONCHA Barnard, 1964
CYAMIONEMA Melvill & Standen, 1914 ?
CYMATIOA Berry, 1964
pro Crenimargo Berry, 1963
(non Cossmann, 1902)
EPHIPPODONTA Tate, 1889
EPHIPPODONTINA Kuroda, 1945
EPHIPPODONTOANA Habe, 1951
GALEOMMELLA Habe, 1958
LACTEMILES Iredale, 1931
LEIOCHASMEA Dall, Bartsch & Rehder, 1938
ACHASMEA Dall, Bartsch & Rehder, 1938
LEPIRODES Fischer, 1887 (ICZN 566)
Paralepida Dall, 1899 (unnec. n.nov.)
LEVANDERIA Sturany, 1905
LIBRATULA Pease, 1865
Libratuta Cossmann, 1912 (err.)
OHSHIMAIA Habe, 1977
PHASCOLIOPHILA Nordsieck, 1969
PHLYCTAENACHLAMYS Popham, 1939
PSEUDOGALEOMMA Habe, 1964
SAGAMISCINTILLA Habe, 1975
pro Scintillorbis Kuroda & Habe, 1971
(non Dall, 1899)
SCINTILLA Deshayes, 1856
SCINTILLONA Finlay, 1926
Varotoga Iredale, 1931 ?
THYREOPSIS A. Adams, 1868
TRYPHOMYAX Olsson, 1961
TURQUETIA Velain, 1877 ?
Triquetra Velain, 1879
(non Blainville, 1828)
Pro Torquetia Velain, 1876 (nom.nud.)
UNCIDENS Coen, 1934 ?
VASCONIELLA Dall, 1899
pro Vasconia Fischer, 1878 (part)
(non de Folin, 1873)
DIVARISCINTILLA Powell, 1932
YAMAMOTOLEPIDA Habe, 1976

Placement ?
AUTONOE Leach, 1852 (nom.nud.) (as Autonöe)
(non Rafinesque, 1815)
BILOBARIA Pelseneer, 1911

KELLIIDAE
KELLIINAE (7)
KELLIA Turton, 1822
Chironia Deshayes, 1839
Kellea DeKay, 1843 (err.)
Oronthea Leach in Gray, 1852
Kellya Philippi, 1853 (em.)
Kellyia Bucquoy, Dautzenberg
& Dollfus, 1892 (em.)
ALIGENA Lea, 1846 (7)
Aligena Lea, 1843 (nom.nud.)
CICATELLIA Laseron, 1956 (Cicatella, err.)
DIPLODONTINA Stempell, 1899

KELLINAE (Cont'd.)
KANEOHA Dall, Bartsch & Rehder, 1938
MARIKELLIA Iredale, 1936
MICROPOLIA Laseron, 1953
NESOBORNIA Dall, Bartsch & Rehder, 1938
ODONTOGENA Cowan, 1964
PSEUDOKELLYA Pelseneer, 1903
PSEUDOLEPTON Cossmann, 1895
PSEUDOPYTHINA Fischer in Monterosato, 1884
(Fischer, 1878) (61)
Pseudopythinia Locard, 1892 (em.)
Borniopsis Habe, 1959 ?
RADOBORNIA Dall, Bartsch & Rehder, 1938
PARABORNIOLA Habe, 1958
SCINTILLULA Jousseaume, 1888

BORNIINAE (7) (Borniolinae ?)
BORNIA Philippi, 1836
Ceratobornia Dall, 1899 ?
BORNIOLA Iredale, 1924
BYSSOBORNIA Iredale, 1936
LIONELITA Jousseaume, 1888
RHAMPHIDONTA Bernard, 1975
SOLECARDIA Conrad, 1849

LASAEIDAE
LASAEINAE
LASAEA Leach in Brown, 1827
Cycladina Cantraine,1835
(non Latrielle, 1827)
Cycladius auctt. (err.)
Lasea Gray, 1842 (nom.van.)
Anapa Gray, 1847
Poronia Récluz, 1843
Peronia Troschel, 1845 (err.)
(non Fleming, 1822)
ARTHRITICA Finlay, 1926
CLATHROCONCHA Coen, 1934 ?
LITIGIELLA Monterosato, 1909
MELLITERYX Iredale, 1924
MYLLITA Orbigny & Récluz, 1850
ZEMYLLITA Finlay, 1927
MYLLITELLA Finlay, 1927
PARVIKELLIA Laseron, 1956
PYTHINA Hinds, 1844
Phytina, Pythinia auctt. (err.)
SCACCHIA Philippi, 1844
SEMIERYCINA Monterosato in Cossmann, 1911

ERYCININAE
ERYCINA Lamarck, 1805
Erycinigenus Renier, 1807 (invalid)
Migonitis Rafinesque, 1815
(unnec. nom.nov. pro Erycina)
Eryx Swainson, 1840
HEMILEPTON Cossmann, 1911
AMERYCINA Chavan, 1959

Placement ?
HYALOKELLIA Habe, 1960
KAMEKIA Kuroda & Habe, 1971
SAGAMIKELLIA Kuroda & Habe, 1971
SOYOKELLIA Habe, 1958
SQUILLACONCHA Kuroda & Habe, 1971
THRACIOKELLIA Kuroda & Habe, 1971

LEPTONIDAE
LEPTON Turton, 1822
Leptum de Folin, 1873 (err.)
Eupoleme Leach in Gray, 1852
PLATOMYSIA Habe, 1951
PROPERYCINA Cerulli-Irelli, 1908
POTIDOMA Deroux, 1961

LEPTONIDAE (Cont'd.)
SEMELOIDEA Bartrum & Powell, 1928
SCINTILLORBIS Dall, 1899 (61)
VERMITEXTA Laseron, 1956

MONTACUTIDAE
MONTACUTINAE
MONTACUTA Turton, 1822
Sphenalia Wood, 1874 (fossil)
Coriareus Hedley, 1907
Montaguia Bronn, 1848
AXINODON Verrill & Bush, 1898
Kelliola Dall, 1899
BARRIMYSIA Iredale, 1929
BOREACOLA Bernard, 1979 (5)
CALLOMYSIA Habe, 1951
CONCHENTOPYX Bernard, 1964
CURVEMYSELLA Habe, 1959
DECIPULA Jeffreys, 1881 (61)
DEVONIA Winckworth, 1930
Synapticola Malard, 1903
(non Voigt, 1892)
ENTOVALVA Voeltzkow, 1890 ?
Cycladoconcha Spaerck, 1931 ?
FASTIMYSIA Iredale, 1929
FRONSELLA Laseron, 1956
ISSINA Jousseaume, 1898
KELLIOPSIS Verrill & Bush, 1898
KONA Dall, Bartsch & Rehder, 1938
MALVINASIA Cooper & Preston, 1910
MANCIKELLIA Dall, 1899
pro Zoe Monterosato, 1878
(non Philippi, 1840)
MERIGNACIA Cossmann, 1914
MONTACUTONA Yamamoto in Habe, 1959
NIPPONOMONTACUTA Yamamoto & Habe, 1961
NIPPONOMYSELLA Yamamoto & Habe, 1959
PILEATONA Laseron, 1956
ROCHEFORTULA Finlay, 1926
SPHAERUMBONELLA Coen, 1933
TAHUNANUIA Powell, 1952
TELLIMYA Brown, 1827
Tellinomya Agassiz, 1846 (em.)

MYSELLINAE
MYSELLA Angas, 1877 (7)
Petricola Gray, 1825
(non Lamarck, 1801)
ROCHEFORTIA Velain, 1877 (1938 ?)
VIRMYSELLA Iredale, 1930

OROBITELLINAE (7)
OROBITELLA Dall, 1900
ISOROBITELLA Keen, 1962
NEAEROMYA Gabe, 1873
PYTHINELLA Dall, 1899
Phytinella Haas, 1938 (err.)
SCIOBERETIA Bernard, 1895

THECODONTINAE
THECODONTA A. Adams, 1864
PRISTES Carpenter, 1864 (7)
Pristiphora Carpenter, 1866
(non Latreille, 1810)
Serridens Dall, 1899

CYAMIOIDEA
CYAMIIDAE
CYAMIUM Philippi, 1845 (Cyanum, Cyanium, err.)
CYAMIOMACTRA Bernard, 1897
Heteromactra Lamy, 1906
KIDDERIA Dall, 1876 (16)
COSTOKIDDERIA Finlay, 1926 ?

CYAMIIDAE (Cont'd.)
LEGRANDINA Tate & May, 1901
LUTETINA Velain, 1877
PTYCHOCARDIA Thiele, 1912
KINGIELLA Soot-Ryen, 1957 (7)
Reloncavia Soot-Ryen, 1969

PERRIERINIDAE
PERRIERINA Bernard, 1897
CYAMIOCARDIUM Soot-Ryen, 1951

SPORTELLIDAE
SPORTELLA Deshayes, 1858 (fossil)
FABELLA Conrad, 1863 (recent)
ANISODONTA Deshayes, 1858
AUSTROSPORTELLA Ponder, 1971
BASTEROTIA C. Mayer in Hörnes, 1859
Eucharis Récluz, 1850
(non Latreille, 1804)
BASTEROTELLA Olsson & Harbison, 1953
CERULLIA Chavan, 1953
pro Scintillula Cerulli-Irelli, 1909
(non Jousseaume, 1888)
ENSITELLOPS Olsson & Harbison, 1953
FULCRELLA Cossmann, 1886
HITIA Dall, Bartsch & Rehder, 1938
ISOCONCHA Pelseneer, 1911
BENTHOQUETIA Iredale, 1930
Austroturquetia Cotton, 1930

NEOLEPTONIDAE
NEOLEPTON Monterosato, 1875
ARCULUS Monterosato, 1909
CALVITIUM Laseron, 1953
EPILEPTON Dall, 1899
JOUSSEAUMIELLA Bourne, 1907
pro Jousseaumia Bourne, 1906
(non Sacco, 1894)
NEODAVISIA Chavan, 1969
pro Davisia Cooper & Preston, 1910
(non Del Guercio, 1909)
NOTOLEPTON Finlay, 1926
PACHYKELLYA Bernard, 1897
PUYSEGURIA Powell, 1927
WALDO Nicoll, 1966 (fossil ?)

GALATHEAVALVIDAE (35a)
GALATHEAVALVA Knudsen, 1970

BERNARDINIDAE (22)
BERNARDINA Dall, 1910
HALODAKRA Olsson, 1961
STOHLERIA Coan, 1984

CHLAMYDOCONCHOIDEA (7)
CHLAMYDOCONCHIDAE
CHLAMYDOCONCHA Dall, 1884

CARDITOIDEA
CARDITIDAE
CARDITINAE
CARDITA Bruguière, 1792
Cardites Lamarck, 1801;
Dalman, 1825 (inv.)
Arcinella Oken, 1815 (non-binom.)
(non Schumacher, 1817)
Herterocarda Rafinesque, 1815 (em.)
Mytilicardita Anton, 1838
BYSSOMERA Olsson, 1961
JESONIA Gray, 1847
Mytilicardes Blainville, 1825
Jesonia Gray, 1840 (nom.nud.)
Mytilocardia Agassiz, 1846 (err.)

CARDITINAE (Cont'd.)
BEGUINA Röding, 1798
Azarella Gray, 1854

CARDITESINAE
CARDITES Link, 1807
(non Lamarck, 1801- err.)
Cardita Megerle, 1811
(non Bruguière, 1792)
Actinobolus Mörch, 1853
(non Westwood, 1842)
STROPHOCARDIA Olsson, 1961

CARDITAMERINAE
CARDITAMERA Conrad, 1838
Lazaria Gray, 1854
LAZARIELLA Sacco, 1899
ARCTURELLINA Chavan 1951
pro Arcturella Chavan, 1941
(non Sars, 1879)
CARDITELLA E. A. Smith, 1881
CARDITELLONA Iredale, 1936
CARDIOCARDITA Anton, 1838
Cardiocardites Blainville, 1825 (err.)
Agaria Gray, 1840 (nom.nud.)
Azaria Tryon, 1872 (err.- nom.null.)
Divergidens Eames, 1957
BATHYCARDITA Iredale, 1924
CHONIOCARDIA Cossmann, 1904 (fossil)
CARDITELLOPSIS Iredale, 1936 (recent)
CRASSICARDIA Savizky, 1979
CYCLOCARDIA Conrad, 1867
Arcturus Humphrey in Sowerby, 1839
(non Berthold in Latreille, 1817)
Bendeglans Eames, 1957
VIMENTUM Iredale, 1925
GLANS Megerle, 1811
CENTROCARDITA Sacco, 1899
MIODONTISCUS Dall, 1903
pro Miodon Carpenter, 1864
(non Duméril, 1859)
PLEUROMERIS Conrad, 1867
Cycloglans Gorodiski & Freneix, 1959 (1960)

MIODOMERIDINAE (16)
PTEROMERIS Conrad, 1862
(non Conrad, 1865)
CORIPIA de Gregorio, 1885

VENERICARDIINAE (16)
VENERICARDIA Lamarck, 1801 (fossil ?)
MEGACARDITA Sacco, 1899

THECALIINAE
THECALIA H. & A. Adams, 1857
MILNERIA Dall, 1881
pro Ceropsis Dall, 1871
(non Gay & Solier, 1839)

CONDYLOCARDIIDAE
CONDYLOCARDIINAE
CONDYLOCARDIA Bernard, 1896
Hippella Mörch, 1861 (invalid)
AMERICUNA Klappenbach, 1962
BENTHOCARDIELLA Powell, 1930
CARDITOPSIS E. A. Smith, 1881
CONDYLOCUNA Iredale, 1936
CUNANAX Iredale, 1936
ERYCINELLA Conrad, 1845
Triodonta van Koenen, 1893
(non Bory, 1827)

CONDYLOCARDIINAE (Cont'd.)
MICROMERIS Conrad, 1866
Pteromeris Conrad, 1865
(non Conrad, 1862)
MESOCUNA Laseron, 1953
PARTICONDYLA Laseron, 1953
RADIOCONDYLA Iredale, 1936

CUNINAE
CUNA Hedley, 1902
Cuma Bartsch, 1915 (err.)
GONIOCUNA Klappenbach, 1962
HAMACUNA Cotton, 1931
SALTOCUNA Iredale, 1936
OVACUNA Laseron, 1953
PROPECUNA Cotton, 1931
VOLUPICUNA Iredale, 1936

(? Order Hippuritoida - 16)

CHAMOIDEA (7)
CHAMIDAE
CHAMA Linnaeus, 1758
Jataronus Bruguière, 1792 (nom.nud.)
Maceris Modeer, 1793
Lazarus Cuvier, 1800 (nom.nud.)
Chamigenus Renier, 1807 (invalid)
Cameola Rafinesque, 1815 (em.)
Ostreocamites Defrance, 1826 (nom.nud.)
Cama Quoy & Gaimard, (1835)
Lacinea Humphrey in Sowerby II, 1842
Macrophyllum Gray in Gevers, 1847
(non Gray, 1838)
Gryphus Humphrey in Gray, 1847
(non Megerle, 1811)
Licinia Humphrey in Gray, 1847
Chamahippopus Gistl, 1848
(unneccesary nom.nov. pro Chama)
Stola Herrmannsen, 1849 (nom.nud.)
Macerophylla Mörch, 1853
Macerovillia Paetel, 1875 (nom.nud.)
AMPHICHAMA Habe, 1964
PSILOPUS Poli, 1795
Psilopoderma Poli, 1795 (nom.null.)
Psilotus Rafinesque, 1815 (em.)
Psilopododerma Agassiz, 1846 (nom.null.)
Psiloderma Fischer, 1887 (err.)
TRICHICHAMA Tapaswi, 1983
ARCINELLA Schumacher, 1817
Camostraea Deshayes, 1830
Echinochama Fischer, 1887
PSEUDOCHAMA Odhner, 1917
EOPSEUMA Odhner, 1919

CRASSATELLOIDEA (21)
CRASSATELLIDAE
CRASSATELLINAE
CRASSATELLA Lamarck, 1799 (fossil)
Crassatellites Krueger, 1823 (unav.)
RIOSATELLA Vokes, 1973
BATHYTORMUS Stewart, 1930
Crenocrassatella Habe, 1951
CRASSATINA Kobelt, 1881
Crassatina Weinkauff, 1881 (nom.nud.)
CHATTONIA Marwick, 1929
EUCRASSATELLA Iredale, 1924
Eucrassinella Cruz, 1980
HYBOLOPHUS Stewart, 1930 ?
SPISSATELLA Finlay, 1926
INDOCRASSATELLA Chavan, 1952
NIPPONOCRASSATELLA Kuroda & Habe, 1971
SALAPUTIUM Iredale, 1924
TALABRICA Iredale, 1924

SCAMBULINAE [21, 7]
CRASSINELLA Guppy, 1874
Pseuderiphyla Fischer, 1887
Gouldia authors
(non C. B. Adams, 1847)

ASTARTOIDEA [7]
ASTARTIDAE
ASTARTE Sowerby, 1816
Crassina Lamarck, 1818
ISOCRASSINA Chavan, 1949
RICTOCYMA Dall, 1871
Rhectocyma von Martens, 1873 (em.)
DIGITARIA S. Wood, 1853
Woodia Deshayes, 1860
Parvati Semper, 1862 (invalid)
FILATOVELLA Merklin, 1959
pro Astartella Filatova, 1958
(non Hall & Whitney, 1958)
GONILIA Stoliczka, 1871
GOODALLIA Turton, 1822
Mactrina Brown, 1827
Mactroidea Brown, 1827
TRIDONTA Schumacher, 1817
Triodonta Agassiz, 1846 (nom.null.)
(non Bory, 1827)
NICANIA Leach, 1819

CARDINIIDAE
TELLIDORELLA Berry, 1963
Lirotarte Olsson, 1964

CARDIOIDEA
CARDIIDAE [34]
CARDIINAE
CARDIUM Linnaeus, 1758
Bucardites Schlotheim, 1820 (nom.van.)
Cardia, Cardima, Cardinium, Carduim,
Cardum, Cordium auctt. (err.)
Tropidocardium Römer, 1868
Trifaricardium Kuroda
& Habe, 1951 (inv.)
BUCARDIUM Gray, 1853
Ringicardium Fischer, 1887
ACANTHOCARDIA Gray, 1851
Orbis Blainville, 1825
(non Müller, 1767)
Archicardium Sandberger, 1863
Eucardium Fischer, 1887
Sphaerocardium Coen, 1933
Sphoerocardium Coen, 1933 (nom.null.)
Anthocardia Mansfield, 1937 (err.)
EUROPICARDIUM Popov, 1977 (fossil ?)
RUDICARDIUM Monterosato in Coen, 1914
PARVICARDIUM Monterosato, 1884
PLAGIOCARDIUM Cossmann, 1886 (fossil)
MAORICARDIUM Marwick, 1944 (recent)
PAPILLICARDIUM Monterosato
in Sacco, 1899 (recent)
VEPRICARDIUM Iredale, 1929

TRACHYCARDIINAE
TRACHYCARDIUM Mörch, 1853
Tachycardium,
Trachicardium auctt. (err.)
Kathocardia Tucker & Wilson, 1932
DALLOCARDIA Stewart, 1930
MEXICARDIA Stewart, 1930
PHLOGOCARDIA Stewart, 1930
ACROSTERIGMA Dall, 1900 (fossil)
REGOZARA Iredale, 1936 (recent)
PAPYRIDEA Swainson, 1840
VASTICARDIUM Iredale, 1927

FRAGINAE
FRAGUM Röding, 1798
Hemicardia Spengler, 1799 (auctt.)
Isocardia Oken, 1815 (= Hemicardia)
(non Lamarck, 1799)
LUNULICARDIA Gray, 1853
Opisocardium Bayle, 1879
CORCULUM Röding, 1798
Cardissa Megerle, 1811; Swainson, 1840
Hemicardia Cuvier in Fleming, 1818
Hemicardium Cuvier in Schweigger, 1820
CTENOCARDIA H. & A. Adams, 1857
AFROCARDIUM Tomlin, 1931
MICROFRAGUM Habe, 1951
TRIGONIOCARDIA Dall, 1900
Trigonocardia Cossmann, 1901 (err.)
AMERICARDIA Stewart, 1930
APIOCARDIA Olsson, 1961

PROTOCARDIINAE
NEMOCARDIUM Meek, 1876
Awadia Abbass, 1962
FRIGIDOCARDIUM Habe, 1952
Erigidocardium Habe, 1951 (nom.null.)
KEENAEA Habe, 1951
LOPHOCARDIUM Fischer, 1887
LYROCARDIUM Meek, 1876
Amphicardium von Martens, 1880
MICROCARDIUM Thiele, 1934
PRATULUM Iredale, 1924

LAEVICARDIINAE [34]
LAEVICARDIUM Swainson, 1840
Liocardium Agassiz, 1846 (nom.van.)
Exocardium Olsson, 1964
DINOCARDIUM Dall, 1900 ? [33]
FULVIA Gray, 1853
CERASTODERMA Poli, 1795
Cerastes Poli, 1791
(non Laurenti, 1768)
Edulicardium Monterosato, 1923
CILIATOCARDIUM Kafanov, 1974
CLINOCARDIUM Keen, 1936
KEENOCARDIUM Kafanov, 1974
SERRIPES Gould, 1841
Aphrodite Lea, 1834
(non Link, 1807)
Acardo Swainson, 1840
(non Lamarck, 1788)
YAGUDINELLA Kafanov, 1975 ?

HEMIDONACIDAE [48]
HEMIDONAX Mörch, 1871
Donacicardium Vest, 1875

LYMNOCARDIIDAE
ADACNA Eichwald, 1838
DACICARDIUM Papainopol, 1975 ?
DIDACNA Eichwald, 1838
MONODACNA Eichwald, 1838
MYOCARDIA Vest, 1861 [61]
PRODIDACNA Logvinenko
& Starobogatov, 1966 ?

TRIDACNOIDEA
TRIDACNIDAE
TRIDACNA Bruguière, 1797
Tridachnes Röding, 1798
Tridacnodites Krueger, 1823
Gataron Berthold in Latrielle, 1827
Dinodacna Iredale, 1937
Hippopus Martini in Gray, 1847
(non Lamarck, 1799)

TRIDACNIDAE
TRIDACNA (Cont'd.)
CHAMETRACHEA Mörch, 1853
Chamaetrachea Klein in Fischer, 1887 (err)
Flodacna, Sepidacna, Vulgodacna Iredale, 1937
Cametrachea Coates, 1925 (err.)
PERSIKIMA Iredale, 1937
HIPPOPUS Lamarck, 1799
Pelvis Megerle, 1811
Cerceis Gistl, 1848 (nom.van.)

MACTROIDEA
MACTRIDAE
MACTRINAE
MACTRA Linnaeus, 1767
Trigonella Da Costa, 1778
Deikea Mayer, 1872
Colorimactra Iredale, 1929
Telemactra Iredale, 1929
AUSTROMACTRA Iredale, 1930
COELOMACTRA Dall, 1895
CYCLOMACTRA Dall, 1895
DIOPHOROMACTRA Iredale, 1930 ?
ELECTOMACTRA Iredale, 1930
Electromactra Thiele, 1934 (err.)
LONGIMACTRA Finlay, 1928
MAORIMACTRA Finlay, 1928
MACTRINULA Gray, 1853
Mactrella Gray, 1853
Papyrina Mörch, 1853
MACTRODERMA Dall, 1894
MACTROTOMA Dall, 1894
MICROMACTRA Dall, 1894
SIMOMACTRA Dall, 1894
NANNOMACTRA Iredale, 1930
AKTSCHAGYLIA Starobogatov, 1970 ?
HARVELLA Gray, 1853
MACTRELLONA Marks, 1915
Mactrella authors (non Gray, 1853)
MACTROMERIS Conrad, 1868
Mactrodesma Conrad, 1869
MULINIA Gray, 1837
Moulinia, Mulina, Mulinea auctt. (err.)
PSEUDOCARDIUM Gabb, 1866
RANGIA DesMoulins, 1832
Gnathodon Gray in Sowerby, 1832 (non Oken, 1816)
Clathrodon Conrad, 1833 (err.)
Colombia Blainville in Rang, 1834 (n.nd.)
Perissodon Conrad, 1863
Columbia Dall, 1898 (err.)
RANGIANELLA Conrad, 1868
RUGOSOXYPERAS Habe, 1977 ?
SCALPOMACTRA Finlay in Marwick, 1928
SCISSODESMA Gray, 1837
Schizodesma Gray, 1838 (nom.van.)
SPISULA Gray, 1837
Spisulina Fischer, 1887
pro Spisula H. & A. Adams, 1856 (non Gray, 1837)
CRASSULA Marwick, 1948
HEMIMACTRA Swainson, 1840
NOTOSPISULA Iredale, 1930
OXYPERAS Mörch, 1853
PSEUDOXYPERAS Sacco, 1901
STANDELLA Gray, 1853
Leptospisula Dall, 1895
SYMMORPHOMACTRA Dall, 1894
TUMBEZICONCHA Pilsbry & Olsson, 1935

LUTRARIINAE
LUTRARIA Lamarck, 1799
Eustylodon Gistl, 1848 (n.v.)
pro Cacophonia Gistl, 1848 (n.v.)
Lutaria Reichenbach, 1828 (err.)
GONIOMACTRA Mayer, 1867
LUTROMACTRA Iredale, 1929
LUTROPHORA Dall, 1895
PSAMMOPHILA Leach in Brown, 1827
EASTONIA Gray, 1853
HETEROCARDIA Deshayes, 1855
MEROPESTA Iredale, 1929
Merope H. & A. Adams, 1856 (non Newman, 1838)
Standella authors (non Gray, 1853)
TRESUS Gray, 1853
Tresus Walckenaer, 1833 (nom.nud.)
Schizothaerus Conrad, 1853
pro Cryptodon Conrad, 1837 (non Turton, 1822)

PTEROPSELLINAE
ANATINA Schumacher, 1817
Labiosa Schmidt in Möller, 1832 (n.v.)
Cypricia Gray, 1847
Leucoparia Mayer, 1867 (nom.van.)
RAETA Gray, 1853
Lovellia Mayer, 1867
RAETELLA Dall, 1895
RAETELLOPS Habe, 1952
RAETINA Dall, 1895

ZENATIINAE
ZENATIA Gray, 1853
Metabola Meyer, 1867 (nom.van.)
ZENATIOPSIS Tate, 1879
DARINA Gray, 1853
RESANIA Gray, 1853 (Jan.)
Vanganella Gray, 1853 (June)
Laminaria, Myomactra Mayer, 1867 (n.v.)
ZENATINA Gill & Darragh, 1963 (61)

ANATINELLIDAE
ANATINELLA Sowerby, 1833

CARDILIIDAE
CARDILIA Deshayes, 1835
Leptina Pictet, 1855 (non Meigen, 1830)
HEMICYCLONOSTA Deshayes in Blainville, 1825 (fossil ?)

MESODESMATIDAE (Paphidae) (8)
MESODESMATINAE
PAPHIES Lesson, 1831
Machaena Leach in Gray, 1843
Taria Gray, 1853
AMESODESMA Iredale, 1930
ATACTODEA Dall, 1895
Paphia Lamarck, 1799 (non Röding, 1798)
MESODESMA Deshayes, 1831
Ceronia Gray, 1853
REGTERENIA De Rooij-Schuiling, 1972

DAVILINAE
DAVILA Gray, 1853
ANAPELLA Dall, 1895
pro Anapa Gray, 1853 (non Gray, 1847)
MONTEROSATUS Beu, 1971
pro Nesus Monterosato in Locard, 1899

ERVILIINAE
ERVILIA Turton, 1822
Rochefortina Dall, 1924
Spondervilia Iredale, 1930
ARGYRODONAX Dall, 1911
COECELLA Gray, 1853
Caecella auctt. (err.)

SOLENOIDEA
SOLENIDAE
SOLEN Linnaeus, 1758
Salen, Solenaria, Solenarius, Solenia, Solex auctt. (err.)
Vagina Megerle, 1811
Listera Leach in Gray, 1852 (non Turton, 1822)
Fistula Martini in Mörch, 1853
SOLENA Mörch, 1853
Hypogella Gray, 1854
Solena Ghosh, 1920 (non Mörch, 1853)

CULTELLIDAE
CULTELLUS Schumacher, 1817
CULTRENSIS Coen, 1933
ENSIS Schumacher, 1817
Ensatella Swainson, 1840
ENSISOLON Habe, 1977
NEOSOLEN Ghosh, 1920
PHARELLA Gray, 1854
PHAXAS Leach in Gray, 1852
Subcultellus Ghosh, 1920
ENSICULUS H. Adams, 1860
Cultellus auctt. (non Schumacher, 1817)
SILIQUA Megerle, 1811
Aulus Oken, 1815 (invalid)
Leguminaria Schumacher, 1817
Solecurtoides DesMoulins, 1832
Machaera Gould, 1841 (non Cuvier, 1832)
NEOSILIQUA Habe, 1965
TANYSIPHON Benson, 1858 ?

TELLINOIDEA
TELLINIDAE (9)
TELLININAE
TELLINA Linnaeus, 1758
Tellinarius Duméril, 1806 (em.)
Liotellina Fischer, 1887
pro Musculus Martini in Mörch, 1853 (Inv.) (non Röding, 1798)
Tellinarius Froriep in Salisbury, 1934 (non Duméril, 1806)
ABRANDA Iredale, 1924
Punipagia Iredale, 1930 (47)
Punigapia Thiele, 1934 (err.)
ACORYLUS Olsson & Harbison, 1953
ANGULUS Megerle, 1811
Oudardia Monterosato, 1884
Angelus Aldrich, 1911 (err.)
Moerella Fischer in Gardner, 1928 (non Fischer, 1887)
Anbulus Megerle in McLean, 1951 (err.)
ARCOPAGINULA Lamy, 1918
ARCOPELLA Monterosato in Thiele, 1934
CADELLA Dall, Bartsch & Rehder, 1938
CLATHROTELLINA Thiele, 1934
CYCLOTELLINA Cossmann, 1887
ELLIPTOTELLINA Cossmann, 1886
ELPIDOLLINA Olsson, 1961

TELLINA (Cont'd.)
EURYTELLINA Fischer, 1887
Peronaeoderma Poli in Stoliczka, 1870 (non Poli, 1795)
Tellinota Iredale, 1936
FABULINA Gray, 1851
Tellinula Mörch, 1853
HEMIMETIS Thiele, 1934
HERTELLINA Olsson, 1961
HOMALINA Stoliczka, 1870
Hornalina Afshar, 1969 (err.)
IRAQUITELLINA Dance & Eames, 1966
LACIOLINA Iredale, 1937
Bosempra Leach in Brown, 1844 (inv.)
Scrobiculina Dall, 1900 (16)
MACOMONA Finlay, 1926
MEGANGULUS Afshar, 1969
MERISCA Dall, 1900
Lyratellina Olsson, 1961
MOERELLA Fischer, 1887
pro Moera H. & A. Adams, 1856 (non Hübner, 1819)
Donacilla Gray, 1851 (non Blainville, 1819)
Bathytellina Kuroda & Habe, 1958
Asbjoernsenia Friele,1886 (Asbjörnsenia)
NITIDOTELLINA Scarlato, 1965
OBTELLINA Iredale, 1929
OMALA Schumacher, 1817
Homala Agassiz, 1846 (err.)
OUDARIA Monterosato, 1884
PERONAEA Poli, 1791
Peronaeoderma Poli, 1795
PERONIDIA Dall, 1900
PHARAONELLA Lamy, 1918
PHYLLODA Schumacher, 1817 (7)
PHYLLODELLA Hertlein & Strong, 1949 (61)
PHYLLODINA Dall, 1900
PINGUITELLINA Iredale, 1927
PISTRIS Thiele, 1934
pro Pristis Jousseaume in Lamy, 1918 (non Link, 1790)
Pristipagia Iredale, 1936
PSEUDARCOPAGIA Bertin, 1878
QUADRANS Bertin, 1878
QUIDNIPAGUS Iredale, 1929
SCISSULA Dall, 1900
Cirsula Dall in Cary & Spaulding, 1909
SCUTARCOPAGIA Pilsbry, 1918
SEMELANGULUS Iredale, 1924
SERRATINA Pallary, 1922
Striotellina Thiele, 1934
TELLINANGULUS Thiele, 1934
TELLINELLA Gray in Mörch, 1853
Capsa Lamarck, 1799 (non Bruguière, 1797)
Eutellina Fischer, 1887
Maoritellina Finlay, 1926
TELLINIDELLA Hertlein & Strong, 1949
TELLINIDES Lamarck, 1818
ZEARCOPAGIA Finlay, 1926
ARCOPAGIA Leach in Brown, 1827
Cydippe Leach, 1852 (non Eschscholtz, 1821)
STRIGILLA Turton, 1822
Strigella Turton in Gray, 1840 (err.)
Strigula Turton in Pfeiffer,1861(err.) (non Perry, 1811)
Limicola Leach in Gray, 1857 (61) (non Koch, 1816)
Strigillina Stoliczka, 1870 (err.)
AERETICA Dall, 1900
PISOSTRIGILLA Olsson, 1961

TELLININAE
STRIGILLA (Cont'd.)
ROMBERGIA Dall, 1900
SIMPLICISTRIGILLA Olsson, 1961
TELLIDORA Mörch in H. & A. Adams, 1856
Tellinodora Paetel, 1875 (err.)
Tellipiura Olsson, 1944

Placement ? (61)
BARTSCHICOMA Afshar, 1969
DALLITELLINA Afshar, 1969
JOHNSONELLA Afshar, 1969
LIOTRIBELLA Afshar, 1969
PILSBRYMETIS Afshar, 1969
ROEMERILLA Afshar, 1969
SMITHARCOPAGIA Afshar, 1969
SMITHSONELLA Afshar, 1969
MACROMANGULUS F. Nordsieck, 1969

MACOMINAE
MACOMA Leach, 1819
Macroma, Macrotoma auctt. (err.)
Limecola Leach in Brown, 1844
Macomopsis Sacco, 1901
Pulvinus Scarlato, 1965 ?
AUSTROMACOMA Olsson, 1961
HETEROMACOMA Habe, 1952
Gastrana authors (part)
MACOPLOMA Pilsbry & Olsson, 1941
PINGUIMACOMA Iredale, 1936
PSAMMACOMA Dall, 1900
REXITHAERUS Conrad in Tryon, 1869
ROSTRIMACOMA Salisbury, 1934
SALMACOMA Iredale, 1929
SCISSULINA Dall, 1924
APOLYMETIS Salisbury, 1929
pro Polymetus Salisbury, 1929
(non Walsingham, 1908)
pro Metis H. & A. Adams, 1856
(non Philippi, 1843)
CYMATOICA Dall, 1889 (7)
EXOTICA Jousseaume in Lamy, 1918
JACTELLINA Iredale, 1929
LOXOGLYPTA Dall, Bartsch & Rehder, 1938
FLORIMETIS Olsson & Harbison, 1953
Apolymetis auctt. (part)
GASTRANA Schumacher, 1817
Diodonta Deshayes, 1846
(non Hartmann, 1843)
Fragilia Deshayes, 1848
LEPORIMETIS Iredale, 1930
MACALIA H. Adams, 1860
Tellinungula Römer, 1873
Macalina Habe, 1952 (nom.null.)
PSAMMOTRETA Dall, 1900
Cydippina Dall, 1900
Schumacheria Cossmann, 1902 (nom.van.)
ARDEAMYA Olsson, 1961
PSEUDOMETIS Lamy, 1918
TELLINOMACTRA Lamy, 1918
SINOMACOMA Yamamoto & Habe, 1959
TEMNOCONCHA Dall, 1921
Psammothalia Olsson, 1961

SEMELIDAE (Scrobiculariidae) (24a)
SEMELE Schumacher, 1817
AMPHIDESMA Lamarck, 1818
ELEGANTULA de Gregorio, 1884
ABRA Leach in Lamarck, 1818
Dorvillia Leach, 1852
Habra Agassiz, 1846 (em.- nom.null.)
Orixa Leach in Gray, 1852
Lutricularia Monterosato, 1884

ABRA (Cont'd.)
Syndesmyella Sacco, 1901
Abrina Habe, 1952
IACRA H. & A. Adams, 1856
Strigillina Dunker, 1861
SYNDOSMYA Récluz, 1843
Sinodesmia, Sinodesmya, Syndesmya,
Syndomya auctt. (err.)
CUMINGIA Sowerby, 1833 (Cummingia - err.)
Harpax Gistl, 1848
(non Parkinson, 1811)
(unnec. nom. nov. pro Cumingia)
Mikrola Meyer, 1887
LEPTOMYA A. Adams, 1864
LEPTOMYARIA Habe, 1960
MONTROUZIERIA Souverbie, 1863
Montrouziera Souverbie, 1863
(non Bigot, 1860)
Eumontrouziera Hedley, 1915
(unnec.nom.nov.)
LONOA Dall, Bartsch & Rehder, 1938
SCROBICULARIA Schumacher, 1815
Ligula Montagu, 1808
(non Bloch, 1782)
Arenaria Megerle, 1811
(non Brisson, 1760)
Lavignonus Cuvier,1817 in Férussac, 1821
Lavigno, Lavignon, Lavignona auctt.(err.)
Listera Turton, 1822
Lutricola Blainville, 1824
Calcinella Deshayes, 1830
Carinella Sowerby, 1839
(non Johnston, 1833)
Martinea Da Costa in Bucquoy,
Dautzenberg & Dollfus, 1898
SEMELINA Dall, 1900
SOULEYETIA Récluz, 1869
THEORA H. & A. Adams, 1856
ENDOPLEURA A. Adams, 1864
THYELLISCA Vokes, 1956
pro Thyella H. Adams, 1866
(non Wallengren, 1858)

PSAMMOBIIDAE
PSAMMOBIINAE
GARI Schumacher, 1817
Psammotaea Lamarck, 1818
Capsella Deshayes, 1855
(non Gray, 1851)
Milligaretta Iredale, 1936
DYSMEA Dall, Bartsch & Rehder, 1938
GOBRAEUS Leach in Brown, 1844
Psammocola Blainville, 1825 (authors)
(non Blainville, 1824)
GRAMMATOMYA Dall, 1898
KERMADYSMEA Powell, 1958
PSAMMOBELLA Gray, 1851
PSAMMOBIA Lamarck, 1818
Haplomochlia Gistl, 1848 (nom.van.)
ASAPHIS Modeer, 1793
Capsa Bruguière, 1797;
Lamarck, 1801 (non 1799)
Corbula Röding, 1798
(non Bruguière, 1797)
Capsula Schumacher, 1817
Psammocola Blainville, 1824
Procos Gistl, 1848 (nom.van.)
Acaphis Paetel, 1875 (nom.null.)
HETEROGLYPTA von Martens, 1880
ASCITELLINA Marwick, 1928
HETERODONAX Mörch, 1853
ORBICULARIA Deshayes, 1850
Elizia Gray, 1854

SANGUINOLARIINAE
SANGUINOLARIA Lamarck, 1799
Lobaria Schumacher, 1817
(non Müller, 1776)
Isarcha Gistl, 1848 (nom.van.)
HAINANIA Scarlato, 1965
PSAMMOSPHAERICA Jousseaume, 1894
Psammosphaerita auctt. (err.)
PSAMMOTELLA Herrmannsen, 1852
PSAMMOTELLINA Fischer, 1887
pro Psammotella H. & A. Adams, 1856 (n.nd.)
(non Herrmannsen, 1852)
Flavomala Iredale, 1936
SOLETELLINA Blainville, 1824
Psammotaea Lamarck of authors
(non Lamarck, 1818)
Florisarka Iredale, 1936
Solenotelina, Solenotallina,
Soletellaria auctt. (err.)
NUTTALLIA Dall, 1898
Nuttalia Fischer, 1889 (err.)
Nuttalina Clark, 1925 (err.)

SOLECURTINAE
SOLECURTUS Blainville, 1824
Psammosolen Risso, 1826
Macha Oken, 1835
Silex Quoy & Gaimard, 1835
Cyrtosolen Herrmannsen, 1848 (n.van.)
Adasius Leach in Gray, 1852
Solecurtellus Ghosh, 1920
Solenicurtus, Solencurtus, Solecurtis,
Solenocurtis auctt. (err.)
AZORINUS Récluz, 1869
pro Azor Leach in Brown, 1844
(non Sowerby, 1824)
Zozia Winckworth, 1930 (pro Azor)
PHARUS Leach in Brown, 1844
Pharus Gray, 1840 (nom.nud.)
Polia Orbigny, 1845
(non Ochsenheimer, 1816)
Ceratisolen Forbes & Hanley, 1848
Erratisolen, Seratisolen auctt. (n.null.)
Artusius Leach in Gray, 1852
PROTAGELUS Tapaswi, 1983
TAGELUS Gray, 1847
Siliquaria Schumacher, 1817
(non Bruguière, 1789)
Cultellus Conrad, 1837
(non Schumacher, 1817)
CLUNACULUM Dall, 1899 ?
MESOPLEURA Conrad, 1868
Subtagelus Ghosh, 1920

NOVACULININAE
NOVACULINA Benson, 1830
Loncosilla Rafinesque, 1831 (nom.dub.)
Navaculina Orbigny, 1846 (nom.null.)
SINONOVACULA Prashad, 1924

DONACIDAE
DONAX Linnaeus, 1758
Donaciarius Duméril, 1806 (em.)
Donacites Schlotheim, 1813 (Invalid)
Donacina Férussac, 1821 (nom.nud.)
Danax, Domax auctt. (err.)
CAPSELLA Gray, 1851
Capisteria Gray, 1847 (nom.nud.)
Peronaeoderma Mörch, 1853
(non Poli, 1795)
CHION Scopoli, 1777
CUNEUS Da Costa, 1778
Cunerus Paetel, 1890 (err.)

DONAX (Cont'd.)
DELTACHION Iredale, 1930
GRAMMATODONAX Dall, 1900
HECUBA Schumacher, 1817
LATONA Schumacher, 1817
MACHAERODONAX Römer, 1870
Platydonax Dall, 1900 ?
PARADONAX Cossmann in Cossmann
& Peyrot, 1911
PLEBIDONAX Iredale, 1930
SERRULA Mörch, 1853
TENTIDONAX Iredale, 1930
AMPHICHAENA Philippi, 1847 (7)
Amphidona Mörch, 1858 (nom.null.)
GALATEA Bruguière, 1797
Egeria Roissy, 1805 (nom.van.)
Potamophila Sowerby, 1822
Megadesma Bowdich, 1822
Galateola Fleming, 1828 (nom.nud.)
Galataea, Galathea,
Galateia auctt.(err.)
IPHIGENIA Schumacher, 1817
Profischeria Dall, 1903
pro Fischeria Bernardi, 1860
(non Robineau-Desvoidy, 1830)
Iphegenia, Ephigenia auctt. (err.)

DREISSENOIDEA
DREISSENIDAE
DREISSENA Beneden, 1835
pro Driessena, Driessenia
Beneden,1835 (Inv.)
Tichogonia Rossmassler, 1835
Dithalmia Jay, 1835 (nom.nud.)
Mytilina Cantraine, 1837
(non Bory, 1824)
Coelogonia Bronn, 1837
Mytilomya Bronn, 1838
Mytolimax Schaufuss, 1869
Dreisena, Dreistena etc. authors (n.nl.)
CARINODREISSENA Lvova & Starobogatov, 1982
MYTILOPSIS Conrad, 1857
Praxis H. & A. Adams, 1857
(non Guenée, 1852)
Mytiloides Conrad, 1874 (err.)
PONTODREISSENA Logvinenko &
Starobogatov, 1966 (61)

GAIMARDIOIDEA (16) (Cyamacea ?)
GAIMARDIIDAE
GAIMARDIA Gould, 1852
Modiolarca Gray, 1847
(non Gray, 1843)
Phaseolicama Rousseau, 1854
Phascolicama Gray, 1855
Gaimarda Gray, 1855 (err.)
NEOGAIMARDIA Odhner, 1924
PROGAIMARDIA Ponder, 1971
EUGAIMARDIA Cotton, 1931 ?
pro Neogaimardia Cotton, 1931
(non Odhner, 1924)

ARCTICOIDEA
ARCTICIDAE
ARCTICA Schumacher, 1817
Arctica Mohring, 1758 (rej.)
Armida Gistl, 1848
pro Cyprina Lamarck, 1818
Asmidia Mörch, 1853 (nom.null.)
Nympha Mörch, 1853
(non Mörch, 1853, p.25)
Cypriniadea Rovereto, 1900 (nom.van.)
Asmida Vokes, 1967 (err.)

TRAPEZIIDAE
TRAPEZIUM Megerle, 1811
Trapezium Hwass, 1797 (rej.)
Libitina Schumacher, 1817
Cypricardia Lamarck, 1819
GLOSSOCARDIA Stoliczka, 1870
NEOTRAPEZIUM Habe, 1951
CORALLIOPHAGA Blainville, 1824
Lithophagella Gray, 1854
FLUVIOLANATUS Iredale, 1924 ?
ISORROPODON Sturany, 1896 ?

KELLIELLIDAE
KELLIELLA M. Sars, 1870
Keliella authors, 1865-1868 (nom.nud.)
Kellyella Fischer, 1887 (nom.null.)
PAULIELLA Munier-Chalmas, 1895
WARRANA Laseron, 1953

GLOSSOIDEA

GLOSSIDAE
GLOSSUS Poli, 1795
Cuculla Seba, 1781 (non-binom.)
Glossoderma Poli, 1795
Isocardia Lamarck, 1799
Buccardium Megerle, 1811
Diceratia Oken, 1815 (Invalid)
Bucardia Schumacher, 1817
Buccardites Schlotheim, 1830
Tychocardia Römer, 1869
MEIOCARDIA H. & A. Adams, 1857
Miocardia Fischer, 1887 (nom.null.)

VESICOMYIDAE
VESICOMYA Dall, 1886
Versicomya, Vesicomia auctt. (err.)
CALLOGONIA Dall, 1889
VENERIGLOSSA Dall, 1886
WAISIUCONCHA Beets, 1942
CALYPTOGENA Dall, 1891
ARCHIVESICA Dall, 1908
ECTENAGENA Woodring, 1938
AKEBICONCHA Kuroda, 1943 ?

CORBICULOIDEA

CORBICULIDAE
CORBICULA Megerle, 1811
Cyrena Lamarck, 1818
Eucorbicula Crosse & Fischer, 1894
Serrilaminula Lindholm, 1933
CORBICULELLA Ihering, 1907 (61)
CORBICULINA Dall, 1903
CYANOCYCLAS Blainville, 1818
Neocorbicula Fischer, 1887
CYRENODONAX Dall, 1903
BATISSA Gray, 1853 (Batista, Bautisa auctt.- err.)
CYRENOBATISSA Suzuki & Oyama, 1943
POLYMESODA Rafinesque, 1820
Egetaria Mörch, 1861
Diodus Gabb, 1868
pro Cyprinella Gabb, 1864
(non Girard, 1856)
Leptosiphon Fischer, 1872
Americana Clessin, 1879
EGETA H. & A. Adams, 1858
pro Anomala Deshayes, 1854
(non von Block, 1799)
GELOINA Gray, 1842
Isodoma Deshayes, 1858
Indica Clessin, 1879
NEOCYRENA Crosse & Fischer, 1894
PSEUDOCYRENA Bourguignat, 1854
Cyrenocapsa Fischer, 1872

CORBICULIDAE (Cont'd.)
SOLEILLETIA Bourguignat, 1885
VILLORITA Gray in Griffith & Pidgeon, 1834
Velorita, Villosita,
Velarita auctt.(err.)

PISIDIIDAE (Sphaeriidae)
PISIDIUM L. Pfeiffer, 1821
Euglesa Gray, 1840 (auctt.) (nom.inv.)
Galileja Costa, 1839 (authors)
Pera Sowerby, 1842 in Alder, 1931
Pisum Megerle in Gray, 1847 & authors
(non Megerle, 1811) (nom.dub.)
Cordula Leach in Gray, 1852
Corrugaria Conrad, 1853 ?
Cycladina Clessin, 1871
(non Latreille, 1827)
Euglesa Leach in Jenyns, 1832 (61)
Euglesia Gray in Turton, 1840
Fossarina Clessin in Westerlund, 1873
(non Adams & Angas, 1864)
Fluminina Westerlund, 1873
Fluminea, Flumininea, Flumina auctt.(n.n.)
Rivulina Clessin in Westerlund, 1873
(non de Belleville, 1820)
Amnicana, Casertiana, Henslowiana,
Pusillana, Roseana Fagot, 1892 (inv.)
Cymatocyclas, Tropidocylas Dall, 1903
Pseudeupera Germain, 1913
Fontinalina, Lacustrina Sterki, 1918
Eupisidium Odhner, 1921
Cletella Strand, 1928
pro Clessinia Piaget, 1913 (pro Fossarina)
(non Döring, 1874)
Australpera, Glacipisum Iredale, 1943
Speleopisidium Zhadin, 1952
AFROPISIDIUM Kuiper, 1962
CYCLOCALYX Dall, 1903 (19)
EUROPISIDIUM Stadnichenko, 1984
KUIPERIPISIDIUM Izzatulaev & Starobogatov,1986
NEOPISIDIUM Odhner, 1921
ODHNERPISIDIUM Kuiper, 1962 (61)
PARAPISIDIUM Kuiper, 1966
TUVAPISIDIUM Izzatulaev & Starobogatov, 1986
USSURIPISIDIUM Izzatulaev & Starobogatov,1986
BYSSANODONTA Orbigny, 1846
EUPERA Bourguignat, 1854
Limosina Clessin, 1872
(non Macquart, 1835)
Clessinella Waagen, 1905
MICRANODONTA Kuiper, 1965 ?
SPHAERIUM Scopoli, 1777
Cyclas Bruguière, 1797; Lamarck, 1799
Cornea Megerle, 1811
Corneocyclas Blainville, 1818
Amesoda Rafinesque, 1820
Cycladites Krueger, 1823
Cyrenastrum Bourguignat, 1854
Sphariastrum Bourguignat, 1854
Cycladella Carpenter, 1865
Corneola Westerlund, 1873
(non Held, 1837)
Serratisphoerium Germain, 1909
Trigonosphaerium Kobelt, 1913
Sphaerinova Iredale, 1943
Spaeridium Jakushina, 1973
MUSCULIUM Link, 1807
Phymesoda Rafinesque, 1820
Calyculina Clessin, 1872
Carneola Westerlund, 1873
Primella Cooper, 1890
PSEUDOCORBICULA Dautzenberg, 1908
SULCASTRUM Sterki, 1930

Placement ?
ARCTEUGLESA Pirogov & Starobogatov, 1974
CINGULIPISIDIUM Pirogov & Starobogatov, 1974
CONVENTUS Pirogov & Starobogatov, 1974
COSTOPISIDIUM Pirogov & Starobogatov, 1974
HIBERNEUGLESA Starobogatov, 1984
PARASPHAERIUM Alimov & Starobogatov, 1968
PULCHELEUGLESA Starobogatov, 1984
STRIATICYCLAS Starobogatov, 1968
TETRAGONOCYCLAS Pirogov & Starobogatov, 1974
ZECYCLAS Starobogatov, 1968

VENEROIDEA
VENERIDAE
VENERINAE
VENUS Linnaeus, 1758
Venusarius Duméril, 1805
Clausina Brown, 1827
Ventricola Römer, 1867
ANTIGONA Schumacher, 1817
Omphaloclathrum Mörch, 1853
VENTRICOLOIDEA Sacco, 1900
AMEGHINOMYA Ihering, 1907 [fossil ?]
CIRCOMPHALUS Klein in Mörch, 1853
Mioclausinella Kautsky, 1936 [inv.]
DOSINA Gray, 1835
Dorsina Gray, 1840 [nom.null.]
Dosinula Finlay, 1926
PLURIGENS Finlay, 1930
GLOBIVENUS Coen, 1934
PERIGLYPTA Jukes-Browne, 1914
pro Cytherea Röding, 1798
[non Fabricius, 1794]
Proxichione Iredale, 1929
Tigammona Iredale, 1930
VENTRICOLARIA Keen, 1954
Ventricola Authors
[non Römer, 1867]

CIRCINAE
CIRCE Schumacher, 1817
CIRCENITA Jousseaume, 1888 [61]
Circentia Keen, 1969 [err.]
FLUCTIGER Iredale, 1924
LAEVICIRCE Habe, 1951
MICROCIRCE Habe, 1951
PARMULOPHORA Dall, 1905 [61]
pro Parmulina Dall, 1902
[non Penard, 1902]
PRIVIGNA Dall, Bartsch & Rehder, 1938
REDICIRCE Iredale, 1936
GAFRARIUM Röding, 1798
Crista Römer, 1857
GOULDIA C. B. Adams, 1847
Thetis C. B. Adams, [1845]
[non Sowerby, 1826]
CRENOCIRCE Habe, 1960
DORISCA Dall, Bartsch & Rehder, 1938
GOULDIOPA Iredale, 1924

SUNETTINAE
SUNETTA Link, 1807
Cuneus Megerle, 1811
[non Da Costa, 1778]
Meroe Schumacher, 1817
Sunemeroe Iredale, 1930
CYCLOSUNETTA Fischer-Piette, 1939
pro Sunettina Jousseaume, 1891
[non L. Pfeiffer, 1869] ?
SUNETTINA Pfeiffer, 1869
Solanderina Dall, 1902

CHIONINAE
CHIONE Megerle, 1811
AUSTROVENUS Finlay, 1926
CHIONISTA Keen, 1958
CHIONOPSIS Olsson, 1932
Gnidiella Parker, 1949
ILIOCHIONE Olsson, 1961
Anomalocardia authors
[non Schumacher, 1817]
LIROPHORA Conrad, 1863
ANOMALOCARDIA Schumacher, 1817
Triquetra Blainville, 1828
Cryptogramma Mörch, 1853
Murcia Römer, 1857
[non Koch, 1835]
ANOMALODISCUS Dall, 1902
CRYPTONEMA Jukes-Browne, 1914
Cryptonemella Kuroda & Habe, 1951
BASSINA Jukes-Browne, 1914
CALLANAITIS Iredale, 1917
Callanatis auctt. [error]
Eusalacia Cossmann, 1920
Euralacia auctt. [err.]
pro Salacia Jukes-Browne, 1914
[non Lamouroux, 1816]
CHAMELEA Mörch, 1853
Ortygia Leach in Brown, 1827
[non Boie, 1826]
Hermione Leach in Gray, 1852
[non Meigen, 1800]
CLAUSINELLA Gray, 1851
Zucleica Leach in Gray, 1852
Anaitis Römer, 1857
[non Duponchel, 1829]
HUMILARIA Grant & Gale, 1931
MERCENARIA Schumacher, 1817
Crassivenus Perkins, 1869
PLACAMEN Iredale, 1925
PROTOTHACA Dall, 1902
ANTINIOCHE Olsson, 1961
CALLITHACA Dall, 1902
Gallithaca Fischer-Piette
& Vukadinovic, 1977 [err.]
COLONCHE Olsson, 1961
LEUKOMA Römer, 1857
Nioche Hertlein & Strong, 1948
Leucoma Mörch, 1861 [err.]
[non Hübner, 1822]
Granithaca Fischer-Piette
& Vukadinovic, 1977
NOTOCHIONE Hertlein & Strong, 1948
NOVATHACA Habe, 1951
PROTOCALLITHACA Nomura, 1937
TROPITHACA Olsson, 1961
TUANGIA Marwick, 1927
PUBERELLA Fischer-Piette & Vukadinovic, 1977 ?
TAWERA Marwick, 1927
TIMOCLEA Brown, 1827 ? [7]
Timoclea Leach in Gray, 1847 [Valid, 61]
Pasiphae Leach in Gray, 1852
[non Latreille, 1819]
Parvivenus Sacco, 1900
Vermolpa Iredale, 1930
CHIONERYX Iredale, 1924
GLYCYDONTA Cotton, 1936

MERETRICINAE
MERETRIX Lamarck, 1799
Cytherea Lamarck, 1806
[non Fabricius, 1794]
Cythereites Krüger, 1823 [err.]
Nympha Mörch, 1853
[non Fitzinger, 1826]

MERITRICINAE (Cont'd.)
TIVELA Link, 1807
Trigona Megerle, 1811
(non Jurine, 1807)
COMUS Cox, 1930
EUTIVELA Dall, 1891
PACHYDESMA Conrad, 1854
pro Trigonella Conrad, 1837
(non Da Costa, 1778)
PLANITIVELA Olsson, 1961
TRANSENNELLA Dall, 1883
Transenella Fischer, 1887 (err.)

PITARINAE
PITAR Römer, 1857
Caryatis Römer, 1862
(non Hübner, 1819)
Pitaria Dall, 1902 (nom.van.)
COSTELLIPITAR Habe, 1951
HYPHANTOSOMA Dall, 1902
HYSTEROCONCHA Dall, 1902
pro Dione Gray, 1847
(non Hübner, 1819)
KATHERINELLA Tegland, 1929
LAMELLICONCHA Dall, 1902
NANOPITAR Rehder, 1943
PITARELLA Palmer, 1927
PITARENUS Rehder & Abbott, 1951
PITARINA Jukes-Browne, 1913
TINCTORA Jukes-Browne, 1914
pro Callizona Jukes-Browne, 1913
(non Doubleday, 1846-50)
Jukesbrownia Cossmann, 1920
(pro Callizona)
Callizonata Strand, 1926
AMIANTIS Carpenter, 1864
EUCALLISTA Dall, 1902
CALLISTA Poli, 1791
Callistoderma Poli, 1795
Chione Gray, 1838; Leach, 1852
(non Megerle, 1811)
COSTACALLISTA Palmer, 1927
MACROCALLISTA Meek, 1876
CALLOCARDIA A. Adams, 1864
Callocallista Weaver,
1916 (nom.null.)
AGRIOPOMA Dall, 1902
Agriodesma Dall, 1916 (nom.null.)
(non Dall, 1909)
APHRODORA Jukes-Browne, 1914
pro Leucothea Jukes-Browne, 1913
(non von Mertens, 1833)
LEPIDOCARDIA Dall, 1902
LIOCONCHA Mörch, 1853
SULCILIOCONCHA Habe, 1951
MEGAPITARIA Grant & Gale, 1931
NOTOCALLISTA Iredale, 1924
STRIACALLISTA Marwick, 1938
PELECYORA Dall, 1902
Sinodia Jukes-Browne, 1908
SAXIDOMUS Conrad, 1837
Ezocallista Kira in
Kamada, 1962 (inv.)
SINODIELLA Fischer-Piette & Delmar, 1967
TRANSENPITAR Fischer-Piette
& Testud, 1967 (61)

SAMARANGIINAE
SAMARANGIA Dall, 1902
Smarangia Cossmann, 1903 (err.)
Petroderma Kuroda, 1945

TAPETINAE
TAPES Megerle, 1811
Parembola Römer, 1857
RUDITAPES Chiamenti, 1900
Amygdala Römer, 1857 (non Gray, 1825)
EUMARCIA Iredale, 1924
EURHOMALEA Cossmann, 1920
pro Rhomalea Jukes-Browne, 1914
(non Koch, 1873)
GOMPHINA Mörch, 1853
GOMPHINELLA Marwick, 1927
JUKESENA Iredale, 1915
pro Acolus Jukes-Browne, 1913
(non Foerster, 1856)
MACRIDISCUS Dall, 1902
IRUS Schmidt, 1818 (61)
Irus Oken, 1815 (Invalid - ICZN)
IRUSELLA Hertlein & Grant, 1972
NOTIRUS Finlay, 1928
pro Irona Finlay, 1926
(non Mabille, 1883) (61)
NOTOPAPHIA Oliver, 1923
PAPHONOTIA Hertlein & Strong, 1948
KATELYSIA Römer, 1857
Catelysia, Katalysia,
Katelisea auctt. (err.)
LIOCYMA Dall, 1870
Lyozymia F. Nordsieck, 1969 (err.)
Lyocima, Lyacyma auctt. (err.)
MARCIA H. &. A. Adams, 1857
Hemitapes Römer, 1864
Levimarcia Cox, 1927
Nipponomarcia Ikebe, 1941 (fossil ?)
Nipponomarucia Habe, 1977 (err.)
GRANICORIUM Hedley, 1906
PAPHIA Röding, 1798
Paratapes Stoliczka, 1870
pro Textrix Römer, 1857
(non Sundevall, 1883)
Eutapes Chiamenti, 1900
Arcritopaphia Iredale, 1936
CALLISTOTAPES Sacco, 1900
Hemitapes authors (non Römer)
NEOTAPES Kuroda & Habe, 1971
PROTAPES Dall, 1902
PSEPHIDIA Dall, 1902
(non Pomel, 1872 - nom.nud.)
pro Psephis Carpenter, 1864
(non Guenée, 1854)
VENERUPIS Lamarck, 1818
Pullastra Sowerby, 1826
Myrsus H. & A. Adams, 1858
Pro Metis H. & A. Adams, 1857 (non 1856)
(non Philippi, 1843)
Aureitapes, Cyaneitapes,
Polititapes Chiamenti, 1900
Venerirupes, Venerirupis, Venerupes,
Venerupsis, Veniorupis auctt. (err.)
PAPHIRUS Finlay, 1927

DOSINIINAE
DOSINIA Scopoli, 1777
Dosinidia Dall, 1902
Bonartemis, Pardosinia Iredale, 1929
Fallartemis, Meridosinia,
Semelartemis Iredale, 1930
ASA Basterot, 1825
Arctoe Risso, 1826 (as Arctoë)
Arctoa Herrmannsen, 1846 (em.)
Dosinia ss. authors (non Scopoli, 1777)
AUSTRODOSINIA Dall, 1902
DOSINELLA Dall, 1902
DOSINISCA Dall, 1902

DOSINIINAE
DOSINIA (Cont'd.)
DOSINOBIA Finlay & Marwick, 1937 ?
DOSINORBIS Dall, 1902
KEREIA Marwick, 1927
LAMELLIDOSINIA Zhuang Qi-Qian, 1964
PECTUNCULUS Da Costa, 1778 (ICZN 1414)
Arthemis Poli, 1791
Arthemiderma Poli, 1795
Artemis, Artimis authors (nom.van.)
Orbiculus Megerle, 1811
Exoleta Brown, 1827
Circompholos Gray, 1847
Cerana Gistl, 1848
Ampithaea Leach in Gray, 1852
PHACOSOMA Jukes-Brown, 1912

CLEMENTINIINAE
CLEMENTIA Gray, 1842
Blainvillia Hupe, 1854
(non Robineau-Desvoidy, 1830)
EGESTA Conrad, 1845
TERENTIA Jukes-Browne, 1914
Euterentia Cossmann, 1920
COMPSOMYAX Stewart, 1930

CYCLININAE
CYCLINA Deshayes, 1850
Eocyclina Dall, 1908
CYCLINELLA Dall, 1902
CYCLINORBIS Makiyama, 1926 (?)
FRIGICHIONE Fletcher, 1936 (?)
PALEOMARCIA Fletcher, 1936 (?)

GEMMINAE
GEMMA Deshayes, 1853
Totteniana Perkins, 1869
NUTRICOLA Bernard, 1982 (6)
PARASTARTE Conrad, 1862
Callicistronia Dall, 1883
ROHINI Semper, 1862 (Fossil ?)

GLAUCONOMIDAE
GLAUCONOME Gray, 1828
Glaucomya, Glauconoma, Glauconomella,
Glauconomya auctt. (err.)
Glauconometta Iredale, 1936

TURTONIIDAE (16)
TURTONIA Alder, 1848

COOPERELLIDAE
COOPERELLA Carpenter, 1864
Oedalina Carpenter, 1865 (unnec.nom.nov.)
pro Oedalia Carpenter, 1864

PETRICOLIDAE
PETRICOLA Lamarck, 1801
Naranio Gray, 1853
CLAUDICONCHA Fischer, 1887
LAJONKAIRIA Deshayes, 1854
Lajonkairea, Lajonkaireira auctt. (n.nl.)
PETRICOLARIA Stoliczka, 1870
Gastranella Verrill, 1872
PETRICOLIRUS Habe, 1951
PSEDOIRUS Habe, 1951
RUPELLARIA de Bellevue, 1802
Choristodon Jonas, 1844
Ropellaria Richards, 1962 (err.)
Roxellaria Agassiz, 1845 (em.)
VELARGILLA Iredale, 1931
MYSIA Lamarck, 1818
Lucinopsis Forbes & Hanley, 1848

Order Myoida
Suborder Myina

MYOIDEA
MYIDAE
MYINAE
MYA Linnaeus, 1758
Hiatula Modeer, 1793
Myarius Duméril, 1806 (nom.van.)
Myes, May, Mega, Myarina,
Myia auctt. (err.)
ARENOMYA Winckworth, 1930
PARAMYA Conrad, 1860 (fossil ?)
pro Myalina Conrad, 1845
(non de Koninck, 1842)
PLATYODON Conrad, 1837
Cryptodonta Nuttall in
Carpenter, 1864
AUSTROPLATYODON Olsson, 1961
TUGONIA Gray, 1842 (as Tagonia)
Tugonia Récluz, 1846 (inv.)
DISTUGONIA Iredale, 1936
TUGONELLA Jousseaume, 1891

CRYPTOMYINAE
CRYPTOMYA Conrad, 1848
VENATOMYA Iredale, 1930

SPHENIINAE
SPHENIA Turton, 1822
Sphaena, Spaeria, Spenia,
Sphaenia, Sphena,
Sphenica, Syphonia
authors (err.)
Tyleria H. & A. Adams, 1854

CORBULIDAE
CORBULINAE
CORBULA Bruguière, 1797
Aloidis Megerle, 1811
ANISOCORBULA Iredale, 1930
BICORBULA Fischer, 1887
CARYOCORBULA Gardner, 1926
HEXACORBULA Olsson, 1932
JULIACORBULA Olsson & Harbison, 1953
NOTOCORBULA Iredale, 1930
PANAMICORBULA Pilsbry, 1932
SERRACORBULA Olsson, 1961
SOLIDICORBULA Habe, 1949
TENUICORBULA Olsson, 1932
VARICORBULA Grant & Gale, 1931
Corbula authors
(non Bruguière)
HERRINGTONIUM Clarke, 1973 (61) (?)
Sphaeridium Jakushina, 1973 ? (fossil)
(non Fabricius, 1775)
MINICORBULA Habe, 1977 ?
PHYSOIDA Pallary, 1900
POTAMOCORBULA Habe, 1955

LENTIDIINAE
LENTIDIUM Cristofori & Jan, 1832
Corbulomya Nyst, 1844

ERODONIDAE
ERODONA Daudin in Bosc, 1801
Potamomya Sowerby, 1839 (nom.nud.)
Azara Orbigny, 1842
Pacyodon Gray, 1847
Phaenomya Weaver & Palmer, 1922

SPHENIOPSIDAE
GRIPPINA Dall, 1912

GASTROCHAENOIDEA
GASTROCHAENIDAE
GASTROCHAENA Spengler, 1783
Chaena Retzius in Philipsson, 1788 (n.v.)
Gastrochoena Brown, 1827 (err.)
Rocellaria Blainville, 1828
Gastrochaenolites Leymerie, 1842 (n.van.)
Gastrochaenoecium Bronn, 1848 (nom.van.)
Gastrochaenites Bronn, 1848 (nom.van.)
Cucurbitula Gould, 1861
Roxania Paetel, 1875 (61)
(non Grey, 1847)
Dufoichaena Jousseaume in Lamy, 1925
SPENGLERIA Tryon, 1861
EUFISTULANA Eames, 1951
pro Fistulana Bruguière, 1789
(non Müller, 1776)
Fistularia auctt. (err.)
LAMYCHAENA Freneix, 1979

HIATELLOIDEA
HIATELLIDAE
HIATELLA Daudin in Bosc, 1801 ?
Saxicava de Bellevue, 1802
Laxicava Conrad, 1855 (nom.null.)
Haicana Sacco, 1901 (nom.null.)
(pro Hiatella)
Hiacana Sacco, 1901 (err.)
Clotho de Saint Fond, 1808 (nom.dub.)
Didonta Schumacher, 1817
Byssonia Blainville, 1816
Byssomya Oken, 1817
Biapholius Leach in Lamarck, 1818
Biapholus, Biaphleus Leach, 1847 (n.nl.)
Pholeobia Leach, 1819
Agina Turton, 1822
Rhomboides Blainville, 1824
(non Goldfuss, 1820)
Coramya Brown, 1844
Spongyophylla Nardo in Brusina, 1866
Turneria Glibert & Van de Poel, 1966 ?
(non Forel, 1895)
CYRTODARIA Reuss, 1801 (ICZN 1414)
Sertodaria Oken, 1817 (err.)
Cyrtodoria, Cytodaria auctt. (err.)
Cyrtodera Blainville, 1825 (nom.null.)
Glycimeris Lamarck, 1801
(non Lamarck, 1799)
PANOPEA Ménard, 1807 (ICZN 1414)
Glycimeris Lamarck, 1799 (Suppr.)
Panope Ménard, 1807 (nom.null.)
Panodea Oken, 1817 (nom.null.)
Panopaea Lamarck, 1818 (nom.null.)
Myopsis Agassiz, 1840
Heteromya Mayer, 1884
PANOMYA Gray, 1857
Chaenopaea Mayer, 1884
SAXICAVELLA Fischer, 1878 (61)
Arcinella Philippi, 1844
(non Schumacher, 1817)
Arcinellina Bronn, 1849 (err.)

Suborder Pholadina

PHOLADOIDEA
PHOLADIDAE
PHOLADINAE
PHOLAS Linnaeus, 1758
Pholadites Davila, 1767
Polas Gronovius, 1781 (nom.null.)
Hypogaea Poli, 1791
Pholas Lamarck, 1799
(non Lamarck, 1801)

PHOLAS (Cont'd.)
Hypogaeoderma Poli, 1795
Phloas Turton, 1806 (nom.null.)
Pholadites Schlotheim, 1813 (invalid)
Pholalites Schlaepfer, 1821 (nom.null.)
Xylotrya Leach in Menke, 1830
Dactylina Gray, 1847
(non Zborzewski, 1843)
Pholadarius Herrmannsen, 1852
Pholalithes Paetel, 1875
Pragmopholas Fischer, 1887
MONOPLAX Thang, Tsi & Li, 1960 (fossil ?)
MONOTHYRA Tryon, 1862
THOVANA Leach in Gray, 1847
Gitocentrum Tryon, 1862
Gitoventrum Paetel, 1875 (nom.null.)
BARNEA Leach in Risso, 1826
Barnia Gray, 1840 (nom.null.)
Bernia Leach, 1852 (err.)
Holopholas Fischer, 1887
ANCHOMASA Leach in Gray, 1852
Anchomosa Lamy, 1925 (nom.null.)
UMITAKEA Habe, 1952
Barnea Habe, 1952 (non Leach)
CYRTOPLEURA Tryon, 1862
SCOBINOPHOLAS Grant & Gale, 1931
pro Scobina Bayle, 1880
(non Lepeltier & Serville, 1828)
pro Pholas Lamarck, 1801
(non Linnaeus, 1758)
Pholadigenus Renier, 1807 (rej.)
Pholas H. & A. Adams, 1856
(non Linnaeus, 1758)
Leuconyx H. & A. Adams, 1863 (nom.obl.)
Leuconix Crosse, 1868 (nom.null.)
TALONA Gray, 1842 (Talonia Iredale, 1913; err.)
Talona Gray, 1842 (nom.nud.)
Talena Sowerby, 1849 (nom.null.)
ZIRFAEA Gray, 1842
Zirfaea, Zirphaea, Zyrphaea auctt.(n.n.)
Thurlosia Leach in Catlow & Reeve, 1845

MARTESIINAE
MARTESIA Sowerby, 1824
Mactresia Leach in Gray, 1840
Mactesia Paetel, 1875 (nom.null.)
Martesiella Verrill & Bush, 1898
Hiata Zetek & McLean, 1936
Mesopholas Taki & Habe, 1945
Diploplax Bartsch & Rehder, 1945
PARTICOMA Bartsch & Rehder, 1945
ASPIDOPHOLAS Fischer, 1887
Scutigera Cossmann, 1886
(non Lamarck, 1801)
Calyptopholas Lamy, 1927
CHACEIA Turner, 1953
DIPLOTHYRA Tryon, 1862
Schroeteria Tryon, 1862 (nom.oblit.)
LIGNOPHOLAS Turner, 1955
PARAPHOLAS Conrad, 1848
Paraphola Paetel, 1875 (nom.null.)
PENITELLA Valenciennes, 1846
Penicilla Conrad, 1854 (nom.null.)
Navea Gray, 1851
PHOLADIDEA Turton, 1819
Pholadidoidea Blainville, 1826
Pholidea Swainson, 1835
Talonella Gray, 1851
Cadmusia Leach in Gray, 1852
Pholameria Conrad, 1865
Pholididae Paetel, 1875 (nom.null.)
HATASIA Gray, 1851
Hatosia, Hastasia auctt. (nom. null.)

JOUANNETIINAE
JOUANNETIA DesMoulins, 1828
Iouannetia Gray, 1840 [nom.van.]
Juanetia Orbigny, 1846 [nom.null.]
PHOLADOPSIS Conrad, 1849
Triomphalia Sowerby, 1849
Triumphala Paetel, 1890 [nom.null.]
NETASTOMA Carpenter, 1864
Nettastomella Carpenter, 1865
[Unnec. nom.nov. pro Netastoma]
Netastomella Paetel, 1875 [nom.null.]
Nettastoma Lamy, 1926 [nom.null.]
SCYPHOMYA Dall, 1898 ?

XYLOPHAGAIDAE [7]
XYLOPHAGA Turton, 1822
Xilophaga Geinitz, 1845 [nom.null.]
Xylotomea Dall, 1898 [unnec. em.]
pro Xylophaga Turton, 1822
[non Xylophagus Meuschen, 1778]
Xilotoma Dall in Gigoux,
1934 [nom.null.]
Protoxylophaga Taki & Habe, 1945
Neoxylophaga Taki & Habe, 1950
Metaxylophaga Taki & Habe, 1950
MESOXYLOPHAGA Habe, 1977 ?
XYLOPHOLAS Turner, 1972
XYLOREDO Turner, 1972

TEREDINIDAE
TEREDININAE
TEREDO Linnaeus, 1758
Teredigenus Renier, 1807 [inv.]
Toredo May, 1929 [nom.null.]
Terebro auctt. [err.]
Teredarius auctt. [err.]
Zopoteredo Bartsch, 1923
Dactyloteredo Moll, 1941 [invalid]
Austroteredo Habe, 1952
Rochia Munari, 1975 ?
[non Gray, 1857]
COELOTEREDO Bartsch, 1923
PINGOTEREDO Iredale, 1932
Pinguiteredo Habe, 1952 [nom.null.]
BACTRONOPHORUS Tapparone-Canefri, 1877
pro Calobates Gould, 1862
[non Kaup, 1829]
DICYATHIFER Iredale, 1932
Pseudodicyathifer Tchang,
Tsi & Li, 1958
LYRODUS Gould in Binney, 1870
Teredops Bartsch, 1921
Cornuteredo Dall, Bartsch
& Rehder, 1938
NEOTEREDO Bartsch, 1920
PSILOTEREDO Bartsch, 1922
TEREDORA Bartsch, 1921
Malleolus Gray, 1847
[non Ehrenberg, 1838]
TEREDOTHYRA Bartsch, 1921
Ungoteredo Bartsch, 1927
Idioteredo Taki & Habe, 1945
UPEROTUS Guettard, 1770
Fistulana Lamarck, 1799
[non Müller, 1776]
Fistulanigenus Renier, 1807 [rej.]
Guetera Gray, 1847
Hyperotus Herrmannsen, 1849 [nom.van.]
Hyperotis Paetel, 1875 [nom.null.]
Guettera H. & A. Adams, 1856 [nom.null.]
Glumebra Iredale, 1936
ZACHSIA Bulatov & Rzhavschikov, 1933
Zachia Turner, 1966 [err.]

BANKIINAE
BANKIA Gray, 1842
Xylotrya Leach in Gray, 1847
[non Menke, 1830]
Xylotyra, Xylotria, Xylotya,
Hylotrya auctt. [nom.null.]
Spathoteredo Moll & Roch, 1937 [61]
[non Moll, 1928]
Lyrodobankia Moll, 1941 [invalid]
Bankiopsis Clench & Turner, 1946
Bankiura Moll, 1952
BANKIELLA Bartsch, 1921
CLUPIBANKIA Moll, 1952
LILIOBANKIA Clench & Turner, 1946
NEOBANKIA Bartsch, 1921
Deviobankia Iredale, 1932
PLUMULELLA Clench & Turner, 1946
NAUSITORA Wright, 1864
Nausitoria, Naucitora auctt. [nom.null.]
Inequarista Iredale, 1932
Nausitorella Moll, 1952
NOTOTEREDO Bartsch, 1923
Phylloteredo Moll & Roch, 1937
SPATHOTEREDO Moll, 1928

KUPHINAE
KUPHUS Guettard, 1770
Furcella Lamarck, 1801
Septaria Lamarck, 1818
[non Férussac, 1807]
Clossonnaria Férussac, 1821
Clausaria Menke, 1828
Cuphus Agassiz, 1846 [nom. van.]
Cloisonnaria Adams, 1858 [nom. van.]
Clossonaria Paetel, 1875 [nom. van.]
Cyphus Fischer, 1887 [nom. van.]
Kyphus Herrmannsen, 1847 [nom. van.]
Bicornia May, 1929 [inv.]
Microvexillum May, 1929 [inv.]
Proteredo May, 1929 [inv.]
EOTEREDO Bartsch, 1923

Subclass Anomalodesmata [42]
Order Pholadomyoida

PHOLADOMYOIDEA
PHOLADOMYIDAE
PHOLADOMYA G. B. Sowerby, 1823
Pholadomyaea, Pholadomia, Phaladomya,
Pholdomya, Pholadonya,
auctt. [nom.null.]
Flabellomya Rollier, 1911

PARILIMYIDAE
PARILIMYA Melvill & Standen, 1899
PANACCA Dall, 1905
pro Aporema Dall, 1903
[non Scudder, 1890]
Notomya Cotton, 1931
[non M'Coy, 1847]
Panacea Salisbury, 1932 [nom.null.]
NIPPONOPANACCA Habe, 1977

PANDOROIDEA [16] [Thracioidea]
THRACIIDAE
THRACIA Leach in Blainville, 1824 [61]
Thracia Leach in Sowerby, 1823 [nom.nud.]
Throna Carpenter, 1859 [nom.null.]
Osteodesma Deshayes in Blainville, 1825
Odoncineta Da Costa, 1829
Odontodineta, Cinetodonta auctt. [n.van.]
Corymya Agassiz, 1843
[non Donovan, 1802]

THRACIIDAE
THRACIA (Cont'd.)
Corimya authors (nom.null.)
Coromya authors (nom.null.)
Homoeodesma Fischer, 1887
Eximiothracia Iredale, 1924
Cetothrax Iredale, 1949
CRASSITHRACIA Soot-Ryen, 1941
IXARTIA Leach in Gray, 1852
Pelopia H. Adams, 1868
(non Meigen, 1800)
Rupicilla Schaufuss, 1869
pro Rupicola de Bellevue, 1802
(non Brisson, 1760)
LAMPEIA MacGinitie, 1959
TRIGONOTHRACIA Yamamoto & Habe, 1959
ASTHENOTHAERUS Carpenter, 1864
BUSHIA Dall, 1886 ?
CYATHODONTA Conrad, 1849
PARVITHRACIA Finlay, 1926
PHRAGMORISMA Tate, 1893
THRACIDORA Iredale, 1924
THRACIOPSIS Tate & May, 1900
pro Alicia Angas, (1867) 1868
(non Johnson, 1861)
Thracidentula Garrard, 1961 ?

LATERNULIDAE
LATERNULA Röding, 1798
Auriscalpium Megerle, 1811
Butorella Strand, 1928
pro Butor Gistl, 1848
(non Forester, 1827)
pro Anatina Lamarck, 1818
(non Schumacher, 1817)
LATERNULINA Habe, 1952
CLISTOCONCHA Smith, 1910
EXOLATERNULA Habe, 1977

PERIPLOMATIDAE
PERIPLOMA Schumacher, 1817
ALBIMANUS Pilsbry & Olsson, 1935
OFFADESMA Iredale, 1930
PENDALOMA Iredale, 1930
COCHLODESMA Couthouy, 1839
Aperiploma Habe, 1952
BONTAEA Leach in Brown, 1844
Galaxura Leach in Gray, 1852
Calcaraea Récluz, 1868
HALISTREPTA Dall, 1904 ?

LYONSIIDAE
LYONSIA Turton, 1822
Hiatella Brown, 1827
(non Bosc, 1801)
Magdala Leach in Brown, 1827
Osteodesma Deshayes, 1830 of authors
(non Blainville, 1825)
Myatella Brown, 1833
Pandorina Scacchi, 1833
(non Bory, 1827)
Tetragonostea Deshayes in
Herrmannsen, 1849 (61)
BENTHOLYONSIA Habe, 1952
AGRIODESMA Dall, 1909
ALLOGRAMMA Dall, 1903
ENTODESMA Philippi, 1845
Philippina Dall & Simpson, 1901
PHLYCTICONCHA Bartsch & Rehder, 1940
pro Phlyctiderma Bartsch & Rehder, 1939
(non Dall, 1899)
MYTILIMERIA Conrad, 1837

LYONSIDAE (Cont'd.)
OSTOMYA Conrad, 1874
Anticorbula Dall, 1898
pro Himella H. Adams, 1860
(non Dallas, 1852)
Guianadesma Morrison, 1943

PANDORIDAE (16a)
PANDORA Bruguière, 1797
Pandora Hwass in Chemnitz, 1795 (inv.)
Calopodium Röding, 1798
Irus Oken, 1821
(non Oken, 1815)
Trutina Brown, 1827
Pandorina Scacchi, 1836
(non Scacchi, 1833)
CLIDIOPHORA Carpenter, 1864
Clidiphora Carpenter in
Johnson, 1934 (error)
FOVEADENS Dall, 1915
FRENAMYA Iredale, 1930
Coelodon Carpenter, 1865
(non Audinet-Serville, 1882)
HETEROCLIDUS Dall, 1903
LASERONELLA Whitley, 1959
pro Pandorella Laseron, 1951
(non Conrad, 1862)
PANDORELLA Conrad, 1863
Kennerlia Carpenter, 1864
Kennerleya, Kennerleyia, Kenerlia,
Kennerlyia authors (err.)

MYOCHAMIDAE
MYOCHAMA Stutchbury, 1830
MYADORA Gray, 1840 (Myodora auctt. - err.)
HUNKYDORA Fleming, 1948
MYADOROPSIS Habe, 1960

CLEIDOTHAERIDAE
CLEIDOTHAERUS Stutchbury, 1830
Chamostrea Blainville in Gray, 1840
Chamostraea Herrmannsen, 1846 (em.)

CLAVAGELLOIDEA

CLAVAGELLIDAE
CLAVAGELLA Lamarck, 1818 (Clavigella auctt.- err.)
Bacilia Valenciennes in Gray, 1858
BRYOPA Gray, 1847
Tiria de Gregorio, 1886
DACOSTA Gray, 1858
STIRPULINA Stoliczka, 1870
Styrpulina auctt. (err.)
Stirpuliniola Kuroda & Habe, 1971
HUMPHREYIA Gray, 1858
Humphreysia Paetel, 1875 (err.)
NIPPONOCLAVA Smith, 1976
PENICILLUS Bruguière, 1789
Brechites Guettard, 1770 (non-binom.)
Bunodus Guettard, 1770 (non-binom.)
Penecilli Da Costa, 1776
Penecillus, Penicellus auctt. (n.null.)
Verpa Röding, 1798
Aquaria Perry, 1811
Arytene Oken, 1815 (inv.)
Bunodus Blainville, 1817 (nom.nud.)
Arytene Gray, 1858
Arytaena, Arythaena authors (err.)
Aspergillum Lamarck, 1818
Adspergillum, Aspergillium,
Aspergillus authors (err.)
Clepsydra Schumacher, 1817
FOEGIA Gray, 1842
WARNEA Gray, 1858

POROMYOIDEA

POROMYIDAE

POROMYA Forbes, 1844
Embla Lovén, 1847
Thetis H. & A. Adams, 1856
(non Sowerby, 1826)
Ectorisma Tate, 1892
Questimya Iredale, 1930
Poromia, Proamya auctt. (err.)
CETOCONCHA Dall, 1886
Silenia Smith, 1885
(non Mulsant, 1873)
CETOMYA Dall, 1889
DERMATOMYA Dall, 1889

CUSPIDARIIDAE

CUSPIDARIA Nardo, 1840
Neaera Gray in Griffith and Pidgeon,1834
(non Robineau-Desvoidy, 1830)
Aulacophora Jeffreys, 1882
(non Chevrolat, 1842)
Vulcanomya Dall, 1886
HALONYMPHA Dall & Smith, 1886
NORDONEAERA Okutani, 1985
PSEUDONEAERA Sturany, 1901
Bendonaeera Cossmann, 1904 (nom.van.)
RHINOCLAMA Dall & Smith, 1886 ?
pro Rhinomya A. Adams, 1864
(non Robineau-Desvoidy, 1830)
Luzonia Dall & Smith, 1890 ?
SOYOMYA Okutani, 1985
TROPIDOMYA Dall & Smith, 1886
pro Tropidophora Jeffreys, 1882
(non Troschel, 1847)
Goniophora Jeffreys, 1883
(pro Tropidophora)
(non Phillips, 1848)
AUSTRONEAERA Powell, 1937
CARDIOMYA A. Adams, 1864
Spathophora Jeffreys, 1882
(non Amyot & Serville, 1843)

CUSPIDARIIDAE (Cont'd.)

JEFFREYSOMYA Nordsieck, 1969 ?
LEIOMYA A. Adams, 1864 (16)
MYONERA Dall & Smith, 1886
PLECTODON Carpenter, 1864
PROTOCUSPIDARIA Allen, 1981
BIDENTARIA Allen, 1981
EDENTARIA Allen, 1981
RENGEA Kuroda & Habe, 1971 ?

VERTICORDIIDAE

VERTICORDIA Sowerby, 1844 (Gray, 1842 - 61)
Trigonulina Orbigny, 1846
Iphigenia Da Costa, 1850
(non Schumacher, 1817)
Hippella authors
(non Mörch, 1861)
Hippagus Philippi, 1844
(non Lea, 1833)
SPINOSIPELLA Iredale, 1930
VERTAMBITUS Iredale, 1930
VERTISPHAERA Iredale, 1930
EUCIROA Dall, 1881
Eucoria Kobelt, 1896 (err.)
ACREUCIROA Thiele & Jaeckel, 1931
HALICARDIA Dall, 1895
Haloconcha authors (err.)
HALICARDISSA Dall, 1913
HALIRIS Dall, 1886
SETALIRIS Iredale, 1930
LAEVICORDIA Seguenza, 1876
Laevicardia Fischer, 1887 (err.)
LYONSIELLA M. Sars in G. Sars, 1872
Lyonsiela Thiele, 1912 (err.)
PROAGORINA Iredale, 1930
POLICORDIA Dall, Bartsch & Rehder, 1938
Laevicordia authors
(non Seguenza, 1876)
ANGUSTEBRANCHIA Ivanova, 1977
LATEBRANCHIA Ivanova, 1977
SIMPLICICORDIA Kuroda & Habe, 1971

References:

Allen, J. A. & F. J. Hannah "Reclassification of Protobranchia" J.Conch.32:225-249 - 1986 (1)

Bergmans, W. "Validity of genera Lissarca, Austrosarepta etc.: Philobryidae" Jour.Mal.Soc.Aust.2(1):35-38 - 1970 (2)
—— "Taxonomic Revision of recent Australian Nuculidae" Rec.Aust.Mus.31:673-719 - 1978 (3)

Bernard, F. R. "Nutricola, new genus: Transenella tantilla from the Northeastern Pacific" Venus 41(2):146-149 - 1982 (6)
—— "Catalogue of Living Bivalvia of the Eastern Pacific Ocean" - 1983 Can.Sep.Pub.Fish.Aquatic Sci.61:102 pp. (7)

Beu, A. G. "Genera of the Mesodesmatidae" Jour.Mal.Soc.Aust. 2(2):113-131 - 1971 (8)

Boss, K. J. "Tellininae: Western Atlantic" (9)
I. Tellina: Johnsonia 4(45):217-272 - 1966
II. Tellina, Tellidora:Ibid 4(46):273-344.1968
III. Strigilla: Ibid.4(47):345-366 - 1969
—— "Taxonomic concepts and superfluity in bivalve nomenclature" Phil.Trans.Roy.Soc.London (B).284:417-424 - 1978 (9a)

—— & A. S. Merrill "Pandoridae in the Western Atlantic" Johnsonia 4(44):181-215 1965(16a)

Clark, A.H. "Potamilus Rafinesque vs. Proptera" Malac.Data.Net(Eco.Serv.)1(3):58-65.1986(20)

Coan, E. "Recent Crassatellinae of the Eastern Pacific" Veliger 26(3):153-169 - 1984 (21)
—— "The Bernardinidae of the Eastern Pacific" Veliger 27(2):227-237 - 1984 (22)
—— "Additional taxonomic units; J. G. Cooper" Nautilus 100(1):30-32 - 1986 (24)
—— "Recent Eastern Pacific Semele" Veliger 31(1-2):1-42 - 1988 (24a)

Davis, G.M. & S.L.H. Fuller"Genetic relationships among recent Unionacea of North America" Malacologia 20(2):217-253 - 1981 (25)

Dell, R. K. "Mollusca of the family Mytilidae (Review of Genera Adipicola, Idasola)" Nat.Mus.N.Z.Rec.3(3):17-36 - 1987 (26)

Haas, F. "Unionacea" in "Das Tierreich" Lief.88 Gruyter & Co., Berlin, 1969

Harry, H.W. "Supraspecific classification; living oysters" Veliger 28(2):121-158 - 1985 (27)
—— J.C. Britton & R. Nunley "New sp. Caribbean, Honduras" Veliger 22(4):318-322 - 1980 (28)

Heard, W.H. & R.H.Guckert"Reevaluation: rec.N.Am. Unionacea"Malacologia 10(2):333-355.1971(30)

References: (Bivalvia, Cont'd.)
Ivanova, V. L. "New data: genus Policordia" Trudy.Inst.Okeanol.Akad.Nauk.SSSR:108:173-197 (Veliger Review) 1977 (32)

Kafanov, A. I. "Systematics of Clinocardiinae" Malacologia 19(2):297-328 - 1980 (33)

Keen, A. M. "The Pelecypod family Cardiidae" Tulane Stud.Geol.& Paleon.16(1):1 -1980 (34)

Kenk, V. C. & Wilson, B. R. "New mussel from hydrothermal vents; the Galapagos Rift Zone" Malacologia 26(1-2):253-271 - 1985 (35)

Knudsen, J. "Systematics & biology of abyssal & halal Bivalvia" Galathea Report, Vol.2 Copenhagen, 1970 (35a)

Krivosheina, L. V. "Sphaerastrinae" Veliger 27(3):344 (Review, 1978) (36)

Lubinsky, I. "Marine Bivalve Mollusks of the Canadian Central & Eastern Arctic" Can.Bull.Fish.& Aquatic Sci.207 - 1980 (37)

McMichael, D. R. & I. D. Hiscock "A monograph of freshwater mussels of the Australian region" J.Mar.Freshw.Res.9:372-508 - 1958 (39)

Moore, R. C.(Ed.) "Treatise on Invertebrate Paleontology" N (Bivalvia) 1-3 - 1969-1971 Geological Soc.Amer.& U. of Kansas Press (41)

Morton, B. "The Anomalodesmata" Malacologia 21(1-2):35-60 - 1981 (42)

Oliver, P.G."Loripes" Basteria 50:47-64 1986 (43)
—— "Estellacar, Rectangularca and systematic position of Barbatia pectunculiformis" Journ.Conch.Lond.32(5):279-288 - 1987 (43a)

Osorio, C. & P.M. Arnaud "Gaimardia sp.nov.: Cent. Chile" Veliger 26(4):311-315 - 1984 (45)

Parodiz, J. J. & A. A. Bonetto "Taxonomy and Zoogeographic relationships: S.Am. Naids" Malacologia 1(2):179-213 - 1963 (46)

Ponder, W.F."On the synonomy of Austr. tellinids" J.Aust.Mal.Soc.3(2):111-119 - 1977 (47)
—— P. N. Colman, C. M. Yonge & M. H. Colman "The taxonomic position of Hemidonax- " J.Malac.Soc.Aust.5(1-2):4-64 - 1981 (48)

Salvini-Plawen, L. Von "A reconsideration of Systematics in the Mollusca: (Phylogeny and higher classification)" Malacologia 19(2):249-278 -1980 (p.264) (51)
—— & G. Haszprunar"The affinities:Septibranchia" Veliger 25(1):83-84 - 1982 (52)

Saunders, H.L. & J.A. Allen "Studies on deep-sea Protobranchia: Prologue and Pristoglomidae" Bull.Mus.Comp.Zool.145(5):237-261 1973 (53)
—— "Siliculidae and Lametilidae" Ibid. 145(6):263-310 - 1973 (54)
—— "Tindariidae & Pseudotindaria" Ibid. 148(2):23-59 - 1977 (55)

Scarlato, O. A & Ya. Starobogatov "Morphology, Systematics and Phylogeny of Molluscs" Trans.Zool.Inst.Acad.Sci.USSR 80:5-58 - 1979 Mus.Comp.Zool.Harv.U.Sp. Pub.5 - 1985 (56)

Smith, B. J. "Revision of recent Clavagellidae" J.Malac.Soc.Aust.3(3-4):187-209 - 1976 (57)
—— & W. P. Wall "Margaritiferidae reinstated" Occ.Pap.Moll.Harvard U.4(69):321-331.1984(58)

Vokes, H. E. "Genera of the Bivalvia" Revised,1980 Paleontological Research Inst.,N.Y.307pp.(61)

Wagner, H. P. "Notes on type material: Pectinidae" Basteria 49(4-6):81-84 - 1985 (61x)
—— "Status of four species, scallops; new genus" Ibid.52:41-44 - 1988 (61xa)

Waller, T. R. "Morphology, morphoclines and new classification of the Pteromorpha" 1978 Phil.Trans.Roy.Soc.London(B).284:345-365(62)
—— "New genus and species of scallop from Somalia: new tribe "Decatopectinini" Nautilus 100(20):39-46 - 1986 (62a)

Zhenru, W.& Qi Zhongyan"Study: Chinese Mytilidae" Inst.Ocean.Acad.Sinica (China) - 1984 (63)

Class Scaphopoda
Order Dentaliida

DENTALIIDAE

DENTALIUM Linnaeus, 1758
- Dentalia Perry, 1811
- Dentalites Schlotheim, 1813
- Paradentalium Cotton & Godfrey, 1933
- Lentigodentalium Habe, 1963

TESSERACME Pilsbry & Sharp, 1898

ANTALIS H.& A. Adams, 1854
- Dentale da Costa, 1778 (Suppressed - ICZN)
- Antalis Herrmannsen, 1846 (Suppressed - ICZN)
- Entaliopsis Newton & Harris, 1894 pro Entalis Gray, 1847 (non Sowerby, 1839)
- Antale Sacco, 1896

GRAPTACME Pilsbry & Sharp, 1897

DENTALIIDAE (Cont'd.)

STRIODENTALIUM Habe, 1964

COCCODENTALIUM Sacco, 1896

SPADENTALINA Habe, 1963

FISSIDENTALIUM Fischer, 1885
- Schizodentalium Sowerby, 1894
- COMPRESSIDENTALIUM Habe, 1963

PICTODENTALIUM Kira, 1959

FUSTIARIA Stoliczka, 1868
- Pseudantalis Monterosato, 1884

LAEVIDENTALIIDAE

LAEVIDENTALIUM Cossmann, 1888
- Calliodentalium Habe, 1964

RHABDUS Pilsbry & Sharp, 1897

EBOREIDENS Chistikov, 1975 (2)

GADILINIDAE

GADILINA Foresti, 1895

EPISIPHON Pilsbry & Sharp, 1897

ANULIDENTALIUM Chistikov, 1975

OMNIGLYPTIDAE
OMNIGLYPTA Kuroda & Habe, 1953

Order Gadilida

ENTALINIDAE
ENTALINA Monterosato, 1872
Eudentalium Cotton & Godfrey, 1933
MEGAENTALINA Habe, 1963
BATHOXIPHUS Pilsbry & Sharp, 1897
RHOMBOXIPHUS Chistikov, 1983
SOLENOXIPHUS Chistikov, 1983
ENTALINOPSIS Habe, 1957
HETEROSCHISMOIDES Ludbrook in Knight et al, 1960 (3a)
Pro Heteroschisma Simroth in Bronn, 1895 (non Wachsmuth, 1883; nor Kofoid & Skogsberg, 1928)
COSTENTALINA Chistikov, 1982
PERTUSICONCHA Chistikov, 1982

PULSELLIDAE
PULSELLUM Stoliczka, 1868
Siphonentalis G. O. Sars, 1878
Siphonodentalis Clessin, 1896 (error)
FISSIPULSELLUM Scarabino, 1979
ANNULIPULSELLUM Scarabino, 1986 (11)
STRIOPULSELLUM Scarabino, 1979 (in press) (9)
COMPRESSIDENS Pilsbry & Sharp, 1897

SIPHONODENTALIIDAE
SIPHONODENTALIUM M. Sars, 1859
Siphonodontum Locard, 1887 (err.)
Tubidentalium Locard, 1887
Pulchellum Morlet, 1888
PLATYSCHIDES Henderson, 1920
STRIOCADULUS Emerson, 1962 (5)
SAGAMICADULUS Sakuri & Shimazo, 1963
POLYSCHIDES Pilsbry & Sharp, 1898
? Gadus "Rang" Deshayes, 1864 (non Rang, 1829; nor Linnaeus, 1758)
DISCHIDES Jeffreys, 1867
Dicides Sacco, 1897 (err.)

GADILIDAE
GADILA Gray, 1834
Helonyx Stimpson, 1865
Loxoporus Jeffreys, 1883
Gadilopsis Woodring, 1925
CADULUS Philippi, 1844
BATHYCADULUS Scarabino, 1979 (9)
(Validation in press, 1988)

WEMERSONIELLIDAE (10)
WEMERSONIELLA Scarabino, 1986
pro Emersoniella Scarabino, 1986 (10) (non Tendeiro, 1965) (nom.nud.)

References:

Chistikov, S. D. "Some problems in the classification of the order Dentaliida" in Linkharev & Starobogatov "Molluscs, etc." Eng.Trans. Malac.Rev.11:71-73 - 1975 (2)

—— "The modern Entalinidae: 1. The Subfamily Heterochismoidinae" Zoologicheskii Zhurnal 61(5):671-682 (Russian) 1982a (3)

—— 2. Ibid. 61(9):1309-1321 (Russian) 1982b (3a)

—— 4. "The Subfamily Bathoxiphinae" Ibid. 62(2):181-190 - 1983 (4)

Emerson, W. K. "A classification of the Scaphopod mollusks" Jour. Paleo. 36(3):461-482 - 1962 (5)

—— "Two new eastern Pacific species of Cadulus, remarks on the classification: Scaphopod mollusks" Nautilus 92:117-123 - 1978 (6) (Contains summary of Starobogatov {1974} and Chistikov {1975} revisions).

Ludbrook, N. H. in R. M. Moore, Ed. - 1969 "Treatise on Invertebrate Paleontology" I.Mollusca.1.Scaphopoda:137-141 (7)

Palmer, C.P. "Supraspecific: Scaphopod Mollusca" Veliger 17(2):115-123 - 1974 (8)

Scarabino, V. "Scaphopodes bathyaux et abyssaux" Thesis; U. de Aix-Marseille - 1979 (9)

—— "Nuevos taxa Abisales: Scaphopoda"1986 (10) Com.Zool.Mus.Hist.Nat.Montevideo XI(155):1-14

—— "Systematics; Scaphopoda I:Order Gadilida" Ibid. XI.161:1-15 - 1986 (11)

A. Taxa proposed by Iredale (1933-1945) in various families of the Pulmonata (Zilch, 1959-60) and not listed in this Outline. These have not been accepted or assigned by subsequent workers to various families and sub-families; some may be nomina nuda.

Allenella Iredale, 1944 (Paralaomidae)
Allocharopa Iredale, 1937 (Charopidae)
Amenixesta Iredale, 1941 (Xestidae)
Annacharis Iredale, 1944 (Microcystidae)
Ballena Iredale, 1944 (Pseudocharopidae)
Bischoffena Iredale, 1937 (Charopidae)
Buffetia Iredale, 1945 (Microcystidae)
Charopella Iredale, 1944 (Charopidae)
Charopinesta Iredale, 1944 (Charopidae)
Chloritisanax Iredale, 1933 (Chloritidae)
Christianoconcha Iredale, 1945 (Paralaomidae)
Chronoceryx Iredale, 1941 (Durgellinidae)
Claudena Iredale, 1941 (Papuinidae)
Corneoxesta Iredale, 1941 (Xestidae)
Cralopa Iredale, 1941
Deceptrena Iredale, 1944 (Pseudocharopidae)
Deliciola Iredale, 1944 (Microcystidae)
Dentherona Iredale, 1933 (Charopidae)
Dignamoconcha Iredale, 1944 (Microcystidae)
Dolapex Iredale, 1945 (Fanulidae)
Egilodonta Iredale, 1937 (Charopidae)
Egilomen Iredale, 1937 (Charopidae)
Ellarion Iredale, 1941 (Helicarionidae)
Elsothera Iredale, 1933 (Charopidae)
Epinicium Iredale, 1939 (Charopidae)
Excellaoma Iredale, 1937 (Laomidae)
Fanulena Iredale, 1945 (Fanulidae)
Forsancyclus Iredale, 1944 (Ancylidae)
Frustropa Iredale, 1945 (Charopidae)
Geodiscus Iredale, 191941 (Geotrochidae)
Glacilimnea Iredale, 1943 (Lymnaeidae)
Goweroconcha Iredale, 1944 (Charopidae)
Gratilaoma Iredale, 1939 (Laomidae)
Gudeoconcha Iredale, 1944 (Gudeoconchidae)
Gyropena Iredale, 1944 (Charopidae)
Howearion Iredale, 1944 (Helicarionidae)
Hunsteinia Iredale, 1941 (Ariophantidae)
Innesoconcha Iredale, 1944 (Nitoridae)
Insullaoma Iredale, 1937 (Laomidae)
Iotula Iredale, 1941 (Paralaomidae)
Kannaropa Iredale, 1937 (Charopidae)
Kermarion Iredale, 1944 (Helicarionidae)
Kermodon Iredale, 1945 (Charopidae)
Laomavix Iredale, 1933 (Laomidae)
Laomopa Iredale, 1945 (Paralaomidae)
Letimola Iredale, 1941
Lidgbirdia Iredale, 1944 (Pseudocharopidae)
Luinodiscus Iredale, 1937 (Charopidae)
Lutilodix Iredale, 1945 (Fanulidae)
Magilaoma Iredale, 1937 (Laomidae)
Melloconcha Iredale, 1944 (Microcystidae)
Miselaoma Iredale, 1933 (Laomidae)
Missioclivus Iredale, 1941 (Charopidae)
Mistarion Iredale, 1941 (Helicarionidae)
Mutalena Iredale, 1944 (Bullinidae)
Mystivagor Iredale, 1944 (Pseudocharopidae)
Nancibella Iredale, 1945 (Microcystidae)
Oreomava Iredale, 1933 (Charopidae)
Papualbinula Iredale, 1941 (Gastrocoptidae)
Papulaoma Iredale, 1941 (Paralaomidae)
Papusuccinea Iredale, 1941 (Succineidae)
Parcolena Iredale, 1945 (Fanulidae)
Parmellops Iredale, 1944 (Helicarionidae)
Penecosta Iredale, 1944 (Charopidae)
Peplimnea Iredale, 1943 (Lymnaeidae)
Pernagera Iredale, 1933 (Charopidae)
Pernastela Iredale, 1944 (Charopidae)
Pillomena Iredale, 1933 (Charopidae)
Planilaoma Iredale, 1937 (Laomidae)
Problancyclus Iredale, 1943 (Ancylidae)
Pulcharopa Iredale, 1944 (Charopidae)
Pygmanisus Iredale, 1943 (Planorbidae)
Roblinella Iredale, 1937 (Charopidae)
Rosselidena Iredale, 1941 (Trochomorphidae)
Scilistylus Iredale, 1941 (Partulidae)
Serostena Iredale, 1933 (Durgellinidae)
Setomedea Iredale, 1933 (Charopidae)
Simlimnea Iredale, 1943 (Lymnaeidae)
Spirancinea Iredale, 1945 (Succineidae)
Stimulator Iredale, 1944 (Ancylidae)
Tribocystis Iredale, 1944 (Microcystidae)
Trocholaoma Iredale, 1937 (Laomidae)
Turbolaoma Iredale, 1937 (Laomidae)
Westralaoma Iredale, 1939 (Laomidae)

B. Taxa validly proposed as "group names" in the Order Unionoida. Although rejected by workers, these must rank as subgenera under the regulations of the Code of Zoological Nomenclature. (H. E. Vokes, 1980)

UNIO Philipsson, 1788
AEGERICIANA Locard, 1889
ALPECANUSIANA (Bourguignat MS) Locard, 1889
AMNICUSIANA (Bourguignat MS) Locard, 1889
ASTICUSIANA (Bourguignat MS) Locard, 1889
ATERIANA (Bourguignat MS) Locard, 1889
BARYANA (Bourguignat MS) Locard, 1889
BATAVUSIANA (Bourguignat MS) Locard, 1889
BERENGUIERIANA (Bourguignat MS) Locard, 1889
Berenguiereana Caziot, 1910 (err.)
Berenguieriane Caziot, 1910 (err.)
BREBISSONIANA Locard, 1889
COURQUINIANA (Bourguignat MS) Locard, 1889
CRASSIANA (Bourguignat MS) Locard, 1889
CYRTUSIANA Locard, 1889
DESFONTAINIANA (Bourguignat MS) Locard, 1889
ELLIPSOPSISIANA (Bourguignat MS) Locard, 1889
ELONGATULUSIANA (Bourguignat MS) Locard, 1889
EYDYUSIANA (Bourguignat MS) Locard, 1889
FALSUSIANA (Bourguignat MS) Locard, 1889
FUSCULUSIANA (Bourguignat MS) Locard, 1889

UNIO (Cont'd.)
GALLICUSIANA (Bourguignat MS) Locard, 1889
GRANIGERIANA (Bourguignat MS) Locard, 1889
HECKINGIANA (Bourguignat MS) Locard, 1889
HISPANIANA (Bourguignat MS) Locard, 1889
HOLANDRIANA Locard, 1889
(non Bourguignat, 1884)
Hollandriana Germain, 1904 (err.)
JACQUEMINIANA (Bourguignat MS) Locard, 1889
JOANNISIANA (Bourguignat MS) Locard, 1889
JOUSSEAUMEANA (Bourguignat MS) Locard, 1889
LEMOTHEUXIANA (Bourguignat MS) Locard, 1889
LOCARDIANA (Bourguigmat MS) Locard, 1889
(non Bourguignat, 1884)
MANCUSIANA (Bourguignat MS) Locard, 1889
MARIANA (Bourguignat MS) Locard, 1889
(non Quoy & Gaimard, 1825)
MELASIANA Locard, 1889
MERETRICISIANA (Bourguignat MS) Locard, 1889
MOQUINIANA (Bourguignat MS) Locard, 1889
MORELETIANA (Bourguignat MS) Locard, 1889

UNIO (Cont'd.)
MUCIDUSIANA (Bourguignat MS) Locard, 1889
NANUSIANA (Bourguignat MS) Locard, 1889
NUBILUSIANA Locard, 1889
PENCHINATIANA (Bourguignat MS) Locard, 1889
PISANIANA (Bourguignat MS) Locard, 1889
PLATYRHYNCHOIDIANA (Bourguignat MS) Locard,1889
PORNIANA (Bourguignat MS) Locard, 1889
REQUIENIANA (Bourguignat MS) Locard, 1889
RHOMBOIDIANA (Bourguignat MS) Locard, 1889
Rhomboideana Fagot, 1892 (err.)
ROSTRATIANA Locard, 1889
(non Bourguignat, 1881)
SANDRIANA (Bourguignat MS) Locard, 1889
SIMONISIANA (Bourguignat MS) Locard, 1889
SOCARDIANA (Bourguignat MS) Locard, 1889
SPERCHINUSIANA (Bourguignat MS) Locard, 1889
TUMIDUSIANA (Bourguignat MS) Locard, 1889
TURTONIANA (Bourguignat MS) Locard, 1889
VILLAEANA (Bourguignat MS) Locard, 1889
VILLANANA Fagot, 1892
VINCELEUSIANA (Bourguignat MS) Locard, 1889
Vincelotiana Germain, 1904 (emend.)

ANODONTA Lamarck, 1799
ABBREVIATIANA Bourguignat, 1881
ACALLIANA Bourguignat, 1881
ADAMIANA Bourguignat, 1881
ANATINIANA Bourguignat, 1881
AREALIANA Bourguignat, 1881
ARNOULDIANA (Bourguignat MS) Locard, 1890
ARROSIANA (Bourguignat MS) Locard, 1890
AVONIANA Bourguignat, 1881
BECCARIANA (Bourguignat MS) Locard, 1890
BRIANDIANA Bourguignat, 1881
BROTIANA Bourguignat, 1881
CAMURIANA (Bourguignat MS) Locard, 1890
CARVALHOIANA Locard, 1890
COLLOBIANA (Bourguignat MS) Locard, 1890
COMPLANATIANA Locard, 1890
CYGNAEANA Bourguignat, 1881
(= Anodonta Lamarck)
DEPRESSIANA Bourguignat, 1881
ELLIPSOPSIANA Bourguignat, 1881
ELONGATIANA Locard, 1890 (nom. nud.)
EMBIANA Locard, 1890
GALLANDIANA Bourguignat, 1881
(non Bourguignat, 1880)
GASTRODIANA Bourguignat, 1881
GESTROIANA (Bourguignat MS) Locard, 1890
GLYCIANA Bourguignat, 1881

ANODONTA Cont'd.)
GRAVIDIANA Locard, 1890
HUMBERTIANA (Bourguignat MS) Locard, 1890
IDRINIANA (Bourguignat MS) Locard, 1890
ILLUVIOSIANA Fagot, 1885
INCRASSATIANA (Bourguignat MS) Locard, 1890
INTERMEDIANA (Bourguignat MS) Locard, 1890
JOURDHEUILIANA Bourguignat, 1881
LACANNICIANA Locard, 1890
LETOURNEUXIANA (Bourguignat MS) Locard, 1890
(non Bourguignat, 1880)
LUSITANIANA Bourguignat, 1881
MACHADOIANA Locard, 1890
MACILENTIANA Bourguignat, 1881
Macilentina Locard, 1890 (err.)
MACROSTENIANA (Bourguignat MS) Locard, 1890
MARIONIANA (Bourguignat MS) Locard, 1890
MERETRIXIANA (Bourguignat MS) Locard, 1890
MILLETIANA Bourguignat, 1881
OGERIENIANA (Bourguignat MS) Locard, 1890
OVULIANA (Bourguignat MS) Locard, 1890
PAMMEGALIANA Bourguignat, 1881
PICARDIANA Bourguignat, 1881
PISCINALIANA Bourguignat, 1881
Piscimaliana Locard, 1890 (err.)
PLATTENICIANA Bourguignat, 1881
PONDEROSIANA Bourguignat, 1881
Ponderosina Locard, 1884 (err.)
PSEUDOGLYCIANA Locard, 1890
RAYANA Locard, 1890
ROSSMAESSLERIANA Bourguignat, 1881
Rossmaessleriana Locard, 1890
(non Bourguignat, 1881
ROSTRATIANA Bourguignat, 1881
RUMANICANA (Bourguignat MS) Locard, 1890
SCRUPEANA Locard, 1890 (nom.nud.)
SPENGLERIANA Bourguignat, 1881
SPONDAEANA (Bourguignat MS) Locard, 1890
Spondeana Germain, 1904 (err.)
STURMIANA (Bourguignat MS) Locard, 1890
TRICASSINIANA (Bourguignat MS) Locard, 1890
UNIONIFORMIANA Locard, 1890
VENTRICOSIANA Bourguignat, 1881
VIETULIANA Locard, 1890
WESTERLUNDIANA Bourguignat, 1881

ETHERIA Lamarck, 1807
CAILLAUDIANA Bourguignat, 1880
CHAMBARDIANA Bourguignat, 1880
LETOURNEUXIANA Bourguignat, 1880
NILOTICIANA Bourguignat, 1880

- B -

NOTES

[Field notes and new literature information.]

NOTES

[Field notes and new literature information.]

NOTES

[Field notes and new literature information.]

NOTES

[Field notes and new literature information.]

NOTES

[Field notes and new literature information.]

NOTES

[Field notes and new literature information.]

american malacologists